Pollution Impacts on Marine Biotic Communities

Marine Science Series

The CRC Marine Science Series is dedicated to providing state-of-the-art coverage of important topics in marine biology, marine chemistry, marine geology, and physical oceanography. The series includes volumes that focus on the synthesis of recent advances in marine science.

CRC MARINE SCIENCE SERIES

SERIES EDITOR

Michael J. Kennish, Ph.D.

PUBLISHED TITLES

Artificial Reef Evaluation with Application to Natural Marine Habitats, William Seaman, Jr.

The Biology of Sea Turtles, Volume I, Peter L. Lutz and John A. Musick

Chemical Oceanography, Second Edition, Frank J. Millero

Coastal Ecosystem Processes, Daniel M. Alongi

Ecology of Estuaries: Anthropogenic Effects, Michael J. Kennish

Ecology of Marine Bivalves: An Ecosystem Approach, Richard F. Dame

Ecology of Marine Invertebrate Larvae, Larry McEdward

Ecology of Seashores, George A. Knox

Environmental Oceanography, Second Edition, Tom Beer

Estuary Restoration and Maintenance: The National Estuary Program, Michael J. Kennish

Eutrophication Processes in Coastal Systems: Origin and Succession of Plankton Blooms and Effects on Secondary Production in Gulf Coast Estuaries, Robert J. Livingston

Handbook of Marine Mineral Deposits, David S. Cronan

Handbook for Restoring Tidal Wetlands, Joy B. Zedler

Intertidal Deposits: River Mouths, Tidal Flats, and Coastal Lagoons, Doeke Eisma

Marine Chemical Ecology, James B. McClintock and Bill J. Baker

Morphodynamics of Inner Continental Shelves, L. Donelson Wright

Ocean Pollution: Effects on Living Resources and Humans, Carl J. Sindermann

Physical Oceanographic Processes of the Great Barrier Reef, Eric Wolanski

The Physiology of Fishes, Second Edition, David H. Evans

Pollution Impacts on Marine Biotic Communities, Michael J. Kennish

Practical Handbook of Estuarine and Marine Pollution, Michael J. Kennish

Practical Handbook of Marine Science, Third Edition, Michael J. Kennish

Seagrasses: Monitoring, Ecology, Physiology, and Management, Stephen A. Bortone

Pollution Impacts on Marine Biotic Communities

Michael J. Kennish

Institute of Marine and Coastal Sciences
Rutgers University
New Brunswick, New Jersey

CRC Press
Taylor & Francis Group
Boca Raton London New York

CRC Press is an imprint of the
Taylor & Francis Group, an **informa** business

CRC Press
Taylor & Francis Group
6000 Broken Sound Parkway NW, Suite 300
Boca Raton, FL 33487-2742

© 1998 by Taylor & Francis Group, LLC
CRC Press is an imprint of Taylor & Francis Group, an Informa business

First issued in paperback 2019

No claim to original U.S. Government works

ISBN-13: 978-0-367-44807-3 (pbk)
ISBN-13: 978-0-8493-8428-8 (hbk)

Visit the Taylor & Francis Web site at
http://www.taylorandfrancis.com

and the CRC Press Web site at
http://www.crcpress.com

Preface

The rapid population growth and uncontrolled development in the coastal zone during the 20th Century have led to major pollution impacts on estuarine and coastal marine biotic communities. The increase in chemical contaminants from land-based sources has caused acute as well as insidious alteration of these communities, particularly near urban and industrial centers. Because estuarine and coastal marine environments have served as repositories for dredged spoils, sewage sludge, industrial and municipal effluents, and other wastes, biotic communities in these environments are commonly exposed to a multitude of contaminants, including petroleum hydrocarbons, halogenated hydrocarbons, volatile organic compounds, heavy metals, and radioactivity. Perhaps of even greater significance are increasing nutrient and organic loading problems which have contributed substantially to eutrophication and occasionally to severe deficiencies in dissolved oxygen concentrations (i.e., hypoxia and anoxia). Eutrophic conditions in estuaries and coastal marine waters often produce shifts in the structure of faunal communities, with pollution-tolerant species tending to predominate. Commercially and recreationally important species may be eliminated from these impacted systems or greatly reduced in numbers, thereby directly affecting fisheries important to man.

The principal objective of this book is to examine in detail pollution impacts on estuarine and coastal marine biotic communities. The approach is to provide case histories of well-known, heavily contaminated systems. I have selected case histories that reflect a broad spectrum of pollution problems and that present a challenge to contemporary marine biologists, marine chemists, and marine geologists.

This publication has been prepared primarily as a reference for marine scientists and administrators engaged in the research and management of estuarine and coastal marine environments. It should also be of value to individuals involved in bio-monitoring and assessment of pollution problems in these environments. In addition, the volume has been prepared in a well-organized framework so that it can be used as a text for senior undergraduate and graduate students taking advanced courses in marine pollution.

Author

Michael J. Kennish, Ph.D., is a research marine scientist in the Institute of Marine and Coastal Sciences and a member of the graduate faculty at Rutgers University, New Brunswick, New Jersey.

He graduated in 1972 from Rutgers University, Camden, New Jersey, with a B.A. degree in geology and obtained his M.S. and Ph.D. degrees in the same discipline from Rutgers University, New Brunswick, in 1974 and 1977, respectively.

Dr. Kennish's professional affiliations include the American Fisheries Society (Mid-Atlantic Chapter), American Geophysical Union, American Institute of Physics, New England Estuarine Research Society, Atlantic Estuarine Research Society, Southeastern Estuarine Research Society, Gulf Research Estuarine Research Society, Pacific Estuarine Research Society, New Jersey Academy of Science, and Sigma Xi. He is also a member of the Center for Deep-Sea Ecology and Biotechnology of Rutgers University.

While maintaining research interests in broad areas of marine ecology and marine geology, Dr. Kennish has been most actively involved in studies of anthropogenic effects on estuarine and coastal marine ecosystems and with investigations of deep-sea hydrothermal vents and seafloor spreading centers. He is the author and editor of eight books dealing with various aspects of estuarine and marine science. In addition to these books, Dr. Kennish has published more than 80 research articles and book chapters and has presented papers at numerous conferences. Currently, he is the marine science editor of the journal, *Bulletin of the New Jersey Academy of Science*, and series editor of Marine Science books for CRC Press. His biogeographical profile appears in *Who's Who in Frontiers of Science and Technology, Who's Who Among Rising Young Americans, Who's Who in Science and Engineering*, and *American Men and Women of Science*.

Acknowledgments

I am grateful to my colleagues at Rutgers University who have worked with me on various investigations of estuarine and coastal marine environments. In the Institute of Marine and Coastal Sciences, I thank K. W. Able, M. P. DeLuca, J. F. Grassle, R. A. Lutz, J. R. Miller, and N. P. Psuty. Special thanks are extended to R. A. Lutz for his collaboration on many marine science projects. In the Department of Biological Sciences, I acknowledge R. E. Loveland for innumerable exchanges of ideas on estuarine and coastal marine pollution problems. At CRC Press, I am indebted to P. Petralia for coordinating all editorial and production matters on the book. Finally, I express my love and devotion to my wife, Jo-Ann, and sons, Shawn and Michael, for their support and understanding during the preparation of the volume.

Michael J. Kennish

Contents

Introduction

STATEMENT OF THE PROBLEM

The effects of pollution on estuarine and coastal marine biotic communities continue to be the focus of numerous academic, industrial, and government investigations worldwide. Burgeoning population growth in the coastal zone has led to escalating pressures on living resources in adjacent waters. Many pollution impacts on marine communities can be traced directly to industrialized centers which release an extensive array of chemical contaminants to influent systems. Others are more difficult to delineate because they largely derive from contaminants supplied by nonpoint source runoff from land and atmospheric fallout. Of even greater concern have been the adverse environmental effects associated with waste disposal activities, particularly sewage sludge and dredged-spoil dumping as well as municipal and industrial wastewater discharges. Oil spills and leakages have also accounted for significant biotic impacts. Because of multiple pollutant sources in watershed areas, estuarine and coastal marine waters receive the greatest concentration of pollutants introduced into marine environments.

Biotic impacts of waste disposal and chemical contamination in coastal environments are well chronicled. For example, eutrophication of estuarine waters in particular often culminates in anaerobiosis, toxic algal blooms, mass kills of benthic and epibenthic organisms, and changes in abundance and diversity patterns of fish.[1-5] The accumulation of toxic chemicals in bottom sediments frequently contaminates shellfish and demersal finfish populations, generating a wide range of diseases.[6] Abnormalities in fish and shellfish are manifested most conspicuously in heavily degraded estuarine and coastal marine systems (e.g., Puget Sound). The occurrence of elevated fecal pathogens has resulted in the closure of valuable shellfish beds to harvesting. Oil spills have decimated or completely eradicated shallow water communities.[7] The devastating effects of the *Florida* oil spill in Buzzards Bay, Massachusetts, in September 1969 provide an example. Dredging and dredged-spoil disposal have impacted benthic communities by disrupting bottom habitats and altering water quality. Dredging increases mortality of the benthos not only by the physical disturbance associated with the removal and relocation of bottom sediments, but also by the mechanical action of the dredge itself and by smothering of the organisms with sediment when they are picked up or dumped. Chemical contaminants in dredged spoils may hinder repopulation of the dumpsites and later development of the benthic community.

In addition to chemical contamination, other anthropogenic factors can be equally detrimental to coastal marine environments and communities. Coastal engineering projects, for instance, often destroy or severely impair valuable habitats used by many marine organisms for breeding and as nursery areas. Land reclamation in estuarine and wetland zones, construction of embankments, bulkheading, ditching, harbor dredging, construction and operation of large electric generating plants, as well as other anthropogenic activities all impact to some degree estuarine and coastal marine ecosystems. In nearshore waters, engineering structures mainly constructed to improve navigation, to diminish coastal erosion, and to prevent property damage may upset natural readjustment processes, sometimes with disastrous consequences for habitats, marine organisms, and coastal communities. Included among the more important engineering structures in the nearshore are groins and seawalls constructed to prevent erosion and protect property damage; breakwaters to protect a portion of the shoreline area and to provide a harbor for anchoring boats; and jetties at river mouths and tidal inlets to deepen and stabilize channels, to preclude shoaling by littoral drift, and to shield channel entrances from storm waves.[8,9] Shore protection features require considerable management strategies to minimize environmental impacts.[10]

Because multiple anthropogenic factors occur in many coastal environments, the understanding of the effect of a specific pollutant or individual anthropogenic activity on estuarine or coastal marine biotic communities is far from complete. Pollution data are often not fully analyzed and integrated. The lack of statistically accurate and understandable information impedes the development of models to predict the environmental and ecological consequences of a given type of pollutant or anthropogenic impact. Hence, it is necessary to implement carefully planned, comprehensive monitoring and assessment programs to collect essential data.

Long-term monitoring of marine populations and communities is necessary to isolate and identify the effects of pollution and other anthropogenic activities.[6] Both short- and long-term monitoring of biotic communities, in turn, is critical to devise sound solutions to acute and insidious pollution problems. Ideally, both pre- and post-impact surveys of estuarine and coastal marine communities should be conducted to accurately predict pollution effects and to formulate effective remedial action. An ultimate goal is to develop quantitative and predictive models (relying on field and laboratory data) not only of biological, chemical, and physical processes, but also of the anthropogenic stresses imposed on natural systems.

Biomonitoring in estuarine and coastal marine environments has increased dramatically in recent years to support national and regional environmental policies.[11] In the United States, for example, two federal programs are underway to monitor and assess the environmental quality of estuarine and coastal marine waters nationwide: (1) the National Status and Trends Program (NS&T) of the National Oceanic and Atmospheric Administration and; (2) the Environmental Monitoring and Assessment Program (EMAP) of the U.S. Environmental Protection Agency (USEPA). Through its NS&T Program, NOAA has monitored the concentration of toxic organic compounds and trace metals in bottom-dwelling fish, shellfish, and sediments of estuarine and coastal marine waters since 1984.[12] The EMAP-Estuaries (E) Program initiated by the USEPA in 1990 provides a quantitative assessment of the regional extent of coastal environmental problems by measuring status and change in selected indicators of ecological condition. EMAP-E focuses on four major environmental problems in near-coastal systems including eutrophication, hypoxia, sediment contamination, and habitat loss. Together, the NS&T and EMAP-E Programs are identifying ongoing and emerging environmental and biotic problems in U.S. estuarine and coastal marine systems.

Estuaries and coastal marine waters rank among the most important aquatic systems on earth in terms of ecologic and economic significance. As noted by the U.S. Office of Technology Assessment,[13] these critically valuable environments are susceptible to anthropogenic problems for three principal reasons: (1) They (particularly estuaries) bear the brunt of marine disposal activities and nonpoint pollution; (2) the physical and chemical features of many estuaries (circulation patterns, semi-enclosed configuration, shallow depth, mixing of fresh and saltwater) cause pollutants to be flushed relatively slowly from the system or to actually become trapped; and (3) many marine organisms use estuaries and coastal waters during critical parts of their life cycles (e.g., for spawning or nursery habitat). Hence, it is imperative to continue to delineate, ameliorate, and solve the problems that threaten their function.

SOURCES OF POLLUTION

This book identifies and describes the effects of significant environmental contaminants and other anthropogenic disturbances on estuarine and coastal marine biotic communities. Changes in the structure and dynamics of these communities may arise when contaminants released into the environment impair feeding, growth, development energetics, and recruitment of populations.[14] As noted above, numerous sources of contaminants exist in the coastal zone due to multiple human activities. Among the most important are municipal wastewater discharges, chemical industry wastes, oil spills and leakages, dredged spoils, agriculture and mariculture wastes, food processing wastes, paper pulp mill wastes, and metallurgical industry wastes. However, an often unquantified, yet significant source of contaminants in estuaries and coastal marine waters is nonpoint pollution. An estimated 80–90% of the pollutants that enter the marine environment derive from land-based sources through pipeline discharges, urban and rural runoff, and atmospheric fallout.[15]

Many estuaries and coastal waters are impacted to varying degrees by toxic substances released from land-based sources. Elevated levels of some toxins have been frequently detected in estuarine and coastal marine organisms including recreationally and commercially important species. However, the direct effects of specific toxic chemicals on these organisms in the environment are generally not well documented.

Bioaccumulation is most conspicuous in urbanized systems (e.g., Boston Harbor, New York Bight Apex, Baltimore Harbor, San Francisco Bay, Puget Sound, etc.). Toxic substances typically found in water, sediment, and biotic samples from these systems include polycyclic aromatic hydrocarbons, halogenated hydrocarbons, and heavy metals. Volatile organic chemicals (e.g., benzene and toluene ethylene chloride) are of major concern because they can be highly toxic to organisms as well as potentially carcinogenic, teratogenic, and mutagenic.

Toxic substances incorporated into the food chain at lower trophic levels may result in harmful effects to organisms at higher trophic levels, including man. Eggshell thinning and/or embryo toxicosis in some bird populations (e.g., brown pelicans, *Pelecanus occidentalis*; herring gulls, *Larus argentatus*; double-crested cormorants, *Phalacrocorax auritus*; peregrine falcons, *Falco peregrinus*; and bald eagles, *Haliaeetus leucocephalus*) have been attributed to environmental contaminants (DDT, PCBs, chlorinated pesticides, and heavy metals). Fish-eating birds, such as the brown pelican and herring gull, may have levels of DDT about 30- to 100-fold greater than those of their prey due to the effects of biomagnification.[16]

Humans are at risk of developing serious health problems when they regularly consume fish and shellfish tainted with high levels of toxic substances. This risk has prompted state and federal governmental agencies to establish consumption advisories. For example, in 1994 the New Jersey Department of Environmental Protection announced a prohibition on the harvesting of blue crabs (*Callinectes sapidus*) in the Newark Bay complex due to dioxin and other chemical contamination. A statewide limited consumption advisory was also authorized on large bluefish ($\geq$ 60 cm in length) caught in the state primarily because of PCB contamination.[17] In addition, DDT and chlordane levels have been monitored in finfish and shellfish of New Jersey as a result of potential health effects in man.[18,19]

Aside from toxic chemicals, water quality in estuaries is often degraded by both point and nonpoint sources of nutrients and organic carbon wastes originating from nearby watershed areas. Nutrient and organic carbon loadings from faulty sewage treatment and septic systems, urban runoff, marinas, and boats periodically stimulate excessive algal growth (e.g., red and brown tides) that can be detrimental to marine life. When the nutrient levels overwhelm the capacity of a system to assimilate them, eutrophic conditions can become so severe that the trophic structure is altered. Mass mortality of finfish and shellfish may occur when dissolved oxygen concentrations drop to dangerously low levels.

Pathogens can be a problem in some shallow coastal bays receiving large amounts of organic pollutants. Coliform bacteria concentrations are used to establish standards that regulate water quality, shellfishing, and swimming in estuaries and coastal waters. Fecal and total coliform bacteria levels commonly exceed government standards after periods of heavy rainfall during summer in some shallow estuaries with poor circulation and flushing. In addition, marinas where boats are concentrated may have consistently elevated fecal and total coliform levels. As a result, restrictions on shellfish harvesting and swimming are enforced in affected areas when standards are exceeded. Although little direct information is usually available on estuary pathogens (protozoa and viruses), areas are designated as nonswimmable largely on the basis of high fecal coliform counts and the risk of associated pathogens.

Watershed modifications such as deforestation and construction, landscape partitioning and paving, marsh filling and bulkheading, diking and lagoon construction, and dredging and dredged-spoil disposal have increased turbidity and siltation, destroyed natural habitats, and directly impacted biotic communities. The gradual transformation of natural cover to impervious surfaces by development has substantially increased storm water pollutant loads. Construction of impoundments, ditching for mosquito control, salt hay farming, and other human uses has extensively altered coastal marsh habitats. The need for dredged material disposal areas in the past has led to the filling in of tidal marshes, as has the demand for urban/commercial development. Many land-use planning efforts have been poorly conceived and hastily implemented, resulting in considerable environmental degradation of coastal ecosystems. Effective watershed-based land planning therefore is critically important for the protection of sensitive habitats and the control of nonpoint source pollution and stormwater runoff that threaten aquatic communities.

PLAN OF THE VOLUME

Chapter 1 provides comprehensive coverage of the sources, forms, and fates of pollutants occurring in estuarine and marine environments. Focus is placed on priority pollutants occurring in these environments and their effects on biotic communities. Seven main categories of pollution are examined: eutrophication, organic loading, oil, polycyclic aromatic hydrocarbons, halogenated hydrocarbons, heavy metals, and radioactivity. Other anthropogenic influences, including coastal development impacts, litter, dredging and dredged-spoil disposal, and the effects of thermal discharges, impingement, entrainment, and toxic chemical releases from electric generating stations, are also discussed. In addition, ecological surveys as well as toxicological assessment and monitoring programs used in the investigation of biological impacts of waste disposal and chemical contaminant inputs in the sea are addressed.

The most common anthropogenic wastes in estuarine and coastal marine waters worldwide are dredged spoils, sewage, and industrial and municipal wastewaters.[20] These wastes often contain a wide range of contaminants such as petroleum hydrocarbons, chlorinated hydrocarbons, and heavy metals. They have accounted for numerous adverse effects on marine organisms, especially in highly degraded systems (e.g., Commencement Bay, Tokyo Harbor, the Thames estuary, and the Baltic Sea).

To demonstrate the harmful consequences and full impact of contaminants on estuarine and coastal marine environments and biotic communities, seven case studies are presented. Chapter 2 details the effects of more than six decades of sewage sludge and dredged-spoil dumping in the New York Bight Apex. Waste dumping in the apex has caused major changes in the benthic communities and habitats. Shifts in species composition and abundance of the benthos have been observed in response to the pollutant stresses. Some commercially and recreationally important finfish and shellfish populations were significantly impacted during the period of active waste dumping. The termination of sewage sludge disposal in 1987 resulted in gradual improvement of the benthic community.

Chapter 3 recounts pollution problems in Chesapeake Bay. This system has been plagued for decades by hypoxia and anoxia in deeper channels at depths below the pycnocline during periods of increased stratification of the water column. Eutrophication and the depletion of dissolved oxygen levels in portions of Chesapeake Bay during summer have raised concern regarding the capability of the estuary to support living resources. The bay also receives an array of chemical contaminants (e.g., polycyclic aromatic hydrocarbons, organochlorine compounds, heavy metals, etc.) that are potentially toxic to estuarine organisms and a threat to sensitive habitats. Since the fisheries of Chesapeake Bay are among the most productive of all estuaries on earth, remedial action to limit chemical contaminant inputs as well as eutrophication remains a high priority.

The highest concentrations of chemical contaminants in Chesapeake Bay occur at Baltimore Harbor and Hampton Roads, two major industrial ports. Much of the contaminant input at these locations is attributable to shipping activities. Elsewhere in the system, a large fraction of contaminants enter from the Susquehanna River, from municipal and industrial wastewater discharges, from runoff along the estuarine perimeter, and from atmospheric deposition.

The baywide decline of eelgrass (*Zostera marina*) and other submerged aquatic vegetation over several decades is an urgent problem. The long-term health and vitality of Chesapeake Bay are closely linked to those of sensitive biotopes, such as eelgrass beds. Decreasing abundances of some fish and benthic invertebrate populations have been ascribed to the loss of eelgrass habitat. Many fundamental research and monitoring programs in the estuary have focused on the effects of various pollutants on benthic invertebrate community structure.

Chapter 4 deals with pollution impacts on biota in the Southern California Bight, concentrating on the effects of municipal wastewater discharges and sewage sludge dumping on bottom-dwelling organisms. Four major sanitation districts (i.e., City of Los Angeles, Los Angeles County, Orange County, and City of San Diego) discharge ~90% of all municipal wastewaters in the bight, amounting to a total of more than 1600×10^9 l/yr. Areas receiving large volumes of wastewater discharges include the Santa Monica Bay, Palos Verdes Shelf, San Pedro Shelf, and Point Loma. While substantial impacts of organic enrichment and sediment contamination on shelf biota were recorded during the 1950s, 1960s, and early 1970s, the upgrade in effluent quality and concomitant improvement in water quality of outfall areas during the past 25 years have led to significant recovery of the benthic invertebrate communities and demersal fish fauna. In past years of high biotic impacts, changes in benthic community structure from the deposition of sewage effluent were driven largely by increased abundances of opportunistic and pollution-tolerant species and the elimination of pollution-sensitive forms, as exemplified by the Palos Verdes Shelf community.

Chapter 5 assesses the effects of pollution and anthropogenic activities in the San Francisco Bay ecosystem. Fish fauna and benthic invertebrate communities in San Francisco Bay have been affected by reductions in riverine inflow associated with freshwater diversions for agriculture, municipal demands, and industrial uses. Agricultural drainwater from the San Joaquin Valley that is enriched in selenium has contaminated invertebrates, fish, and waterfowl. In addition, a long history of urban and industrial development and mining along the estuarine shoreline and in watershed areas has induced numerous stresses in estuarine communities. Chemical contaminant inputs derive from many municipal and industrial point sources, as well as nonpoint source runoff from urban and agricultural lands. The most important contaminants in the estuary are chlorinated hydrocarbon compounds (DDT and PCBs), polycyclic aromatic hydrocarbons, and heavy metals.

Pollution impacts on estuarine and marine organisms in Puget Sound are treated in Chapter 6. Urbanized bays in Puget Sound (e.g., Bellingham Bay, Commencement Bay, and Elliott Bay) have accumulated high concentrations of various chemical contaminants. In 1981, the U.S. Environmental Protection Agency designated Commencement Bay as one of the 10 highest priority hazardous waste disposal sites in the United States for remedial investigation under the Superfund Program. Chemical manufacturing plants, smelting facilities, oil refineries, pulp mills, marinas, and other facilities have delivered large concentrations of halogenated hydrocarbon compounds, polycyclic aromatic hydrocarbons, heavy metals, and other chemical contaminants to embayments. The principal contaminants found in the tissues of organisms inhabiting industrialized areas of Puget Sound are aromatic hydrocarbons, chlorinated hydrocarbons (DDT, PCBs, chlorinated butadienes, and chlorinated benzenes), phthalate esters, and heavy metals. These contaminants have been linked to skeletal and genetic abnormalities, physiological malfunctions, metabolic disorders, and some forms of cancer in organisms. Numerous histopathological conditions (e.g., neoplasms, necroses, and lesions) have been documented in crabs, shrimp, demersal fishes, and other fauna in urban embayments. Although there is evidence of chemical contaminant impacts on the benthic community of Commencement Bay, adverse effects of chemical contamination on the benthos of the central basin of Puget Sound have not been clearly established.

Chapter 7 describes anthropogenic effects on biota of the Firth of Clyde on the west coast of Scotland and the Tees River estuary on the northeast coast of England. At Garroch Head in the Firth of Clyde, sewage sludge has been dumped since the early 1900s, with various biotic impacts arising from organic enrichment of bottom sediments. Two sewage sludge dumpsites occur immediately south of Garroch Head. Dumpsite 1 received sewage sludge virtually continuously from 1904 to 1974; dumpsite 2 began receiving sewage sludge in 1974. Major contaminants in the sludge include organochlorine compounds (particularly PCBs) and heavy metals (notably copper, lead, and zinc). Sewage sludge disposal induced changes in the structure of benthic communities at both dumpsites, owing largely to organic enrichment. The benthic communities exhibited a decrease in species diversity and an increase in abundance and biomass of constituent populations. Eleven years after the termination of sewage sludge disposal at dumpsite 1, surveys of the benthic community revealed largescale recovery from the effects of organic enrichment.

In contrast to the problems of organic loading at Garroch Head in the Firth of Clyde, the Tees River estuary has been impacted primarily by chemical contamination. Between 1900 and 1950, population growth and industrial development along the Tees River were rapid and contaminant impacts on benthic flora and fauna, as well as fish populations, were significant. By 1937, salmon were eliminated from the estuary. Many macroalgal species recorded in the system during the 1930s were missing by 1970. The number of species of macroflora and macrofauna also declined upestuary, with most forms concentrated at the estuarine mouth. Since the implementation of environmental remedial programs in the mid–1970s, however, the benthic community has recovered substantially from the earlier pollution impacts. This recovery has been ascribed mainly to improvements in water quality during the past 25 years as a result of tighter controls placed on industrial and sewage wastewater discharges.

The effects of increasing eutrophication in the Dutch Wadden Sea since 1950 are examined in Chapter 8. Nutrient enrichment in the western Wadden Sea during the 1970s and 1980s led to significant increases in primary and secondary production. This escalating production may have caused local depletion of dissolved oxygen levels, leading to greater mortality of the benthos. On the plus side, however, the higher biomass of primary and secondary producers has improved conditions for cockle fisheries and mussel culturing. In addition, the growth of juvenile finfish has been enhanced, which has had a positive effect on commercial and recreational fisheries.

REFERENCES

1. National Research Council, *Managing Wastewater in Coastal Urban Areas*, National Academy Press, Washington, D.C., 1993, 477.
2. Nienhuis, P. H., Nutrient cycling and food webs in Dutch estuaries, *Hydrobiologia*, 265, 15, 1993.
3. Kennish, M. J., Pollution in estuaries and coastal marine waters, *J. Coastal Res.*, Spec. Iss. 12: Coastal Hazards, 1994, 27.
4. Valiela, I., *Marine Ecological Processes*, 2nd ed., Springer-Verlag, New York, 1995.
5. Kennish, M. J., Ed., *Practical Handbook of Estuarine and Marine Pollution*, CRC Press, Boca Raton, FL, 1997.
6. Sindermann, C. J., *Ocean Pollution: Effects on Living Resources and Humans*, CRC Press, Boca Raton, FL, 1996.
7. Kennish, M. J., *Ecology of Estuaries: Anthropogenic Effects*, CRC Press, Boca Raton, FL, 1992.
8. Silvester, R. and Hsu, J. R. C., *Coastal Stabilization: Innovation Concepts*, Prentice-Hall, Englewood Cliffs, NJ, 1993.
9. Beer, T., *Environmental Oceanography*, 2nd ed., CRC Press, Boca Raton, FL, 1997.
10. Clark, J. R., *Coastal Zone Management Handbook*, CRC Press, Boca Raton, FL, 1995.
11. Kramer, K. J. M., Ed., *Biomonitoring of Coastal Waters and Estuaries*, CRC Press, Boca Raton, FL, 1994.
12. O'Connor, T. P. and Beliaeff, B., Recent Trends in Coastal Environmental Quality: Results from the Mussel Watch Project, NOAA Tech. Rept., Department of Commerce, Rockville, MD, 1995.
13. Office of Technology Assessment, Wastes in Marine Environments, U.S. Congress, OTA-O-334, U.S. Government Printing Office, Washington, D.C., 1987.
14. Capuzzo, J. M. and Kester, D. R., Biological effects of waste disposal: experimental results and predictive assessments, in *Oceanic Processes in Marine Pollution*, Vol. 1, Robert E. Krieger Publishing, Malabar, FL, 1987, 3.
15. Curtis, C. E., Protecting the oceans, *Oceanus*, 33, 19, 1990.
16. Blus, L. J., Organochlorine pesticides, in *Handbook of Ecotoxicology*, Hoffman, D. J., Rattner, B. A., Burton, G. A., Jr., and Cairns, J., Jr., Eds., Lewis Publishers, Boca Raton, FL, 1995, 275.
17. Kennish, M. J. and Ruppel, B. E., Polychlorinated biphenyl contamination in selected estuarine and coastal marine finfish and shellfish of New Jersey, *Estuaries*, 19, 288, 1996.
18. Kennish, M. J. and Ruppel, B. E., DDT contamination in selected estuarine and coastal marine finfish and shellfish of New Jersey, *Arch. Environ. Contam. Toxicol.*, 31, 256, 1996.
19. Kennish, M. J. and Ruppel, B. E., Chlordane contamination in selected freshwater finfish of New Jersey, *Bull. Environ. Contam. Toxicol.*, 58, 142, 1997.
20. Clark, R. B., *Marine Pollution*, 3rd ed., Clarendon Press, Oxford, 1992.

1 Pollution in Estuarine and Marine Environments

I. INTRODUCTION

This chapter examines the sources, forms, and fates of pollutants in marine environments. While open ocean waters are a great resource to mankind, estuaries and coastal marine waters rank among the most ecologically sensitive and economically important of all ecosystems. The coastal zone is clearly at greatest risk to anthropogenic activities, especially those associated with waste disposal and the release of toxic pollutants. Thus, much of the discussion herein emphasizes pollution problems in shallow water marine environments, their impacts on biotic communities, and the strategies devised to mollify or control adverse human effects on critically important habitat areas.

Estuarine and coastal marine environments are subject to a multitude of anthropogenic impacts attributable to accelerated population growth and development in the coastal zone; to disposal of agricultural, industrial, and municipal wastes; and to numerous recreational and commercial activities that can compromise their ecological integrity.[1,2] The disposal of wastes by the United States alone exceeds 1 billion metric tons (mt) each year. These wastes fall broadly into four categories: dredged materials (400×10^6 mt), industrial wastes (400×10^6 mt), sewage sludge (300×10^6 mt), and municipal solid wastes (180×10^6 mt). More than 200 million mt of wastes are discharged or dumped in marine waters annually by the United States, with dredged materials, industrial wastes, and sewage sludge accounting for approximately 80, 10, and 9% of this total, respectively. Their subaqueous disposal can release considerable quantities of chemical contaminants and other substances to marine waters. When the concentrations of the contaminants reach sufficiently high levels, measurable adverse effects invariably arise in biotic communities or sensitive habitat areas. The highest concentrations of contaminants occur in estuarine and shallow coastal marine systems, particularly those in close proximity to heavily industrialized metropolitan centers. Less contamination exists in the open ocean, where highest levels typically appear along sea lanes of the world due to inputs from ocean-going vessels (e.g., oil slicks, litter, and other substances). The proximal deep ocean floor has been used for disposal of sewage sludge, dredged spoils, industrial wastes, pharmaceuticals, and low level radioactive wastes. Atmospheric deposition is responsible for more widespread input of contaminants in the open sea, albeit at low concentrations.

Davis[3] proposed that marine pollutants can be divided into three broad categories: (1) those that are more concentrated in the open ocean environment than in coastal waters (e.g., radioactive substances); (2) those that are equally concentrated in open ocean and coastal waters (e.g., DDT and its metabolites); and (3) those that are more concentrated in coastal waters than the open ocean (e.g., heavy metals and most chlorinated hydrocarbons). Compared to other aquatic systems, the open ocean may be subjected to more deliberate dumping of wastes. Although measurably contaminated, the open ocean is not polluted.[4]

Approximately 80% of marine pollution stems from land-based sources that reach estuaries and the ocean via nonpoint runoff, direct disposal of wastes, and atmospheric fallout.[5,6] Public perception tends to equate marine pollution with site-specific ocean dumping, but other forms of waste inputs (e.g., agricultural and urban runoff, municipal and industrial discharges, and oil spills and leakages) are at least as important as ocean dumping in terms of their potential impact on marine resources.[7] Agricultural wastes frequently release a wide range of pollutants to coastal waters, most importantly pesticides, insecticides, high oxygen-demanding substances, and excessive nutrients (particularly nitrogen and phosphorus). Industrial wastes also are highly variable in chemical content and may be assessed on the basis of their biochemical oxygen demand, concentration of suspended solids, and specific toxicity of inorganic and organic constituents. The volume of these wastes can be strictly controlled at the point of generation within a facility or by applying various levels of pretreatment prior to their discharge. Municipal facilities process wastewaters from homes and commercial establishments by employing primary (grit removal, screening, grinding, flocculation, and sedimentation), secondary (oxidation of dissolved organic matter), and tertiary (nutrient removal) methods of treatment. The ultimate goal is to reduce or eliminate suspended solids, oxygen-demanding substances, dissolved inorganic (nitrogen and phosphorus) compounds, and bacteria that can significantly lower the quality of receiving waters. Examples of severely degraded, anthropogenically impacted estuarine/coastal areas include Boston Harbor, New Bedford Harbor, New York Bight Apex, Santa Monica Basin, and Puget Sound (United States); Liverpool Bay and the Tal, Tyne, and Thames estuaries (England); the Rhine estuary (The Netherlands); Masan Bay (Korea); and Tokyo Harbor (Japan).

Marine pollution is closely coupled to human demographics.[8] Because about 60% of the world population lives within 60 km of the coast, poorly planned urban expansion and uncontrolled growth of coastal settlements, together with a wide diversity of land/sea margin activities, have hastened the decimation of nearby marine environments in many regions. With the world population expected to exceed 7 billion people by the turn of the century and the coastal population anticipated to double during the next 25 years, marine pollution problems must be effectively and expeditiously addressed at local, regional, national, and international levels in order to avert further habitat and ecological destruction.

II. CONFLICTING COASTAL RESOURCE USES

The coastal zone provides innumerable recreational, economic, and aesthetic benefits for millions of people. However, various resources of the land/sea margin are the target of conflicting uses. The dredging of waterways in estuaries and harbors to facilitate shipping and transportation, and the disposal of dredged spoils and sewage wastes in shallow waters, disrupt benthic habitats and communities and render extensive areas inhospitable for recreational or commercial fin- and shellfishing.

Accidents associated with the extraction and transport of oil have periodically destroyed entire benthic communities in intertidal and shallow subtidal zones and severely degraded habitats for extensive periods of time, often a decade or more. The filling in of marshes and other wetlands to complete reclamation projects has eliminated valuable spawning and nursery grounds for many marine species, including recreationally and commercially important forms. Comprising 24 million ha of coastal habitat in subtropical and tropical countries, mangrove forests are subject to rapid deterioration from marine pollution, and their loss threatens coastal productivity and biodiversity.[9] The construction of seawalls, which essentially maintain a fixed position on a migrating shoreline, ultimately replaces beaches and dunes that enhance the aesthetic and recreational value of the shore area.[10]

Goldberg[11] identified several competitors for coastal ocean space and resources. Owing to their great economic importance, the tourism and recreation industries play integral roles in the planning and use of coastal ocean space in countries worldwide. Many port and coastal cities support their economies with an array of hotels, restaurants, convention centers, shops, and recreational establishments. Clean beaches and the quality of coastal-zone waters are crucial to the long-term viability of tourism in these cities. Marine parks — notably fish reserves, natural marine reserves, and other protected habitats — augment the cultural, scientific, and aesthetic value of these seacoast regions. In proximity to industrialized and metropolitan centers, improved and expanded harbor facilities enhance transportation by marine vessels which spurs local, state, regional, and national economies and increases world trade. However, the release of chemical contaminants from some of these vessels, as well as industrial wastes discharged or deliberately dumped into marine waters, commonly impacts fishing, shellfishing, and nursery grounds. Municipal sewage waste disposal results in similar effects. The world marine fisheries catch from nearshore areas exceeds 80 million mt annually which represents more than 90% of the total catch from the world's oceans. Hence, the disposal of wastes in coastal marine waters may contend with valuable living resources. Aside from tourism, recreation, marine transportation, dredging, and waste disposal, other contenders for ocean space and resources include: (1) energy development such as electric generation, tidal power, and ocean thermal energy conversion; (2) coastal mining, especially sand and gravel; (3) commercial fishing; and (4) mariculture.

III. DEFINITIONS

A. MARINE ENVIRONMENTS

In this work, marine environments are classified into three broad categories: estuaries, coastal waters, and the open ocean. Estuaries may be defined as semi-enclosed coastal waterbodies that have a free connection to the open ocean, where seawater enters and mixes with freshwater derived from land drainage. They vary greatly in size, shape, and the amount of freshwater inflow. Nevertheless, their productivity is generally high, and they provide critical breeding, spawning, and nursery habitats for fish, shellfish, and many other organisms. Although most estuaries exist as drowned lower reaches of rivers, others occur as lagoon-type (bar-built) waterbodies, fjord-type basins, and tectonically produced systems. Chesapeake Bay, Pamlico Sound, Sogne Fjord, and San Francisco Bay provide examples of each type of estuary, respectively.

Coastal waters overlie the inner continental shelf, typically within 4.8 km of shore (i.e., within the boundary of the territorial sea) and, as such, are less enclosed and more saline than estuaries. In addition, oceanic processes affect coastal waters much more greatly than estuaries, with various oceanic phenomena (e.g., waves, tidal action, longshore currents, coastal upwelling of bottom waters, eddies, and riptides) strongly influencing the movement of water and materials along the inner shelf. Because of the direct link to the open ocean, pollutants in coastal waters tend to be dispersed and diluted more readily than in estuaries, where trapping of substances is of overwhelming importance. Coastal waters, which are moderately productive, often harbor economically important fishing and shellfishing grounds that are susceptible to pollutants from waste disposal activities. Many organisms inhabiting these waters utilize estuaries during portions of their lives. Like estuaries, coastal systems vary markedly in size, shape, and configuration and include bays (e.g., Monterey Bay), sounds (e.g., Puget Sound), and open waters along the shoreline (e.g., the New York Bight and Southern California Bight).

Waters overlying the outer continental shelf, continental slope, and beyond constitute the open ocean. Freshwater input from river discharges does not measurably affect the open ocean, which is more saline and less biologically productive than either coastal waters or estuaries. Ocean currents exert the greatest influence on the movement of waters in the deep sea, with tidal effects being less important. As a consequence of their great depth and distance from land, open ocean environments are considered by many individuals to be more ideal locations for waste dumping than coastal regions. Open ocean systems have considerable capacity to dilute, transport, and disperse wastes and associated pollutants due to their large volume and free exchange of water and, hence, are less vulnerable to the impact of waste disposal than other marine waters. However, those organic chemicals, trace metals, and other contaminants which tend to bioaccumulate, biomagnify, persist in the environment, or cause impacts at low concentrations are of concern even in open ocean waters. Many organisms inhabiting the open sea spend part of their lives in coastal waters and estuaries, where they are subjected to greater concentrations of contamination.

B. MARINE POLLUTION

The United Nations Joint Group of Experts on the Scientific Aspects of Marine Pollution (GESAMP), an international advisory body sponsored by the IMO/FAO/Unesco/WMO/WHO/IAEA/United Nations/UNEP, has formulated perhaps the most widely accepted definition of marine pollution. According to GESAMP,[12] marine pollution is defined as the "introduction by man, directly or indirectly, of substances or energy into the marine environment (including estuaries) resulting in such deleterious effects as harm to living resources, hazards to human health, hindrance to marine activities including fishing, impairment of quality for use of seawater and reduction of amenities." This definition stresses anthropogenic rather than natural waste inputs to the sea and focuses on the impacts of the waste. GESAMP periodically prepares reports on the health of marine waters,[13,14] with the last global report on the state of the marine environment being issued in 1990.[5]

A distinction is made between pollution and contamination in marine environments. Pollution occurs when the concentration of a waste substance exceeds the level at which damaging effects are manifested in the sea. It results in measurable impacts on individual organisms, populations, or biological communities. In contrast, contamination takes place when the concentration of a waste substance in seawater, sediments, or organisms exceeds background levels without causing measurable damaging effects. It is coupled to human activities that modify properties of environmental conditions or the availability and quality of resources over a given space range and time interval.[15]

IV. WASTE INPUTS

A. ESTUARINE AND COASTAL MARINE ENVIRONMENTS

Pollutants enter estuarine and coastal marine waters by numerous routes. Untreated or poorly treated sewage discharged from outfalls and sewage sludge dumped at nearshore sites frequently contribute to nutrient and organic enrichment of coastal waters. They also may release large numbers of pathogenic microorganisms that can significantly degrade water quality and pose a serious health hazard to man. High coliform bacteria levels are common in estuaries, especially in densely populated regions that are unsewered and subjected to high rates of precipation.

Insidious alteration of biotic communities in estuarine and coastal marine environments is closely linked to the accumulation of toxic substances derived from industrial and municipal wastes, dredged-spoils, and nonpoint source pollution, (e.g., halogenated hydrocarbons, polycyclic aromatic hydrocarbons, petroleum hydrocarbons, and heavy metals). Physical, chemical, and biological processes affect the distribution and fate of these substances in the marine environment. Among important physical processes acting on anthropogenic wastes in the water column are advection, diffusion, and sedimentation. Various chemical processes — particularly adsorption, desorption, dissolution, oxidation, reduction, flocculation, volatilization, neutralization, and precipitation — influence the availability, persistence, and degradation of wastes in sediments and overlying waters. Biological responses to the wastes involving uptake, bioaccumulation, and toxicity factors modulate the transfer of pollutants through food chains and the long-term stability of aquatic populations and communities (Figure 1).

Clark[16] identified five main categories of wastes that are sources of pollution in coastal marine environments: degradable wastes, fertilizers, dissipating wastes, particulates, and conservative wastes (Table 1). Degradable wastes are those composed of organic matter and subject to bacterial decomposition (e.g., sewage and oil). Fertilizers constitute a primary source of nutrients (nitrogen and phosphorus) that may promote the eutrophication of estuarine and coastal waters. Dissipating wastes

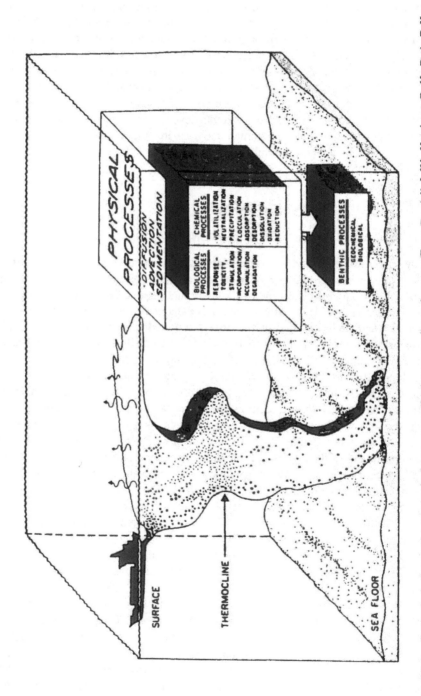

FIGURE 1 Biological, chemical, and physical processes affecting waste disposal at sea. (From Duedall, I. W., Ketchum, B. H., Park, P. K., and Kester, D. R., in *Wastes in the Ocean*, Vol. 1, *Industrial and Sewage Wastes in the Ocean*, Duedall, I. W., Ketchum, B. H., Park, P. K., and Kester, D. R., Eds., John Wiley & Sons, New York, 1983, pp. 3-45. With permission.)

TABLE 1
Sources of Anthropogenic Wastes in
Estuarine and Coastal Marine Waters

Category	Source
Degradable wastes	Sewage
	Oil
	Agricultural organic wastes
	Food processing wastes
	Paper pulp mill wastes
	Brewing and distillery wastes
Fertilizers	Nitrogen-rich fertilizers
	Phosphorus-rich fertilizers
Dissipating wastes	Thermal discharges
	Acids and alkalis
	Cyanide
Particulates	Dredged spoils
	Fly ash
	China clay waste
	Colliery waste
Conservative wastes	Halogenated hydrocarbons
	Heavy metals
	Radioactivity

Compiled from Clark, R. B., *Marine Pollution*, 3rd ed.,
Clarendon Press, Oxford, 1992, 172 pp.

largely refer to industrial discharges that rapidly lose their damaging properties after entering an outfall area. The dissipation of heated effluent discharged from condenser cooling systems of electric generating stations is a prime example. Particulates, such as dredged spoils and fly ash, consist of an array of inert substances that settle to the seafloor when dumped, and they frequently impact benthic habitats and communities. Conservative wastes comprise materials not subject to bacterial attack which are potentially harmful to marine organisms because of their reactive nature. The principal types of conservative wastes are halogenated hydrocarbons, heavy metals, and radioactive substances.

The deterioration of estuarine and coastal marine waters by waste inputs is often severe and accompanied by a number of overt effects such as:

• degraded water quality manifested by hypoxia or anoxia over extensive areas;
• disease, abnormalities, reproductive failure, and mortality of fish and shellfish populations;
• changes in abundance, diversity, and distribution of marine communities;
• loss of submerged aquatic vegetation, wetlands, and other critical habitats;

- closure of shellfish grounds and beaches due to chemical or microbial contamination;
- outbreaks of human disease caused by individuals swimming in contaminated marine waters or consuming contaminated shellfish.

All of these effects can greatly limit resource uses in these shallow coastal systems.

B. DEEP-SEA ENVIRONMENTS

The world ocean is an immense body of water with a volume of 137×10^9 km^3. While waste disposal is concentrated in estuaries or coastal marine environments on the continental shelf, there is growing pressure to use the deep sea as a waste repository, owing to political and logistical constraints being placed on subaqueous waste disposal in nearshore habitats. As mentioned previously, the deep-sea bed has been used for disposal of dredged spoils, sewage sludge, industrial wastes, pharmaceuticals, and low level radioactive wastes. International treaties have been enacted to control the disposal of such wastes. Two of the most important are the Convention on the Prevention of Marine Pollution by Dumping of Wastes and Other Matter of 1972 (better known as the London Dumping Convention and now simply the London Convention), which controls waste dumping in the ocean principally through national legislation and regulations of states contracting to the London Convention, and the International Convention for the Prevention of Pollution from Ships of 1973, which was amended in 1978 (MARPOL of 1973/78). The London Convention, in particular, provides a legal framework (involving more than 90 countries) to regulate dumping in the open ocean and deep sea). Table 2 lists substances controlled by the London Convention.

TABLE 2
Substances Controlled by the London Dumping Convention

Black List

1. Organohalogen compounds
2. Mercury and mercury compounds
3. Cadmium and cadmium compounds
4. Persistent plastics and other persistent synthetic materials, for example, netting and ropes, which may float or may remain in suspension in the sea in such a manner as to interfere materially with fishing, navigation, or other legitimate uses of the sea
5. Crude oil, fuel oil, heavy diesel oil, lubricating oils, hydraulic fluids, and mixtures containing any of these, taken on board for the purpose of dumping
6. High-level radioactive wastes or other high-level radioactive matter, defined on public health, biological, or other grounds, by the competent international body in this field, at present the International Atomic Energy Agency, as unsuitable for dumping at sea
7. Materials in whatever form (such as solids, liquids, semi-liquids, gases, or in a living state) produced for biological and chemical warfare

TABLE 2 (CONTINUED)
Substances Controlled by the London Dumping Convention

8. The preceding paragraphs of this Annex do not apply to substances which are rapidly rendered harmless by physical, chemical, or biological processes in the sea provided they do not: (i) make edible marine organisms unpalatable, or (ii) endanger human health or that of domestic animals. The consultative procedure provided for under Article XIV should be followed by a Party if there is doubt about the harmlessness of the substance

9. This Annex does not apply to wastes or other materials (such as sewage sludges and dredged spoils) containing the matters referred to in paragraphs 1 to 5 above as trace contaminants. Such wastes shall be subject to the provisions of Annexes II and III as appropriate

10. Paragraphs 1 and 5 of the Annex do not apply to the disposal of wastes or other matter referred to in these paragraphs by means of incineration at sea. Incineration of such wastes or other matter at sea requires a prior special permit. In the issue of special permits for incineration the Contracting Parties shall apply the Regulations for the Control of Incineration of Wastes and Other Matter at Sea set forth in the Addendum to this Annex (which shall constitute an integral part of this Annex) and take full account of the Technical Guidelines on the Control of Incineration of Wastes and Other Matter at Sea adopted by the Contracting Parties in consultation

Grey List
The following substances and materials require special permits, issued only according to the articles of the LDC.

A. Wastes containing significant amounts of the matters listed below:
   Arsenic, lead, copper, zinc, and their compounds
   Organosilicon compounds
   Cyanides
   Fluorides
   Pesticides and their byproducts not covered in Annex 1 (Black List)

B. In the issue of permits for the dumping of large quantities of acids and alkalis, consideration shall be given to the possible presence in such wastes of the substances listed in paragraph A and to beryllium, chromium, nickel, vanadium, and their compounds

C. Containers, scrap metal, and other bulky wastes liable to sink to the sea bottom which may present a serious obstacle to fishing or navigation

D. Radioactive wastes or other radioactive matter not included in Annex 1 (Black List) — In the issue of permits for the dumping of this matter, the contracting parties should take full account of the recommendations of the competent international body in this field, at present the International Atomic Energy Agency

E. In the issue of special permits for the inceration of substances and materials listed in this Annex, the Contracting Parties shall apply the Regulations for the Control of Incineration of Wastes and Other Matter at Sea set forth in the addendum to Annex 1 (Black List) and take full account of the Technical Guidelines on the Control of Incineration of Wastes and Other Matter at Sea adopted by the Contracting Parties in consultation, to the extent specified in these Regulations and Guidelines

From Final Act of the London Dumping Convention, Annex I and II, Office of the London Dumping Convention, International Maritime Organization, London, 1972. With permission.

The disposal of wastes into the deep sea effectively isolates pollutants from surface living communities in the open ocean and precludes their impact on coastal environments. Deepwater Dumpsite (DWD)–106 in the northwest Atlantic and the Northeast Atlantic Dumpsite (NEADS) serve as case studies. Consisting of an easterly deepwater municipal sludge site and a westerly deepwater industrial waste site, DWD–106 (now inactive) is located about 200 km southeast of New York Harbor. The easterly deepwater municipal sludge site occupies an area of approximately 258 km², being bounded by coordinates 38°40′N–39°00′N and 72°00′W–72°05′W. The westerly deepwater industrial waste site, located about 18 km from the sludge site, covers an area of approximately 97 km², with center coordinates of 38°45′N–72°20′W. Water depths at the two dumpsites range from 2250–2750 m.

More than 30 million mt (wet) of sewage sludge had been dumped at DWD–106 by the end of 1992, when use of the site was terminated. During dumping, measurable amounts of sludge reached the seafloor immediately west of the site as evidenced by bacterial spores of *Clostridium perfringens* (a human sewage indicator), significant levels of trace metals in sediments, and stable isotope ratios in benthic infauna.[17] Studies have revealed increased density of some benthic populations at the dumpsite, owing to increased concentrations of organic carbon which serves as an additional food source. However, the species diversity of the benthic community has declined in response to the sludge dumping.

Between 1973 and 1983, approximately 4.17×10^6 mt of aqueous industrial wastes were dumped at the industrial dumpsite at DWD–106. Although no detectable impacts of industrial wastes on benthic fauna were demonstrated during this dumping period, some water column effects were observed. Within waste plumes, for example, a potential existed for alterations in the composition of phytoplankton populations and for impaired swimming, respiration, and egg production in zooplankton.[18] However, no major biotic impacts of industrial waste dumping were recorded at DWD–106.

The NEADS dumpsite, which has been in existence since 1967, covers a rectangular area of the seafloor bounded by coordinates 45°59′–46°10′N and 16°–17°30′W. It lies at a mean depth of 4400 m. This dumpsite received large quantities (~2000 mt/yr) of low level radioactive wastes from nuclear power stations and scientific and medical establishments for more than a decade. Prior to being dumped, these wastes were packaged in steel drums or concrete containers to preclude leakage and to provide shielding.

There is little evidence of radioactive impacts on the benthic fauna at NEADS. Only the anemone *Chitoanthus abyssorum* showed increased levels of Sr and Cs between 1967 and 1980, possibly due to waste-container leakage. All other benthic invertebrates sampled at NEADS and a control site have exhibited no significant differences in radionuclide concentrations.[19]

Although contents of radioactive waste dumped at NEADS are expected to corrode and leach through their containment sometime in the future, the delay should allow enough time to ensure the loss of radioactivity via decay of shorter-lived radionuclides. The slow release of the contaminants should also result in great dilution. Theoretical models predict minimal, if any, effect of the radioactive waste on marine communities in and around the NEADS dumpsite. Exposure of marine organisms at NEADS is expected to be at or near the natural background level of radiation and well below the dose rates at which harmful effects are manifested.[16]

V. POINT AND NONPOINT POLLUTION SOURCES

Pollutants in marine environments derive from both point and nonpoint sources (Table 3). Waste disposal operations involving the intentional release of materials to marine waters either via direct dumping or pipeline discharges constitute point sources of pollution. The dumping of municipal sewage sludges, dredged spoils, and industrial wastes (e.g., acid-iron wastes, alkali chemicals, and pharmaceuticals), as well as the discharge of municipal and industrial effluents from outfalls, are the primary point-source categories responsible for the introduction of pollutants to U.S. marine waters. Major contaminant loadings of U.S. marine waters from point sources have decreased substantially during the past two decades due to tighter state and federal government regulations and improved industrial controls of point-source discharges. Consequently, emphasis has shifted more recently to the assessment of detrimental effects ascribable to secondary, less-easily regulated, but recurrent pollutants originating from nonpoint sources.[20] However, the input of nonpoint-source pollutants is more problematical, and acute and insidious biological effects of these pollutants are extremely difficult to assess.[21]

Principal nonpoint sources of pollution from land-based systems include urban and rural runoff, septic tank leakage, groundwater transport, erosion/deposition of contaminated soils, and atmospheric deposition. Nonpoint runoff is a major source

---

TABLE 3
Point and Nonpoint Sources of Pollution in Marine Waters

Sources	Common Pollutant Categories
Point Sources	
Municipal sewage treatment plants	BOD, bacteria, nutrients, ammonia, toxic chemicals
Industrial facilities	Toxic chemicals, BOD
Combined sewer overflows	BOD, bacteria, nutrients, turbidity, total dissolved solids, ammonia, toxic chemicals
Nonpoint Sources	
Agricultural runoff	Nutrients, turbidity, total dissolved solids, toxic chemicals
Urban runoff	Turbidity, bacteria, nutrients, total dissolved solids, toxic chemicals
Construction runoff	Turbidity, nutrients, toxic chemicals
Mining runoff	Turbidity, acids, toxic chemicals, total dissolved solids
Septic systems	Bacteria, nutrients
Landfills/spills	Toxic chemicals, miscellaneous substances
Silvicultural runoff	Nutrients, turbidity, toxic chemicals

From U.S. Environmental Protection Agency, National Water Quality Inventory, Washington, D.C., 1986.

of pollutants to rivers. Nonpoint-source pollutants also originate from human activities at sea associated with accidental releases (e.g., oil spills), marine mining, and the operation of vessels (e.g., bilge waters). Although nonpoint pollution represents a rather diffuse source of contaminants in marine waters, it is quantitatively significant, particularly on developed coastlines. Nonpoint pollution occurs in virtually all estuarine and coastal marine waters along developed shorelines, varying dramatically both spatially and temporally. Because nonpoint runoff is so diffuse, widespread, and variable, it is difficult to accurately quantify. Comprehensive data are available primarily for urban and rural runoff, with large information gaps still constraining the analyses of other nonpoint sources. Many state assessment programs in the United States suffer from inadequate funding, which typically translates into a lack of information-gathering, inadequate systematic analyses of gathered data, and ineffective dissemination of results. Therefore, information needed for accurate national assessment and the determination of the relative inputs of pollutants from point and nonpoint sources is often incomplete.

Marine waste disposal activities continue to be overwhelmingly concentrated in estuarine and coastal marine waters, which receive 80–90% of all wastes released to marine environments worldwide. More than half of all industrial and municipal pipelines discharge directly into estuaries, and more than half of all dredged-spoil dumpsites lie in these coastal ecotones. Dredged material, sewage sludge from municipal treatment plants, liquid industrial wastes, and land runoff are the sources of most pollutants released to coastal marine waters in the United States (Table 4).

TABLE 4
Major Sources of Coastal Marine Pollution in the United States[a,b]

| | Source | | |
Type of Pollutant	Sewage Treatment	Industrial Facilities	Land Runoff
Nutrients (N and P)	41	7	52
Bacteria	16	<1	84
Oil	41	10	47
Toxic Metals	6	46	47

[a] Values in percent.
[b] Excludes dredged spoils.

Data from Natural Resources Defense Council, Washington, D.C.

While the relative contribution of pollutants from pipeline discharges, dumping, and nonpoint sources in estuarine and coastal marine waters varies with the type of pollutant and the location, outfall discharges and runoff generally deliver greater concentrations of pollutants to these coastal systems than dumping. However, in some cases (e.g., Liverpool Bay) dumping is the major source of pollutant entry. Between 1970 and 1990, the general trend in U.S. marine waters was a gradual increase in sewage sludge dumping and a dramatic decrease in industrial waste dumping. The greatest volume of waste material dumped in shallow marine waters of the United States during these two decades consisted by far of dredged spoils. Most (approximately two-thirds) of the dredged material currently dumped in U.S. marine waters is in estuaries, the remainder being divided nearly equally between waters within the 4.8-km territorial boundary and those beyond.

Because of the great variability in composition of marine-dumped wastes and the intermittent and localized nature of dumping operations, it is difficult to compare pollutant inputs derived from sewage sludge and dredged material dumping and those resulting from pipeline discharges and runoff. However, based on a comprehensive national database dealing with relative contributions of pollutants from major sources in the United States,[22] some general observations are possible (Table 4). Nonpoint runoff is the major contributor of fecal coliform bacteria to U.S. marine waters. It also represents the principal source of suspended solids, total phosphorus, and certain heavy metals (e.g., copper, chromium, iron, lead, and zinc). In contrast, municipal point sources are the dominant contributors of pollutants which raise the biochemical oxygen demand, and they also account for large concentrations of total nitrogen, oil, and grease. Industrial discharges are the principal sources of some heavy metals and many organic chemicals. For example, 90% or more of the input of cadmium, mercury, and chlorinated hydrocarbons is attributable to these discharges.

The types of pollutants found either locally or regionally depend on several important factors. For instance, the relative contributions of agricultural and urban runoff, the type and magnitude of industrial discharges, and the size and number of sewage treatment plants that release effluent, all influence the quantity and quality of pollutants in receiving waters. The amount of harbor or port maintenance, shipping, and recreational and commercial activity in marine waters also affects the input of pollutants and their subsequent biotic impacts on both local and regional scales.

VI. FATE OF WASTES AND POLLUTANTS

Biogeochemical processes govern the distribution and fate of anthropogenic wastes and associated pollutants in estuarine and coastal marine environments. Part of the logic of discharging and dumping wastes in estuaries and coastal marine waters is an expected dilution response, with the transport and dispersion of contaminants to offshore areas presumably mitigating impacts on nearshore habitats. Dilution may be defined as "the process by which one constituent is mixed with others, causing the ingredients to spread out over a larger volume than before, with a consequent reduction in concentration."[23] Wastes discharged or dumped in marine waters undergo substantial dilution within a few hours, often by a factor of 5000 or more. Diffusion enhances dilution of waste materials, resulting in lower concentrations of pollutants at points of discharge. Physical transport processes (i.e., currents and mixing) also disperse waste particles and pollutants away from disposal sites. In the coastal zone, hydrographic processes produce more complex circulation patterns than in offshore areas, with water movement being influenced by coastal boundary layer effects, a broad spectra of turbulent eddies, and flow modulated by local bathymetry and shoreline configurations.[24] These factors make accurate prediction of pollutant dispersal extremely tenuous. Estuarine circulation appears to be even more problematic.[25,26] New monitoring techniques and dispersion models have been introduced to study mixing and transport mechanisms; these approaches are valuable in evaluating pollutant dispersal. Hydraulic and numerical models have been employed for many years to examine the spread of pollutants in estuarine and coastal marine waters. However, it is difficult to construct a single model for the movement, transformation, and fate of a pollutant because of the many diverse, unrelated forces acting on it simultaneously when released into marine waters.[23]

Organisms likewise can transport wastes and contaminants from one location to another. Subsequent to the uptake of contaminants, eggs and plankton may be passively carried considerable distances in coastal or open ocean waters. Many fauna, particularly fish, migrate long distances from open ocean and coastal waters into estuaries to breed and to seek nursery habitats for their offspring. Organisms are often exposed to a variety of pollutants when they migrate to or emigrate from estuaries.

The seafloor is the ultimate repository for many pollutants. Flocculation and sedimentation facilitate the flux of wastes and pollutants to the seafloor. Excretory products and the remains of marine organisms, which commonly contain contaminants, also accumulate on the bottom. After deposition, contaminants can be remobilized by bioturbation activities of benthic organisms, by chemical processes (e.g., dissolution, diagenetic alteration of sediments, and desorption of substances from particle surfaces), and by waves and currents which roil bottom sediments. Advective processes often transport these substances long distances from original depositional sites.

VII. BIOTIC EFFECTS OF WASTE INPUTS

Biological concerns of waste loading in marine environments have focused on four main points:

1. The accumulation and transfer of metals and xenobiotic compounds in marine food webs, including accumulation in commercial resources;
2. The toxic effects of such contaminants on the survival and reproduction of marine organisms and the resulting impact on marine ecosystems;
3. The uptake and accumulation of pathogenic organisms in commercially harvested species destined for human consumption; and
4. The release of degradable organic matter and nutrients to the ocean, resulting in localized eutrophication and organic enrichment.

Contaminants of biological concern (e.g., petroleum hydrocarbons, halogenated hydrocarbons, heavy metals, and pathogenic microorganisms) are largely associated with particulate waste materials and sediments. Toxic chemicals and other substances derived from hazardous wastes commonly pose a potential danger to marine communities and man. This is so because they may be acutely lethal to organisms, a cause of detrimental cumulative effects, or persistent in the environment. In addition, their effects may be biomagnified in food chains. Examples include organochlorine compounds (e.g., PCBs and DDT), radionuclides, and many other substances).

The uptake of the contaminants by marine organisms occurs through the ingestion of food and detrital particles, water exchange at feeding and respiratory surfaces, and adsorption of chemicals onto body surfaces.[27] The capacity to store, remove, and detoxify contaminants varies among different taxa. In marine animals, contaminants may be stored in skeletal material, concretions, and soft tissues. The release of excretory and particulate products (e.g., fecal pellets, eggs, and molts) removes contaminants. When contaminant uptake by an organism exceeds the elimination rate, bioaccumulation can take place.

The bioavailability of contaminants, their chemical form, the physiological state of the organism, and its ability to regulate contaminant uptake must be considered in bioaccumulation studies. Toxicologists often employ the terms "bioconcentration" and "biomagnification" when describing bioaccumulation. Bioconcentration refers to an organism's ability to accumulate a contaminant significantly in excess of that in ambient water. Biomagnification, in turn, is the concentration of a pollutant up the food chain, with relatively low levels accumulating in organisms at the base of the chain and higher levels, possibly reaching harmful or lethal amounts, in organisms at the top of the chain. The bioaccumulation of contaminants in the tissues of marine organisms has important implications because it is widely used to delineate the degree of contamination of marine waters.[28]

The responses of estuarine and marine organisms to waste inputs are manifested on four levels of biological organization: cellular, organismal, population, and community levels (Table 5).[28] The earliest detectable changes within the cell in response to toxic environmental chemicals (xenobiotics) involve subcellular organelles (lysosomes, endoplasmic reticulum, and mitochondria). Subcellular damage, such as destabilization of lysosomal membranes, may develop along with the disruption of cellular systems. Some biochemical processes respond specifically to certain types of pollutants (e.g., the cytochrome P450-mediated system of mixed function oxygenation of organic compounds, and the metal-binding proteins). Examples of significant responses to pollutant exposure at the cellular level are the impairment of energy metabolic pathways and the induction of enzymatic detoxification systems.

Individual organisms display multiple responses to pollutant stress. Physiological responses of marine organisms to pollutants depend on the bioavailability, uptake, accumulation, and disposition of contaminants in the body, and on the interactive effects of multiple contaminants.[29] Chief among the negative physiological changes are those that directly impact an organism's growth, reproduction, and survival. For instance, sublethal toxic effects of contaminants commonly alter the energy available for growth and reproduction of marine organisms. Apart from the impairment of these physiological functions, other reproductive and developmental processes may be adversely affected, which can ultimately lower the reproductive and developmental potential of the population and lead to serious long-term problems (e.g., genetic damage). Chronic exposure to chemical contaminants occasionally results in modifications of adaptive and feeding behaviors. Changes in population structure and dynamics may also become evident.[29] In addition, individuals generally exhibit greater susceptibility to disease when exposed to elevated contaminant levels, as

TABLE 5
Marine Organism Responses to Chemical Contaminants

Level	Responses	Effects at the Next Level
Cell	Toxication	Toxic metabolites available
	Metabolic impairment	
	Cellular damage	Disruption in energetics and cellular processes
	Detoxication	Adaptation
Organism	Physiological changes	Reduced population performance
	Behavioral changes	
	Susceptibility to disease	
	Reduced reproductive effort	
	Decreased larval viability	
	Readjustment in rate functions	Population regulation and adaptation
	Altered immunities	
Population	Changes in age/size structure, recruitment, mortality, and biomass	Negative impacts on species productivity as well as coexisting species and communities
	Adjustment of reproductive output and other demographic characteristics	Adaptation of population
Community	Changes in species abundance, species distribution, and biomass	Replacement by more-adaptive competitors
	Altered trophic interactions	Reduced secondary production
	Ecosystem adaptation	No change in community structure and function

From McDowell, J. E., *Oceanus*, 36, 56–61, 1993. With permission.

manifested by histopathological disorders that develop (e.g., high rates of neoplasm formation and tissue inflammation or degeneration).

At the population level, acute changes in population dynamics arise subsequent to episodes of severe pollution stress. Reductions in growth, reproduction rates, biomass, and density as well as increases in mortality and shifts in the distribution of populations can be substantial soon after pollutant exposure. Other changes, such as those related to age/size structure and recruitment of the affected population, may be detected much later. Long-term, pollution-induced variability and natural variability of population parameters are often difficult to differentiate. Natural environmental perturbations frequently have a profound effect on marine populations and, consequently, adequate baseline data must be obtained on populations under pre-pollution conditions or from existing uncontaminated habitats to properly evaluate later impacts of pollutant exposure.

Community responses to pollution may be assessed at both the structural and functional levels. Estimates of species composition, abundance, trophic status, biomass, and diversity, and the spatial and temporal variability in these parameters provide data on structural characteristics of the affected community. Changes in functional aspects due to pollutant exposure usually are apparent in altered trophic interactions and diminished production that develop through time.

Effects of an insidious and cumulative nature on communities include graded responses to pollutants, such as: (1) loss of rare or sensitive species; (2) quantitative changes (e.g., in age structure) of longer-lived species; (3) decreased species diversity; and (4) dominance of opportunistic species.[30] The community-level approach has proven to be effective in the detection and monitoring of the biological effects of pollutants.[31,32] Benthic communities, in particular, undergo dramatic shifts in species composition, abundance, and other parameters because of the variable sensitivities of the constituent species to pollutant exposure. In these communities, acute pollutant stresses cause death in many cases, whereas the effects of chronic pollutant stresses are more subtle, with alterations in community structure often ascribed to changes in growth, fecundity, recruitment, physiological processes, or other factors. Because of their importance to overall ecosystem structure and function, benthic organisms (especially infauna) are effective indicators of impacts at higher levels of biological organization (e.g., the community level).[33]

VIII. TYPES OF MARINE POLLUTANTS

A. PRIORITY POLLUTANTS

Substances that pose the greatest threat to marine environments as a result of their toxicity, persistence, bioaccumulation, or other unusual properties have been placed on priority pollutant lists of a number of countries. The U.S. Environmental Protection Agency (U.S. EPA) designated 127 chemicals as priority pollutants in the United States in the late 1970s. These chemicals are pollutants which have commonly occurred in discharged wastewaters and for which there were existing stocks of chemicals to make standard solutions.[34] Nearly all of the chemicals measured in U.S. estuarine and coastal marine environments by NOAA's National Status and Trends Program are priority pollutants (Table 6). The 74 priority chemicals not measured by this program consist almost entirely of low-molecular-weight, highly soluble, volatile organics.

In England, *The Red List* of highest priority substances was published in 1988.[35] Table 7 shows the initial priority chemicals of *The Red List*. The Department of the Environment (United Kingdom) has conducted monitoring programs for these priority chemicals in British waters. The long-term goal has been to reduce the environmental concentrations of these chemicals as much as possible regardless of their current levels.

B. COMMONLY OCCURRING POLLUTANTS

Common pollutants that have been reported from estuarine and marine environments include the following:

1. Excessive nutrients causing progressive enrichment and periodic eutrophication problems;

2. Sewage and other oxygen-demanding wastes (principally carbonaceous organic matter) which promote anoxia or hypoxia of coastal waters subsequent to microbial degradation;
3. Pathogens (e.g., certain bacteria, viruses, and parasites) and other infectious agents often associated with sewage wastes;
4. Petroleum hydrocarbons originating from oil tanker accidents and other major spillages, routine operations during oil transportation, effluent from nonpetroleum industries, municipal wastes, and nonpoint runoff from land;
5. Polycyclic aromatic hydrocarbons entering estuarine and marine ecosystems from sewage and industrial effluents, oil spills, creosote oil, combustion of fossil fuels, and forest fires;
6. Halogenated hydrocarbon compounds (e.g., organochlorine pesticides) derived principally from agricultural and industrial sources;
7. Heavy metals accumulating from smelting, sewage-sludge dumping, ash and dredged-material disposal, antifouling paints, seed dressings and slimicides, power station corrosion products, oil refinery effluents, and other industrial processes;
8. Radioactive substances generated by uranium mining and milling, nuclear power plants, and industrial, medical, and scientific uses of radioactive materials;
9. Calefaction of natural waters, owing primarily to the discharge of condenser cooling waters from electric generating stations;
10. Litter and munitions introduced by various land-based and marine activities;
11. Fly ash, colliery wastes, flue-gas desulfurization sludges, boiler bottom ash, and mine tailings;
12. Drilling muds and cuttings;
13. Acid-iron and alkali chemicals;
14. Pharmaceuticals; and
15. Suspended solids, turbidity.

McIntyre[7] considers excessive nutrients, sewage, and halogenated hydrocarbons as pollutants of top priority concern in coastal waters, followed by oil, heavy metals, polycyclic aromatic hydrocarbons, and radioactive substances. The status of science and management on pollutants derived from land-based sources is given in Table 8.

1. Nutrient Enrichment

a. Eutrophication

Protecting water quality and the long-term health of marine biotic communities through effective nutrient management strategies is an issue of international concern. Despite the acknowledged problems associated with coastal eutrophication, environmental degradation of coastal waters resulting from excessive nutrient input continues unabated in many regions of the world. These problems often stem from overpopulation, poor planning, and uncontrolled development in nearby watershed areas.

TABLE 6
List of Priority Pollutants Measured by the NOAA National Status and Trends Program in the Estuarine and Coastal Marine Waters of the United States

DDT and Its Metabolites
2,4'-DDD
4,4'-DDD
2,4'-DDE
4,4'-DDE
2,4'-DDT
4,4'-DDT

Chlorinated Pesticides Other Than DDT
Aldrin
cis-Chlordane
trans-Nonachlor
Dieldrin
Heptachlor
Heptachlor epoxide
Hexachlorobenzene
Lindane (-HCH)
Mirex

Polychlorinated Biphenyls
PCB congeners 8, 18, 28, 44, 56, 66,
101, 105, 118, 128, 138, 153, 179,
180, 187, 195, 206, 209

Metals

Major Elements
Aluminum
Iron
Manganese

Trace Elements
Arsenic
Cadmium
Chromium
Copper
Lead
Mercury
Nickel
Selenium
Silver
Tin
Zinc
Tri-, di-, and mono-butyltin

Polycyclic Aromatic Hydrocarbons

2-Ring
Biphenyl
Naphthalene
1-Methylnaphthalene
2-Methylnaphthalene
2,6-Dimethylnaphthalene
1,6,7-Trimethylnaphthalene

3-Ring
Fluorene
Phenanthrene
1-Methylphenanthrene
Anthracene
Acenaphthene
Acenaphthylene

4-Ring
Fluoranthene
Pyrene
Benz[a]anthracene
Chrysene

5-Ring
Benzo[a]pyrene
Benzo[e]pyrene
Perylene
Dibenz[a,h]anthracene
Benzo[b]fluoranthene
Benzo[k]fluoranthene

6-Ring
Benzo[ghi]perylene
Indeno[1,2,3-cd]pyrene

From National Oceanic and Atmospheric Administration (NOAA), Ocean Assessments Division, Rockville, MD.

TABLE 7
Initial "Red List" of Priority
Chemicals in England

Mercury and its compounds
Cadmium and its compounds
Gamma-Hexachlorocyclohexane
DDT
Pentachlorophenol
Hexachlorobenzene
Hexachlorobutadiene
Aldrin
Dieldrin
Endrin
Polychlorinated biphenyls
Dichlorvos
1,2-Dichloroethane
Trichlorobenzene
Atrazine
Simazine
Tributyltin compounds
Triphenyltin compounds
Trifluralin
Fenitrothion
Azinphos-methyl
Malathion
Endosulfan

From Department of the Environment, Inputs
of dangerous substances to water: proposals
for a unified system of control. The government's consultative proposals for tighter controls over the most dangerous substances
entering the aquatic environment, *The Red
List*, July, 1988.

They have been exacerbated by shortsighted political decisions that delay or even avoid the implementation of remedial actions.

The ocean margins trap most of the nutrients originating from land-based sources and transported seaward by rivers and the atmosphere. The anthropogenic flux of nutrients in rivers and the atmosphere in many regions of the world is equal to or greater than the natural flux. Globally, river runoff and atmospheric deposition deliver approximately equal concentrations of nutrients to marine environments. The annual per capita anthropogenic load of nitrogen in U.S. rivers alone is about 8–10 kg.

Eutrophication resulting from enrichment of nutrients is one of the major stresses imposed on coastal ecosystems, often accounting for severe deterioration of water quality and significant impacts on biotic communities. These excess nutrients (particularly nitrogen and phosphorus) — which derive mainly from municipal and

TABLE 8
Status of Pollutants from Land-Based Sources

Substance	Status of Science and Management	Known/Suspected Targets/Effects
Nutrients	Science limited	Eutrophication
	Conservative management possible	Harmful algal blooms
Sewage	Science adequate	Human health
	Management deficient	Pathogens
		Eutrophication
Oil/hydrocarbons	Science adequate	Animal health
	Management deficient	Decreased productivity
		Destruction of amenities
PAHs	Science limited	Human health
	Management deficient	Animal health
		Tainted food sources
Metals	Science adequate	Human health
	Management deficient	Animal health
Synthetic organics	Science limited	Human health
	Conservative management possible	Animal health
Radionuclides	Science adequate	Human health
	Management deficient	Animal health
Thermal	Science adequate	Animal health
	Conservative management possible	
Sediment	Science limited	Decreased productivity
	Conservative management possible	Destruction of amenities (habitats/organisms)
Litter	Science adequate	Animal life
	Management deficient	Destruction of amenities

Modified from UNCED, Preparatory Committee for the United Nations Conference on Environment and Development, Options for Agenda 21: protection of oceans, all kinds of seas including enclosed and semi-enclosed seas, coastal areas and the protection, rational use, and development of their living resources, UNCED, UNEP, Geneva, 1991.

industrial wastewaters, agricultural and urban runoff, dredging and dredged-spoil disposal operations, as well as boats and marinas — stimulate primary production (e.g., algal blooms), culminating in high oxygen demands in bottom waters. Toxic phytoplankton blooms termed "red tides" periodically develop and cause mass mortality of invertebrates and fish. Nitrogen is the primary nutrient responsible for the eutrophication of temperate estuaries and coastal seas, while phosphorus appears to be critically important in many tropical estuarine and coastal systems.

The accumulation of plant and animal remains on the seafloor and their subsequent bacterial decay deplete bottom waters of oxygen, often to seriously low levels (i.e., anoxic conditions with 0 mg/l dissolved oxygen and hypoxic conditions with <2.0 mg/l dissolved oxygen).[36] The highest concentrations of organic carbon occur in waters receiving sewage wastes, where both dissolved and particulate organic carbon levels exceed 100 mg/l. Dissolved organic carbon concentrations in estuarine and coastal marine waters typically range from about 1 to 5 mg/l, and particulate organic carbon levels generally range from approximately 0.5 to 5.0 mg/l in estuaries and 0.1 to 1.0 mg/l in coastal seawater. Organic matter released in sewage wastes can exacerbate anoxia or hypoxia by raising the biochemical oxygen demand (i.e., the oxygen consumed during the microbial decomposition of the waste) and the chemical oxygen demand (i.e., the oxygen consumed through the oxidation of ammonium and other inorganic reduced compounds).[37] Bottom-water hypoxia may persist for months in some estuarine and shallow coastal marine systems where water column stratification inhibits vertical mixing (e.g., parts of Chesapeake Bay, Pamlico River estuary, Potomac River estuary, and Long Island Sound).[38] The intensity of water column stratification is a principal factor controlling the severity of dissolved oxygen depletion over broad areas.

Oxygen-depleted bottom waters impact benthic communities by producing acute changes in the distribution, abundance, and diversity of species. Nuisance organisms and opportunists may become dominant in an area, supplanting more desirable forms, such as commercially and recreationally important finfish and shellfish (e.g., flounder, clams, and lobsters). Eutrophic or hypereutrophic systems characterized by little or no dissolved oxygen in bottom waters, excessive nuisance conditions, and undesirable biological communities, are notable examples.

Eutrophication is an urgent problem that threatens coastal waters of the United States and many other countries. In response, the National Oceanic and Atmospheric Administration (NOAA) is assessing conditions and trends in eutrophic symptoms within U.S. estuaries, compiling data in a National Estuarine Eutrophication Survey for 129 systems. Some alarming observations have been made. A growing number of estuarine and coastal marine systems are showing signs of eutrophication problems — increases in productivity, shifts in community dominants, development of anoxia in bottom waters, and losses of submerged aquatic vegetation.[39] Additional monitoring programs are needed, and a new generation of nutrient management strategies must be formulated to develop effective and intelligent solutions to the eutrophication dilemma.

2. Organic Loading

a. Sewage Discharges

Many eutrophication and organic loading problems in coastal regions throughout the world are linked to the discharge of sewage effluent and dumping of sewage sludge, respectively. Sewage may be discharged untreated or only partly treated near coasts. Raw sewage is mainly water (~99%), with the remainder consisting of solid waste (e.g., sediment, plastics, and floatables), suspended and dissolved organic matter that exerts a biochemical oxygen demand (BOD), oil and grease, nutrients, and pathogens (i.e., bacteria, viruses, protozoa, and helminths) (Figure 2). Since surface water runoff and industrial discharges are combined with domestic sewage

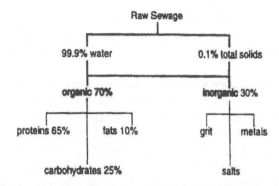

FIGURE 2 Composition of typical raw sewage. (From Lester, J. N., in *Pollution: Causes, Effects, and Control*, 2nd ed., Harrison, R. M., Ed., Royal Society of Chemistry, Cambridge, 1990, pp. 33–62. With permission.)

in many treatment systems, a wide variety of contaminants can be present (e.g., heavy metals and toxic organic compounds), largely sorbed to particulate matter.

The goal of municipal treatment facilities which process domestic wastewaters is to reduce the content of the suspended solids, oxygen-demanding materials, dissolved inorganic compounds (particularly nitrogen and phosphorus compounds), and pathogens. Complete treatment of municipal wastewaters involves a three-stage process (Figure 3).

Stage 1, primary treatment, entails grit removal, screening, grinding, flocculation, and sedimentation, with the end products containing large concentrations of dissolved material and sludge. The liquid waste may then be processed in secondary treatment, and the sludge dumped at designated disposal sites.

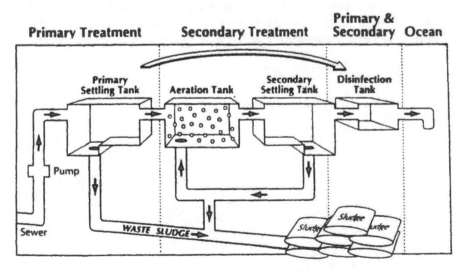

FIGURE 3 Municipal wastewater treatment processes. (From Duedall, I. W., *Oceanus*, 33, 29–38, 1990. With permission.)

Stage 2, secondary treatment, focuses on the oxidation of dissolved organic matter and, hence, the removal of BOD. With 40–60% of the suspended solids and 20–40% of the BOD removed during primary treatment by physical methods, secondary treatment reduces the organic matter remaining in the liquid waste via aerobic bacterial attack.

Stage 3, tertiary treatment, removes nutrients from the effluent. Additional steps may be taken to further improve effluent quality. Specifically, advanced wastewater treatment systems remove refractory pollutants and more than 99% of the suspended solids and BOD. Its high cost, however, is prohibitively expensive for many countries.

The discharge of untreated, primary-treated, or secondary-treated sewage to inland and nearshore waters remains a problem for coastal communities worldwide. While the discharge of raw sewage exerts the greatest impact on water quality, owing to the influx of large amounts of pathogens, nutrients, organic matter, and various toxic chemicals, the release of incompletely treated sewage at municipal outfalls also contributes to declining conditions. In rural areas, faulty septic tank systems can be an additional source of sewage waste — especially nutrients — to rivers and estuaries, which may promote eutrophic conditions in receiving waters.[39] Wastewaters derived from unsewered domestic sources, combined sewer overflows, and short outfalls all have the potential to facilitate disease transmission in shallow marine environments.[40]

Sewage-contaminated marine environments pose a significant health threat to humans. More than 100 human enteric pathogens (i.e., bacteria, viruses, parasites) have been found in treated municipal wastewater and urban stormwater runoff.[37] Although many of these microbial pathogens only survive for hours to several days in seawater, they remain viable for longer periods in fish and shellfish. People who either swim in sewage-contaminated waters or consume contaminated fish and shellfish risk developing serious illnesses. Table 9 lists a diverse number of bacterial, viral, protozoan, and helminthic pathogens capable of initiating waterborne infections. Exposure to these pathogens result in serious diseases, such as dysentary, hepatitis, typhoid, gastroenteritis, and a wide variety of parasitic infections.

TABLE 9
Human Pathogenic Microorganisms Potentially Waterborne

Pathogen	Clinical Syndrome
Bacteria	
Aeromonas hydrophila	Acute diarrhea
Campylobacter spp.	Acute enteritis
Enterotox. Clostridium perfringens	Diarrhea
Enterotox. E. coli	Diarrhea
Francisella tularensis	Mild or influenzal, febrile, typhoidal illness
Klebsiella pneumoniae	Enteritis (occas.)
Plesiomonas shigelloides	Diarrhea
Pseudomonas aeruginosa	Gastroenteritis (occas.)
Salmonella typhi	Typhoid fever
Other salmonellae	Gastroenteritis

TABLE 9 (CONTINUED)
Human Pathogenic Microorganisms Potentially Waterborne

Pathogen	Clinical Syndrome
Shigella spp.	Shigellosis ("bacillary dysentery")
Vibrio cholerae	Cholera dysentery (01 serovars) or cholera-like infection (non-01)
V. fluvialis	Gastroenteritis
Lactose-positive Vibrio	Pneumonia and septicemia
V. parahaemolyticus	Gastroenteritis
Yersinia enterocolitica	Enteritis, ileitis
Cyanobacteria	
Cylindrospermopsis spp.	Hepatoenteritis
Viruses	
Enteroviruses	Aseptic meningitis, respiratory infection, rash, fever
Poliovirus	Paralysis, encephalitis
Coxsackie virus A	Herpangina, paralysis
Coxsackie virus B	Myocarditis, pericarditis, encephalitis, epidemic pleurodynia, transient paralysis
Echovirus	Meningitis, enteritis
Types 68–71	Encephalitis, acute hemorrhagic conjunctivitis
Hepatitis A	Infectious hepatitis type A
Hepatitis non-A, non-B	Hepatitis type non-A, non-B
Influenza A	Influenza
Norwalk and other parovirus-like agents	Epidemic, acute nonbacterial gastroenteritis
Rotavirus	Nonbacterial, endemic, infantile gastroenteritis: epidemic vomiting and diarrhea
Protozoa	
Balantidium coli	Balantidiasis (balantidial dysentery)
Cryptosporidium	Cryptosporidiosis
Entamoeba histolytica	Amoebiasis (amoebic dysentery)
Giardia lamblia	Giardiasis — mild, acute, or chronic diarrhea
Helminths	
Ascaris	Ascariasis (roundworm infection)
Ancylostoma	Hookworm infection
Clonorchis	Clonorchiasis (Chinese liver fluke infection)
Diphyllobothrium	Diphyllobothriasis (broadfish tapeworm infection)
Dracunculus mediensis	Dracontiasis (Guinea worm infection)
Fasciola	Fascioliasis (sheep liver fluke infection)
Fasciolopsis	Fasciolopsiasis (giant intestinal fluke infection)
Paragonimus	Paragonimiasis (Oriental lung fluke infection)
Spirometra mansoni	Sparganosis (plerocercoid tapeworm larvae infection)
Taenia	Taeniasis (tapeworm infection)
Trichostrongylus	Trichostrongyliasis
Trichuris	Trichuriasis (whipworm infection)

From McNeill, A. R., in *Pollution in Tropical Aquatic Systems*, Connell, D. W. and Hawker, D. W., Eds., CRC Press, Boca Raton, FL, 1992, 193. With permission.

b. Sewage Sludge

One of the principal products of municipal wastewater treatment is sewage sludge, a liquid waste containing up to 10% by weight of solid particles. The solid fraction consists of a heterogeneous mixture of particles (e.g., organic detritus, food wastes, inorganic matter, and microbial organisms) ranging from colloidal to macroscopic (>1 cm) in size. In addition, sewage sludge contains trace elements, synthetic organic compounds, oil, grease, and other substances. A thin film typically floats on its surface. With a bulk density of about 1.01 g/cm^3, sewage sludge is slightly less dense than seawater (1.025 g/cm^3) and substantially less dense than marine sediment (1.2–1.6 g/cm^3). However, when dumped in the sea, sewage sludge fractionates in the water column, with the less dense components (e.g., dissolved constituents) mixing rapidly in surface waters and remaining above the pyncnocline and the dense solid particles settling through the water column to the seafloor. Water column stratification, therefore, is an important barrier to the vertical flux of some sewage sludge constituents in marine waters.

Sewage sludge dumping at sea has been practiced mainly by industrialized nations. Controlled sewage sludge disposal in the ocean offers several advantages: (1) particulate organic waste mixes with seafloor sediments and is degraded by natural processes; (2) toxic substances are diluted to low levels and gradually buried under seafloor sediments; and (3) sludge dumping at sea is less expensive and generally safer than on land. The trend is for sewage sludge dumping to be phased out by industrialized nations. For example, countries of the European community expect to cease the disposal of sewage sludge via dumping and outfall discharges by the end of 1998.[41] Currently, England leads all other European countries in the amount of sewage sludge disposal at sea (~210,000 mt dry solids/yr). Of the total volume of sewage sludge generated in England, sea disposal accounts for about 25%, and land disposal and incineration approximately 68% and 7%, respectively.[42]

In the United States, the ocean dumping of sewage sludge commenced in 1924 when New York City began dumping sludge at a single site, the 12 Mile Dumpsite (12-MDS), in the New York Bight Apex located about 22 km southeast of the Hudson-Raritan estuary mouth (Figure 4). Disposal of sewage sludge at this 20-m deep site ceased at the end of 1987. The maximum volume of sewage sludge dumped at 12-MDS was 8.3 million mt in 1983, which exceeded the volumes dumped at all other dumpsites in the world. Other continental-shelf sites off New Jersey that have been used as sewage sludge dumping grounds in the past include the New York Bight alternate sludge site and the Philadelphia sludge site, both located at water depths of approximately 40 m.

Between 1924 and 1987, virtually all sludge dumping in the United States occurred at inner- or mid-continental-shelf sites, which are now closed. Between 1986 and 1992, substantial volumes of sludge were dumped at DWD–106 (Figure 5). As noted previously, more than 30 million mt (wet) of sewage sludge have been dumped at DWD–106, but none since 1992. Alternatives to sewage sludge dumping at sea have been implemented by most U.S. cities. Similar alternative and practicable means of disposal have not been followed by many other countries outside the United States or European community.[40] Thus, it is likely that sewage sludge

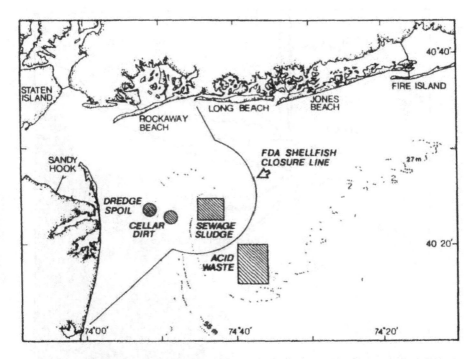

FIGURE 4   Map showing the location of the 12 Mile Dumpsite (Sewage Sludge) in the New York Bight Apex. (From Sindermann, C. J., *Ocean Pollution: Effects on Living Resources and Humans*, CRC Press, Boca Raton, FL, 1995, 275 pp. With permission.)

dumping at sea will continue to be a problem in the foreseeable future in some coastal regions.

The environmental effects of sewage sludge dumping at sea depends on the volume of sludge dumped, the types of pollutants contained in the sludge, and the hydrodynamics at the disposal sites. The U.S. EPA has identified more than 100 chemical compounds of concern in sewage sludge; many of them are halogenated aliphatic and aromatic hydrocarbons, organochlorine pesticides, polychlorinated biphenyls, and phthalate esters. A substantial number of these compounds are not readily amenable to biological or chemical degradation.[43]

c. Sewage Impacted Systems

There are several well-known sewage disposal sites in coastal waters of the United States and England. Aside from the dumping of sewage sludge at the 12-MDS in the New York Bight Apex, the New York Bight alternate sludge site, and the Philadelphia sludge site as mentioned above, Boston Harbor has been used for disposal of sewage wastes for hundreds of years, resulting in a highly degraded benthic environment.[44] On the west coast of the United States, Southern California shelf waters receive billions of liters of municipal wastewaters each day from a series of submarine outfalls. Most of these wastewaters (~90%) derive from the Hyperion Treatment Plant (HTP) of the City of Los Angeles, the Joint Water Pollution Control Plant (JWPCP) of the County Sanitation Districts of Los Angeles County, Wastewater

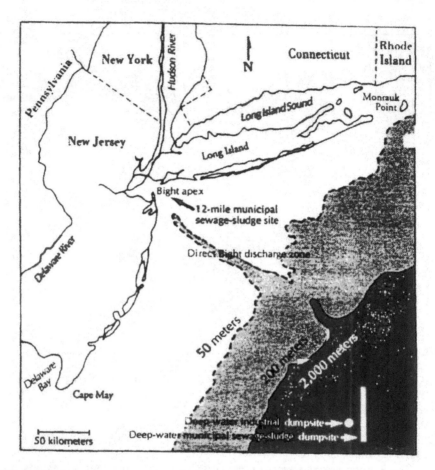

FIGURE 5 Map showing the location of deepwater dumpsites off New Jersey. (From Stegeman, J. J., *Oceanus*, 33, 54–62, 1990. With permission.)

Treatment Plants 1 and 2 of the County Sanitation Districts of Orange County (CSDOC), and the Point Loma Wastewater Treatment Plant (PLWTP) of the City of San Diego. The HTP, JWPCP, and CSDOC release effluent at a depth of ~60 m, while the PLWTP discharges at a depth of 93 m. Over the years, an increase in organic matter and contaminants was recorded in proximity to the outfalls. Adverse effects were observed on benthic communities and fish assemblages, with the greatest impacts observed on benthic infauna and demersal fishes.[45] During the past 25 years, however, effluent quality of the largest municipal treatment plants has markedly improved; although marine communities at waste discharge sites have been revitalized as a result of upgraded conditions, various biota still carry body burdens of some contaminants, notably PCBs and DDT.

At a site about 2 km from Garroch Head, Isle of Bute, Firth of Clyde (Scotland), sewage sludge was dumped virtually continuously from 1904 to 1974. Studies of the site in the early 1970s revealed classical effects of organic enrichment on the benthic fauna (i.e., decreases in species diversity, increases in population abundance and biomass). Later surveys in 1985, however, delineated a marked recovery of the benthic community 11 years after cessation of sewage sludge dumping.[46] The Tees and Tyne estuaries, two heavily degraded systems in England, have also been subject to considerable sewage disposal during the 20th Century. Brown and red seaweeds have decreased in diversity since the early 1930s in the Tees estuary, and during the past 20 years in the Tyne estuary, due in part to this waste disposal. Dramatic changes in the benthic macrofaunal community have also been recorded in areas affected by the sewage disposal. However, cleanup measures undertaken during the past two decades indicate some signs of recovery of the benthos.[47]

d. Other Organic Enrichment Sources

Two other important sources of organic enrichment in some estuarine and coastal marine waters that must be considered are wastes from waterfowl, wildlife, and aquaculture operations. Nutrient enrichment from animal excretions (ammonia and urea) and fecal matter can cause nuisance phytoplankton blooms. Unused foods of fish net pens and cages at aquaculture sites, which are good sources of nitrogen, also increase system productivity. Possible acute effects of open-system aquaculture include the development of hypoxia, anoxia, and hydrogen sulfide, creating conditions inimical to estuarine and marine communities.

e. Effects on Benthic Communities

The impacts of organic enrichment due to sewage disposal and waste input from wildlife and aquaculture operations are occasionally very pronounced in marine benthic communities. Results of numerous investigations have provided consistent correlations between organic enrichment and gross measures of community structure (i.e., species richness, dominance, diversity, and total animal abundance). Structural changes in the benthos subsequent to sewage disposal are the basis of empirical models that describe the impacts of organic enrichment on marine benthic fauna. For example, macrobenthic communities subjected to increased organic loading, either spatially or temporally, will often exhibit: (1) a decrease in species richness and an increase in total number of individuals attributable to high densities of a few opportunistic species; (2) a general reduction in biomass, although there may be an increase in biomass corresponding to a dense assemblage of opportunists; (3) a decrease in body size of the average species or individual; (4) shifts in the relative dominance of trophic guilds; and (5) a shallowing of that portion of the sediment column occupied by infauna.[48] Spatial and temporal gradients commonly develop at outfalls and dumpsites along which infaunal biomass, density, and species diversity change until unaffected conditions are observed.[49] It is conceivable that toxic chemicals associated with sewage wastes also may play a significant role in generating the benthic impacts.

3. Oil

a. General Impacts

The noxious effects of oil pollution on marine communities are well chronicled, as are the damaging impacts on marine and coastal habitats.[2,16,28,38,50] Crude oil consists of thousands of chemical compounds, many of which are toxic to marine life. The aliphatic and polycyclic aromatic hydrocarbon fractions of dissolved oil, for example, can decimate marine populations over extensive areas due to their extreme toxicity, rapid uptake by biota, and persistence in the environment. These fractions tend to bioconcentrate in marine organisms because of their high lipid solubility and are of major concern in commercially and recreationally important finfish and shellfish consumed by man. Petroleum hydrocarbons sorb readily to particulate matter and accumulate in bottom sediments, where they may pose a chronic threat to the benthos. Water soluble compounds, such as benzene, toluene, and xylene, frequently kill meroplankton, ichthyoplankton, or other life stages of organisms exposed to them in the water column, even at concentrations as low as 5 mg/l.

In addition to the toxic action of petroleum hydrocarbons and other chemical components, polluting oil may physically smother marine organisms. Such direct impacts are manifested most conspicuously in intertidal and shallow subtidal habitats that receive heavy coatings of oil washed ashore after major spills. Estuaries are especially sensitive to oil pollution. Benthic communities in these systems usually suffer total or near total decimation immediately after major oil spills; although recovery may commence weeks after these episodic events, communities generally do not return to pre-spill levels for years. Occasionally, the effects of a heavy oil spill on marine communities and habitats are detected more than a decade after the event.[28]

Adverse effects of oil on estuarine and marine biota also occur indirectly via the degradation of critical habitat areas and the decomposition of the oil. Salt marshes, mangroves, and seagrasses are highly sensitive habitats to oil pollution. While the physical presence of polluting oil in these habitats creates inhospitable conditions for successful settlement of many benthic marine populations, decreased pH and dissolved oxygen levels resulting from microbial decomposition and other processes increase mortality and lower food availability for the survivors, thereby amplifying the effects of physical deterioration of the environment.

b. Crude Oil Composition

Crude oil consists of a complex mixture of thousands of gaseous, liquid, and solid organic compounds. Hydrocarbons, which contain molecules ranging in molecular weight from 16 to 20,000, comprise more than 75% by weight of most crude and refined oils. They are divided into four major classes: (1) straight-chain alkanes (n-alkanes or n-paraffins), (2) branched alkanes (isoparaffins), (3) cycloalkanes (cyclo-paraffins), and (4) aromatics (Figure 6). Alkenes also occur in crude oil, but are rare.[28,50] The toxicity of the various classes of petroleum hydrocarbons tends to increase along the series from the alkanes, cycloalkanes, and alkenes to the aromatics. Nonhydrocarbon compounds containing oxygen, nitrogen, sulfur, and metals constitute the remaining 25% of most crude oils, although they may comprise more than 50% by weight of heavy crude oils.

Apart from their chemical composition, crude oils may be classified according to their physical properties, which vary considerably. For example, differences in specific gravity determine whether an oil is classified as light, medium, or heavy. The relative concentrations of certain compounds in various categories have been effectively used to classify oil (e.g., high or low sulfur, naphthenic, and paraffinic).[50]

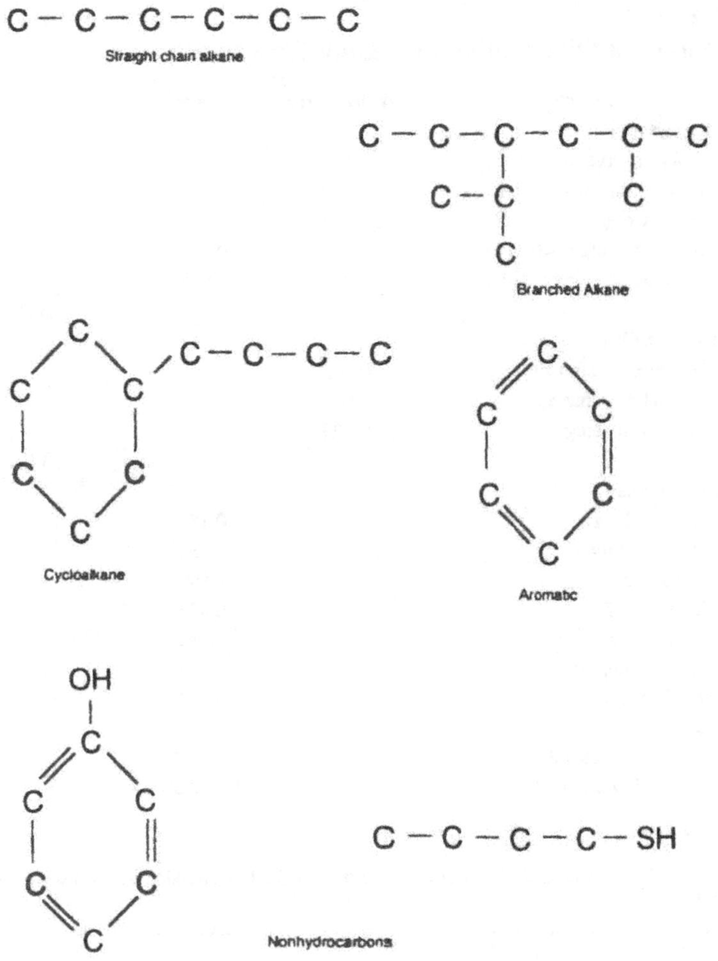

C — C — C — C — C — C

Straight chain alkane

C — C — C — C — C — C
           | |
      C — C   C
           |
           C

Branched Alkane

Cycloalkane

Aromatic

OH

C — C — C — C — SH

Nonhydrocarbons

FIGURE 6   Types of molecular structures found in petroleum. Hydrogen atoms bonded to carbon atoms are omitted. (From Albers, P. H., in *Handbook of Ecotoxicology*, Hoffman, D. J., Rattner, B. A., Burton, G. A., Jr., and Cairns, J., Jr., Eds., Lewis Publishers, Boca Raton, FL, 1995, pp. 330–355. With permission.)

c. Sources of Oil Pollution

Global crude oil production amounts to about 3 billion mt/yr, with approximately half of this total transported by sea. The quantity of oil entering marine waters each year from all sources (excluding biosynthesis) is estimated to be about 2.145 million mt (Table 10). In recent years, there has been a substantial decline in marine oil pollution. The International Maritime Organization reported a 60% reduction in oil pollution at sea between 1981 and 1989, attributable primarily to improvements in tanker operations and fewer accidental spills.[51] Further reductions will depend on future upgrades of shore-based installations (e.g., reception facilities) and decreases in the influx of oil from land-based sources.

TABLE 10
Sources of Oil Pollution in Marine Environments[a]

Source	Oil Industry	Other	Total
Transportation			
Tanker operations	0.143		
Tanker accidents	0.110		
Dry docking	0.004		
Other shipping operations		0.229	
Othering shipping accidents		0.018	
			0.504
Fixed installations			
Offshore oil production	0.045		
Coastal oil refineries	0.091		
Thermal loading	0.027		
			0.163
Other sources			
Industrial waste		0.181	
Municipal waste		0.635	
Urban runoff		0.109	
River runoff		0.036	
Atmospheric fallout		0.272	
Ocean dumping		0.018	1.251
Natural inputs		0.227	0.227
Biosynthesis			
Marine phytoplankton		23,582	
Atmospheric fallout		91–3,630	

[a] Values in million metric tons per year.

Modified from Clark, R. B., *Marine Pollution*, 3rd ed., Clarendon Press, Oxford, 1992, 172 pp.

Oil enters marine waters from both natural and anthropogenic sources including cold-water seeps, atmospheric deposition, municipal and industrial wastewaters, urban and river runoff, ocean dumping of wastes, major tanker spills, losses during marine transportation, and leakages from fixed installations (e.g., coastal refineries, offshore production facilities, and marine terminals). Most (58.3%) of the world input of oil to the sea derives from municipal and industrial wastewaters, urban and river runoff, ocean dumping, and atmospheric fallout. An additional 23.5% is ascribable to marine transportation activities, notably routine tanker operations, tanker accidents, non-tanker accidents, deballasting, and dry docking. Fixed installations (i.e., coastal refineries, offshore production sites, and marine terminals) and natural inputs account for 7.6 and 10.6%, respectively, of all oil entering the sea.

Based on data compiled in Table 10, marine input of petroleum hydrocarbons from anthropogenic sources (1.918 million mt/yr) far exceeds that from natural sources (0.27 million mt/yr) excluding biosynthesis. In addition, oil influx from the users of petroleum products is far greater than that from the oil extraction and transport industries. While the public generally perceives tanker spills as the principal source of oil in the sea, owing in no small part to media coverage, such episodic events actually deliver a relatively small volume of oil to the sea (0.110 ml/yr). In fact, many accidents involving oil tankers yield little or no spillages of oil. It is chronic oil pollution that causes the most insidious damage to estuarine and marine environments. In U.S. coastal waters, for example, such pollution originates primarily from municipal and industrial wastewater discharges, leakages of oil refineries and other shore-based installations, and nonpoint sources (e.g., urban and river runoff).

d. Fate of Polluting Oil

Various physical-chemical processes act within hours of an oil spill on the sea surface to alter its composition and toxicity, most importantly evaporation, dissolution, photochemical oxidation, advection and dispersion, emulsification, and sedimentation (Figure 7). During a spill event, oil spreads across the sea surface as a slick, varying in thickness from micrometers to a centimeter or more. The slick's behavior is a function of the composition of the oil and the prevailing abiotic factors in the area (i.e., water temperature, wind, wave action, tides, and currents) (Figure 8). Currents transport, or advect, the oil from a spill site. Oil slicks travel downwind at 2.5–4.0% of the wind velocity, with light oils spreading faster than heavy oils. The spreading rate is more rapid at higher than lower temperatures and depends on the volume and density of the oil. As wind and wind waves develop, the slick breaks up into distinct patches of oil that drift slowly apart under the influence of horizontal eddy diffusion. Subsurface advection mixes the oil in subsurface waters to depths of about 10 m; vertical diffusion plays a less integral role in subsurface mixing of the oil.

During the first 24–48 hr of an oil spill, evaporation and dissolution produce the greatest change in composition of the oil by causing the rapid loss of the lighter, more toxic and volatile components. The low-molecular-weight aromatics (e.g., toluene and xylene) present at peak concentrations of 10–100 μg/l are lost relatively quickly. The water-soluble components gradually dissolve in seawater and become effectively isolated from the sea-surface slick and incorporated into subsurface waters. Dissolved constituents attain highest concentrations below the oil slick within the first 24 hr of a spill.

Photochemical oxidation acts on oil at or near the sea surface to convert high-molecular-weight aromatic hydrocarbons to polar oxidized compounds. Some of the oxidation products formed by this process include alcohols, ethers, dialkyl peroxides, and aliphatic and aromatic acids.[52] The photochemical oxidation process, which operates most efficiently on thin oil slicks (~ 10^8 cm thick),[53] takes on greater importance after the volatile oil fractions have evaporated.

Agitation provided by waves and currents mixes the oil and seawater leading to turbulence and dispersion of oil slicks. In very rough seas, aerosol formation ensues at the sea surface. Intense agitation and mixing cause the formation of oil-in-water emulsions or water-in-oil emulsions. As the heavier oil fractions mix with seawater, viscosity increases and a water-in-oil emulsion forms. Viscid pancake-like masses called "chocolate mousse," which develop from 50–80% water-in-oil emulsions, may

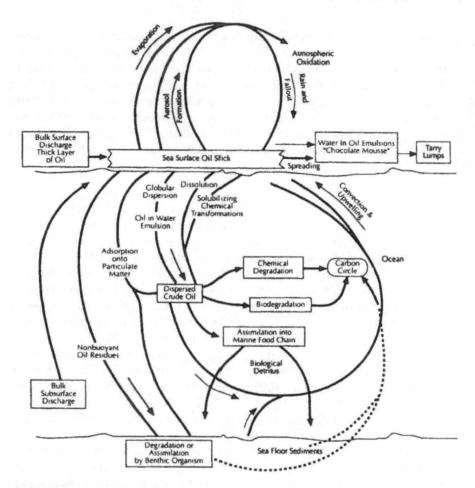

FIGURE 7 Fate and weathering of polluting oil in estuarine and marine waters showing various abiotic and biotic processes that act to alter the oil. (From Burwood, R. and Speers, G. C., *Est. Coastal Mar. Sci.*, 2, 117, 1974. With permission.)

persist for months at sea. The heaviest residues of crude oil — tar balls — measuring 1 mm to 25 cm in diameter are often advected, like chocolate mousse, to remote impact sites. These products degrade very slowly, being extremely stable and persistent in marine waters and, consequently, a major problem for shorelines subsequent to stranding.

Evaporation, dissolution, photooxidation, and the formation of viscous water-in-oil emulsions increase the density of oil on the sea surface. The change in oil density through time is facilitated by the incorporation of particulate matter and the agglomeration of oil particulate mixtures. The sorption of hydrocarbons to particulate matter creates a high specific gravity mixture more than twice that of seawater alone (1.025 g/cm³), which promotes its sedimentation.

Oil is subject to biodegradation by many species of bacteria, fungi, and yeast. Bacteria are the most important biological agents in the breakdown process.

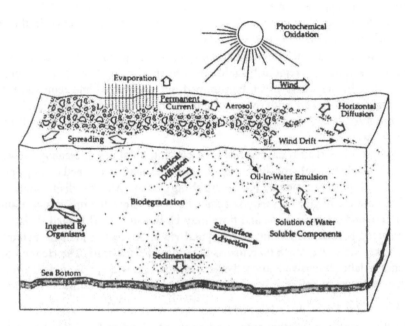

FIGURE 8 Effects of wind and other factors on the movement of polluting oil at sea. (From Bishop, J. M., *Applied Oceanography*, John Wiley & Sons, New York, 1984, 252 pp. With permission.)

Microbes collectively degrade as much as 40–80% of a crude oil spill, with structurally simple hydrocarbon and nonhydrocarbon compounds degrading most readily, especially at higher temperatures. Alkanes and cycloalkanes, for example, breakdown rapidly, whereas the high-molecular-weight species (e.g., polycyclic aromatic hydrocarbons) degrade slowly. The asphaltenes, the highest-molecular-weight components of crude oils, persist with little alteration to form tarry residues that ultimately settle to the seafloor or become stranded along coastlines. Biodegradation rates of oil are influenced most greatly by water temperature, nutrient availability, dissolved oxygen levels, and salinity. Microbial breakdown of oil is active in coastal marine environments where nutrient supply, dissolved oxygen levels, and water temperatures are not limiting. Biodegradation rates of oil decline significantly in high latitudes due to reduced temperatures.

Four main approaches are followed in oil spill cleanup of marine environments: (1) chemical applications; (2) mechanical cleanup; (3) shoreline cleanup; and (4) no spill control. The most appropriate treatment option hinges on conditions existing at the cleanup site. Oil treatment procedures on sandy beaches usually differ from those on rocky shorelines or in salt marshes. In the case of oil spills at sea, the best approach may be no treatment of the oil at all, especially when prevailing winds are offshore.

Chemical treatment of an oil slick usually involves the spraying of dispersants from ships or aircraft to accelerate the emulsification of the oil. Solvents and agents that reduce surface tension are also utilized to remove oil slicks from the surface of pools and enclosed inshore areas. However, dispersants are not effective on heavy or weathered oils, and some of them may be toxic to marine life.[16] Despite these disadvantages, chemical treatment of oil slicks has been instrumental in the preservation of many threatened coastal habitats.

Devices that physically contain floating oil are of great value in cleanup efforts in harbors and inshore regions.[54] Included here are floating booms, which corral oil in a confined zone where it can be pumped out or diverted and removed by other means, and slick lickers with rotating belts of absorbent, which extract oil and pass it onto barges. These devices are most effective in dealing with small oil spills in sheltered waters.[2,16]

When oil is stranded on beaches or rocky intertidal areas, cleanup crews often enact physical removal procedures. Rocky surfaces may be cleaned by high-pressure hoses, high-pressure steam, hand-scrubbing techniques, or dispersants. These methods are ineffective on sandy beaches, however, since the oil simply drains into the beach, and endemic fauna and flora may be threatened. The oiled surface layers of impacted beaches commonly are stripped via bulldozers or manual methods.[16] In some cases, however, much oil remains behind in the substrate. The cleanup of oiled beaches is labor intensive and costly. Hence, the best strategy is to deal with an oil slick expeditiously before it becomes stranded along the coastline.

e. Effects on Biotic Communities

Dramatic changes in benthic communities occur along rocky shorelines, estuaries, and shallow coastal marine environments exposed to polluting oil. Low-energy habitats likely to trap oil, such as salt marshes, mangroves, and seagrasses, are generally teeming with life. Fine-grained sediments in these habitats sorb hydrocarbon compounds and other components of the oil, and may re-release the chemicals for years (especially during storms and other major episodic events), which arrests the development of benthic communities. Once the oil permeates through the bottom sediments, it creates long-term hazardous conditions that threaten the overall stability and health of the benthos. Low oxygen concentrations in deeper sediment layers hinder bacterial degradation of the oil, further delaying recovery of the impacted habitats. Subsequent to severe oiling, marine substrates may remain inhospitable to benthic organisms for a decade or more.

The rate of recovery of an intertidal or shallow subtidal marine habitat is controlled by the number of oiling events and the depth of oil penetration into the substrate. Multiple oilings are much more destructive than single oiling events. When the oil penetrates below the level of the roots and rhizomes of vascular plants, substantial degradation of the biotope can ensue, with recovery being delayed considerably. Shallow-rooted annuals with limited food reserves are much more susceptible to oil pollution than perennials with extensive root and rhizome systems. They are frequently killed by short-term exposure to a single oiling event. Table 11 compares the recovery rates of various aquatic ecosystems to catastrophic oil spills.

The magnitude of oil impacts on estuarine and marine organisms is contingent upon many factors, the most notable being (1) the amount of the oil; (2) composition of the oil; (3) form of the oil (i.e., fresh, weathered, or emulsified); (4) occurrence of the oil (i.e., in solution, suspension, dispersion, or adsorbed onto particulate matter); (5) duration of exposure; (6) involvement of neuston, plankton, nekton, or

TABLE 11
Recovery of Various Aquatic Ecosystems Subjected to Catastrophic Oil Spills

Ecosystem Type	Time Between Major Stresses (in Years)				
	3	5	10	20	100
River					
Headwaters	50–70% of species recovered	Recovery less than 95% of species	Recovered	Recovered	Recovered
Middle reach	50–75% of species recovered	State of constant recovery less than 95% of species	Recovered	Recovered	Recovered
Slow	50–75% of species recovered	Recovery less than 95% of species	Recovered	Recovered	Recovered
Lakes	Most species would not be recovered	Biological integrity not maintained	State of constant recovery	Final state of recovery	Recovered
Estuaries	Principally clams and mollusks are not recovered	Clam and mollusk populations still reduced	Recovered	Recovered	Recovered
Marine					
Beaches	Beaches are in state of final repopulation	Repopulated and probably recovered	Recovered	Recovered	Recovered
Rock shore	Colony communities not recovered	Colony communities generally recovered	Recovered	Recovered	Recovered
Tidal flat	Principally bivalves not recovered	Bivalves still reduced	Recovered	Recovered	Recovered
Marshes	Annual plants and short life span	Long-lived plants not reestablished; most organisms recovered	Final stages of recovery	Recovered for very large systems	Recovery depends upon size of area affected
Open water	Very small area repopulated	Long life-span organisms in recovery	Most species present	Recovered except for very large systems	Recovery depends upon size of area affected

From Cairns, J., Jr., Ed., *Rehabilitating Damaged Ecosystems*, Vol. 2, CRC Press, Boca Raton, FL, 1988. With permission.

benthos in the spill or release; (7) juvenile or adult forms involved; (8) previous history of pollutant exposure of the biota; (9) season of the year; (10) natural environmental stresses associated with fluctuations in temperature, salinity, and other variables; (11) type of habitat affected; and (12) cleanup operations (e.g., physical methods of oil recovery and the use of chemical dispersants).[2,16,28,54,55] Lethal and sublethal effects of oil contamination are manifested in both acute and chronic responses of biota. Organisms which are trapped, smothered, and suffocated by an oil spill, for example, suffer essentially immediate lethal effects. The oil typically interferes with cellular and subcellular processes in these individuals soon after the spill. Those individuals surviving the physical impact of the oil may lose normal physiological or behavioral function if coated, thus predisposing them to greater long-term risk of death. Sublethal effects, such as the impairment of organisms to obtain food or to escape from predators after being coated by the oil, likewise increase mortality of individuals within days or weeks of a spill. Sublethal doses of toxins to eggs and juveniles are responsible for even greater long-term damage to populations. Sublethality can be as devastating as lethal effects to a community by adversely affecting reproduction, growth, distribution, and behavior of many different populations resulting in gradual shifts in species composition, abundance, and diversity that can persist for many years (Table 12).

TABLE 12
Sublethal Effects of Petroleum Exposure on Marine Organisms in the Laboratory[a]

Effect	Species	Exposure Regime and Effective Concentration[b]
Reduced avoidance response	Spotted seatrout larvae (*Cynoscion nebulosus*)	No. 2 fuel WSF for 48 h at 0.01–1.0 $\mu g\ g^{-1}$ (nominal)
50% reduction in defense response to predator	Sea urchin (*Strongylocentrotus droebachiensis*)	Prudhoe Bay Crude WSF for 15 min at 50 ng g^{-1} (measured)
Oil detection, indicated by antennular flicking rate	Blue crabs (*Callinectes sapidus*)	Prudhoe Bay Crude WSF for 2 min at $2 \times 10^{-6}\ \mu g\ g^{-1}$ (median threshold detection limit) Naphthalene for 3 min at $10^{-7}\ \mu g\ g^{-1}$ (median threshold detection limit)
Oil detection, indicated by antennular flicking rate	Dungeness crabs (*Cancer magister*)	Prudhoe Bay Crude WSF for 1 min at 0.1 $\mu g\ g^{-1}$ (median threshold detection limit) Naphthalene for 1 min at 0.01 $\mu g\ g^{-1}$ (median threshold detection limit)
Decreased burying or increased surfacing in sediment	Intertidal clams (*Macoma balthica*)	Prudhoe Bay Crude OWD for 180 d at 0.3 $\mu g\ g^{-1}$ (response threshold)
Resurfacing within 3–5 d	Intertidal clams (*Macoma balthica*)	Prudhoe Bay Crude WSF for 6 d at ~0.3 $\mu g\ g^{-1}$ (median response)
Failure to burrow	Intertidal clams (*Macoma balthica*)	Prudhoe Bay Crude WSF for 6 d at ~2.5 $\mu g\ g^{-1}$

TABLE 12 (CONTINUED)
Sublethal Effects of Petroleum Exposure on Marine Organisms in the Laboratory[a]

Effect	Species	Exposure Regime and Effective Concentration[b]
38% depression of feeding rate	Copepods (*Eurytemora affinis*)	Aromatic heating oil WSF for 24 h at 0.52 µg g⁻¹
Depression of feeding rate	Copepods (*Acartia clausi* and *Acartia tonsa*)	No. 2 fuel oil WAF for 18 h at 250 ng g⁻¹
Depression of feeding rate	Amphipods (*Boeckosimus affinis*)	Prudhoe Bay Crude WSF for 16 wk at ≤0.2 µg g⁻¹
Food detection and finding slowed by one half	Mud snails (*Ilyanassa obsoleta*)	No. 2 fuel oil OWD for 48 h at 0.015 µg g⁻¹
	Mud snails (*Ilyanassa obsoleta*)	Kerosene WSF with a continuous exposure to 4 ng g⁻¹
Feeding rate and growth rate depressed	Rock-crab larvae (*Cancer irroratus*)	No. 2 fuel oil WAF for 27 d at 0.1 µg g⁻¹
Food detection (reduced antennular flicking in response to food extract)	Dungeness crab (*Cancer magister*)	Prudhoe Bay Crude WSF for 24 h at 0.25 µg g⁻¹
Inhibition of attraction to reproductive aggregation	Nudibranchs (*Onchidoris bilamellata*)	Prudhoe Bay Crude WSF for 24 h at 10 ng g⁻¹ Toluene for 24 h at 32 ng g⁻¹
Reproduction (inhibited reproductive aggregation, delayed and reduced egg deposition)	Nudibranchs (*Onchidoris bilamellata*)	Prudhoe Bay Crude WSF for 14 d at 278 ng g–1
Histopathology	Mummichogs (*Fundulus heteroclitus*)	
Lateral line necrosis		Naphthalene for 15 d at 0.02 µg g⁻¹
Ischemia in brain, liver		Naphthalene for 15 d at 0.2 µg g⁻¹
Inclusions in muscle columnar cells	Chinook salmon (*Oncorhynchus tshawytscha*)	Mixed aromatic compounds for 28 d at 5 µg g⁻¹ in food
Delayed overall embryonic development	Pacific herring embryos (*Clupea harengus pallasi*)	Benzene for 24 h at 0.9 ng g⁻¹

[a] The effects documented here were selected based on the degree of realism of exposure conditions relative to chronic discharges and the potential significance of the measured effect to survival and reproduction of the organism. Many other sublethal effects have been documented at higher exposure concentrations.

[b] Exposure concentrations and compositional characteristics depend strongly on the techniques for preparing the oil-water mixtures. This table retains the designations used by the original authors for their preparations. WSF is water-soluble fraction; OWD is oil-water dispersion; and WAF is the water-accommodated fraction.

From Wolfe, D. A., in *Wastes in the Ocean, Vol. 4, Energy Wastes in the Ocean*, Duedall, I. W., Kester, D. R., Park, P. K., and Ketchum, B. H., Eds., John Wiley & Sons, New York, 1985, 45. With permission.

Long-term adverse effects of oil spills on estuarine and marine communities follow major episodes of habitat destruction. In this regard, coastal shoreline and estuarine habitats are assigned the highest priority of biological protection, and offshore areas the lowest priority.[56] The coastal regions of the world are critically important, not only for their fisheries and energy resources, but also for their value in recreation, transportation, and tourism.[57]

Some general observations can be made on the responses of marine biotic groups to oil pollution. For example, planktonic populations, particularly neuston at the sea surface, are highly sensitive to oil spills, but their patchy distribution and short generation times mitigate the overall impacts. Therefore, most adverse effects on the plankton are transient, with populations in an area returning quickly to levels existing prior to a spill.

Benthic communities often incur the greatest damage from oil spills, as evidenced by acute increases in mortality and long-term changes in community structure. Because of the immobility of rooted vegetation and the limited mobility of the infauna, benthic communities are highly prone to oil accumulation in bottom sediments. Thus, the *Florida* oil spill in West Falmouth, Massachusetts, in 1969 caused mass mortality of the benthos and detectable impacts on the community lasting for more than a decade. More recently, the *Exxon Valdez* oil spill in Prince William Sound, Alaska, in 1989 culminated in acute adverse effects on rocky intertidal and shallow subtidal communities that persisted for years.

Fish tend to avoid oil spills because they can swim. Hence, immediate impacts on fish populations may not be apparent. Effects on adults are usually more subtle and long term. Feeding, migration, reproduction, swimming activity, schooling, or burrowing behavior may be altered in response to sublethal concentrations of petroleum hydrocarbons. Early life stages (i.e., ichthyoplankton: eggs and larvae) are much more sensitive and usually experience higher mortality which can translate into long-term reduction in population abundances.

Mammals and seabirds exhibit an array of responses to oil. Sublethal effects chronicled in marine mammals include gastrointestinal and blood disorders, respiratory problems, changes in enzymatic activity in the skin, renal deficiencies, interferences with swimming, eye irritation, and lesions. Oiled plumage of seabirds frequently causes death via drowning or hypothermia. Other casualties arise from the internal uptake of hydrocarbon and nonhydrocarbon compounds subsequent to drinking of contaminated waters, consumption of contaminated food, inhalation of fumes from evaporating oil, and preening of oiled feathers. Depending on the toxicity of the oil, various maladies may develop (e.g., pneumonia, red blood cell damage, immune system depletion, hormonal imbalance, gastrointestinal irritation, growth retardation, and abnormal parental behavior).[2,16,28] Table 13 summarizes the effects of oil pollution on marine communities.

4. Polycyclic Aromatic Hydrocarbons

a. Sources and Concentrations

Among the most ubiquitous organic pollutants in marine environments are polycyclic aromatic hydrocarbons (PAHs), a class of widely distributed organic compounds

TABLE 13
Summary of the Effects of Oil Spills on Marine Communities

Community	Effect	Period of Impact[a]
Plankton		
Phytoplankton biomass and primary production	Increase due to diminished grazing; depression of chlorophyll *a*	Days to weeks, during occurrence of slicks
Zooplankton	Population reduction; contamination	
Fish Eggs	Decreased hatching and survival	
Benthos		
Amphipods, isopods, ostracods	Initial mortality; population decrease	Weeks to years, depending on oil-retentive
Mollusks, especially bivalves	Initial mortality; contamination; histopathology	characteristics of habitat
Opportunistic polychaetes	Population increase	
Overall macrobenthic community	Decreased diversity	
Intertidal and littoral		
Meiofaunal crustaceans, crabs	Initial mortality; population decrease	Weeks to years, depending on oil-retentive
Mollusks	Initial mortality; contamination; histopathology	characteristics of habitat
Opportunistic polychaetes	Population increase	
Overall community	Decreased diversity	
Algae	Decreased biomass; species replacement	
Phanerogams	Initial die-back	
Fish		
Eggs and larvae	Decreased hatching and survival	Weeks to months
Adults	Initial mortality; contamination; histopathology	
Birds		
Adults	Mortality; population decrease	Years

[a] Period of impact depends on the scale and duration of the spill and on the oceanic characteristics of the specific system.

From Wolfe, D. A., in *Wastes in the Ocean*, Vol. 4, *Energy Wastes in the Ocean*, Duedall, I. W., Kester, D. R., Park, P. K., and Ketchum, B. H., Eds., John Wiley & Sons, New York, 1985, 45. With permission.

consisting of hydrogen and carbon arranged in the form of two or more fused benzene rings in linear, angular, or cluster arrangements with substituted groups possibly attached to one or more rings. Compounds range from naphthalene ($C_{10}H_8$, two rings) to coronene ($C_{24}H_{12}$, seven rings) (Figure 9). Many PAH compounds occur in marine environments, all differing in the number and position of aromatic rings. The unsubstituted lower-molecular-weight PAH compounds are noncarcinogenic but more toxic than the high-molecular-weight PAHs which tend to be carcinogenic,

FIGURE 9   Structure of PAH and other aromatic hydrocarbon compounds. (From Futoma, D. J., Smith, S. R., Smith, T. E., and Tanaka, J., *Polycyclic Aromatic Hydrocarbons in Water Systems*, CRC Press, Boca Raton, FL, 1991, 2. With permission.)

Structure	1957 I.U.P.A.C. Name	Other Names	Mol. Weight	Relative Carcino-genicity	Common Abbreviation (if any)
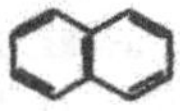	Naphthalene	–	128	–	–
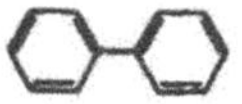	Biphenyl	–	154	–	–
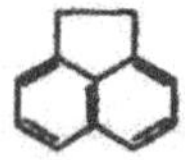	Acenaphthene	–	154	–	–
	Fluorene	–	166	–	–
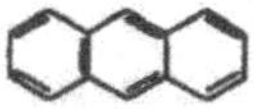	Anthracene	–	178	–	–
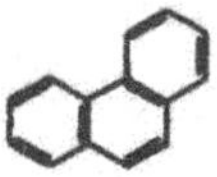	Phenanthrene	–	178	–	–
	Pyrene	–	202	–	–
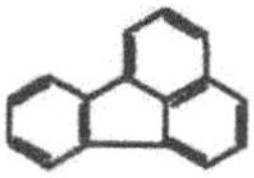	Fluoranthene	–	202	–	–
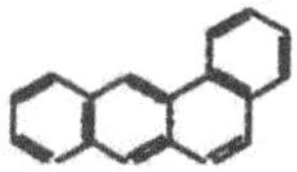	Benzo(a)anthracene	1,2 Benzanthracene	228	< +	B(a)A
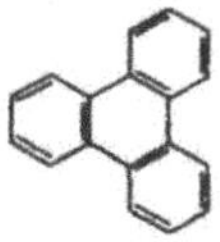	Triphenylene	–	228	–	–

Structure	1957 I.U.P.A.C. Name	Other Names	Mol. Weight	Carcino- genicity	Abbreviation (if any)
	Chrysene	–	228	< +	–
	Naphthacene	Tetracene	228	–	–
	Benzo(b)fluoranthene	3,4 Benzfluoranthene	252	+ +	B(b)F
	Benzo(j)fluoranthene	10,11 Benzfluoranthene	252	+ +	B(j)F
	Benzo(k)fluoranthene	11,12 Benzfluoranthene	252	–	B(k)F
	Benzo(a)pyrene	3,4 Benzopyrene	252	+ + + +	B(a)P
	Benzo(e)pyrene	1,2 Benzopyrene	252	< +	B(e)P
	Perylene	–	252	–	–
	Cholanthrene	–	254	–	–
	7,12 Dimethylbenz-(a)anthracene	7,12 Dimethyl-1,2-benzan-thracene	256	+ + + + +	–
	Benzo(ghi)perylene	1,12 Benzperylene	276	–	–

Structure	1957 I.U.P.A.C. Name	Other Names	Mol. Weight	Relative Carcino- genicity[a]	Common Abbreviation (if any)
	Indeno(1,2,3-cd)pyrene	o-Phenylenepyrene	276	+	IP
	Anthanthrene	–	276	< +	–
	Dibenz(a,h)anthracene	1,2,5,6 Dibenzanthracene	278	+ + +	–
	Dibenz(a,j)anthracene	1,2,7,8 Dibenzanthracene	278	–	–
	Dibenz(a,c)anthracene	1,2,3,4 Dibenzanthracene	278	–	–
	Coronene	–	300	–	–

[a] + + + + + = extremely active; + + + + = very active; + + + = active; + + = moderately active; + = weakly active; < = less than; — = inactive or unknown.

mutagenic, and teratogenic. Marine organisms rapidly metabolize most PAHs, and some of these compounds become carcinogenic, mutagenic, or both after metabolic activation. The ability of these organisms to take up PAHs from contaminated environments is well documented.[58] Benthic invertebrates may be continuously exposed to PAHs in contaminated areas since PAHs are relatively insoluble in water, sorb strongly to particulate matter, and accumulate in bottom sediments.

PAH compounds originate from a variety of anthropogenic sources (e.g., municipal and industrial effluents, creosote, oil spills, urban and agricultural runoff, fossil fuel combustion, asphalt production, and waste incineration). Incomplete combustion of organic matter, especially in the high temperature (500–800°C) range, is a primary mechanism for atmospheric contamination by PAH compounds, many of which enter marine waters via fallout. The process of thermal decomposition of organic molecules and subsequent recombination of the organic particles (pyrolysis) represents a principal pathway of PAH formation.[50] Hence, natural sources of PAHs (e.g., forest and brush fires and volcanic eruptions) also can be important.

Eisler[59] estimated that 2.3×10^5 mt of PAHs enter aquatic environments annually, being derived mainly from oil spills (1.7×10^5 mt) and atmospheric deposition (0.5×10^5 mt) (Table 14). Forest and brush fires (0.19×10^5 mt) and agricultural burning (0.13×10^5 mt) are the major sources of PAHs for the atmosphere. Wastewaters, surface land runoff, and biosynthesis supply relatively small quantities of PAHs to marine environments, with each contributing less than 0.05×10^5 mt/yr.

b. Distribution

i. PAHs in Estuarine and Coastal Marine Environments

The transport of PAHs to marine environments occurs via surface waters and the atmosphere.[60] As PAHs enter the water column, they sorb to particulates and settle

---

TABLE 14
Major Sources of PAHs in Atmospheric and Aquatic Environments[a]

Ecosystem and Sources	Annual Input
Atmosphere	
Total PAHs	
Forest and prairie fires	19,513
Agricultural burning	13,009
Refuse burning	4769
Enclosed incineration	3902
Heating and power	2168
Benzo(a)pyrene	
Heating and power	
Worldwide	2604
United States only	475
Industrial processes	
Worldwide	1045
United States only	198
Refuse and open burning	
Worldwide	1350
United States only	588
Motor vehicles	
Worldwide	45
United States only	22
Aquatic environments	
Total PAHs	
Petroleum spills	170,000
Atmospheric deposition	50,000
Wastewaters	4400
Surface land runoff	2940
Biosynthesis	2700
Total benzo(a)pyrene	70

[a] Concentrations in metric tons.

From Eisler, R., Polycyclic Aromatic Hazards to Fish, Wildlife, and Invertebrates: A Synoptic Review, Biol. Rep 85(1.11), U.S. Fish and Wildlife Service, Washington, D.C. 1987.

---

to the seafloor. Bottom sediments of estuaries and nearshore coastal marine waters located near urban and industrial centers serve as major repositories of PAHs.[1,2,38] However, these compounds may be remobilized by bioturbation activity, bottom currents, and other advective processes. In such dynamic environments, bed shear stresses are commonly elevated, which facilitates particle movement along the benthic boundary layer and the re-entrance of PAHs into the water column.[61]

Because of the strong affinity of PAHs for sediments and other particulate matter, they accumulate to much higher concentrations on the seafloor than in overlying waters. While dissolved PAHs in the water column degrade rapidly by photooxidation, once deposited in bottom sediments, PAHs are more persistent since photochemical or biological oxidation is reduced. In anoxic sediments, PAHs tend to remain unaltered for long periods of time. Guzzella and De Paolis[62] suggest that the analysis of sediment PAHs can serve as an index of PAH input rates to marine environments.

A number of coastal systems contain high PAH levels in bottom sediments. For example, Huggett et al.[63] recorded PAH concentrations in Chesapeake Bay sediments generally ranging from 1–400 µg/kg, with highest levels found in the northern part of the estuary, where human population densities peak. McLeese et al.[64] documented total PAHs in bottom sediments of Boston Harbor amounting to 120 mg/kg. O'Connor et al.[65] tabulated highest PAH concentrations in sediments of the Hudson-Raritan estuary at Newtown Creek (182,000 ng/g). In the New York Bight, they reported relatively high sediment PAH levels (6000 ng/g) in the Christiaensen Basin. Barrick and Prahl[66] estimated total combustion-derived PAH concentrations of 16–2,400 ng/g in sediments of Puget Sound. Readman et al.[67] ascertained total PAH levels of 4900 µg/kg in sediments of the Tamar estuary in England. These systems receive much of their PAH input from urbanized centers.

The concentrations of PAHs in seafloor sediments usually exceed those in the water column by a factor of 1000 or more. Marine waters unaffected by anthropogenic activity generally exhibit PAH levels of less than 1 µg/l. Somewhat higher values (1–5 µg/l) are observed in coastal seawater in proximity to industrialized areas.

c. Effects on Biotic Communities

Polycyclic aromatic hydrocarbons adversely affect marine life as revealed by both laboratory experiments and field observations.[28,50] However, the response of marine organisms to PAH exposure varies widely in nature, owing to variation in bioavailability of the contaminants and different capacities of the organisms to metabolize them. For example, sediment-sorbed PAHs have only limited bioavailability, which ameliorates their toxicity potential and the severity of their impacts. Furthermore, taxonomic groups with poorly developed mixed function oxygenase (MFO) capability (e.g., bivalve mollusks and echinoderms) do not metabolize the compounds efficiently. Hence, PAHs readily accumulate in these organisms, and they are often used as biomonitors of PAH contamination in coastal waters. Other taxa (e.g., fish, annelids, and some crustaceans) have well-developed MFO systems and, therefore, rapidly metabolize PAHs. They accumulate the contaminants only when exposed to heavily polluted environments. Table 15 shows PAH concentrations in selected

TABLE 15
PAH Concentrations in Selected Shellfish and Finfish from
Estuarine and Coastal Marine Waters

Taxonomic Group, Compound, and Other Variables	Concentration[a]
Shellfish	
Rock crab, *Cancer irroratus*	
Edible portions	
New York Bight, 1980	
Total PAHs	1600 FW[b]
Benzo(a)pyrene	1 FW
Long Island Sound, 1980	
Total PAHs	1290 FW
Benzo(a)pyrene	ND[c]
American oyster, *Crassostrea virginica*	
Soft parts	
South Carolina, 1983	
Total PAHs (spring)	
Palmetto Bay	520 FW
Outdoor Resorts	247 FW
Fripp Island	55 FW
Total PAHs (summer)	
Palmetto Bay	269 FW
Outdoor Resorts	134 FW
Fripp Island	21 FW
American lobster, *Homarus americanus*	
Edible portions	
New York Bight, 1980	
Total PAHs	367 FW
Benzo(a)pyrene	15 FW
Long Island Sound, 1980	
Total PAHs	328 FW
Benzo(a)pyrene	15 FW
Softshell clam, *Mya arenaria*	
Soft parts	
Coos Bay, Oregon, 1978–1979	
Contaminated site	
Total PAHs	555 FW
Phenanthrene	155 FW
Fluorene	111 FW
Pyrene	62 FW
Benzo(a)pyrene	55 FW
Benz(a)anthracene	42 FW
Chrysene	27 FW
Benzo(b)fluoranthene	12 FW
Others	<10 FW
Bay mussel, *Mytilus edulis*	
Soft parts	

TABLE 15 (CONTINUED)
PAH Concentrations in Selected Shellfish and Finfish from Estuarine and Coastal Marine Waters

Taxonomic Group, Compound, and Other Variables	Concentration[a]
Oregon, 1979–1980	
Total PAHs	
Near industrialized site	106–986 FW
Remote site	27–274 FW
Sea Scallop, *Placopecten magellanicus*	
Muscle	
Baltimore Canyon	
Benz(a)anthracene	1 FW
Benzo(a)pyrene	<1 FW
Pyrene	4 FW
Edible portions	
New York Bight, 1980	
Total PAHs	127 FW
Benzo(a)pyrene	3 FW
Clam, *Tridacna maxima*	
Soft parts	
Great Barrier Reef, 1980–1982	
Total PAHs	
Pristine areas	0.07 FW
Powerboat areas	Up to 5 FW
Finfish	
Five Species	
Muscle	
Baltimore Canyon	
Benz(a)anthracene	Max 0.3 FW
Benzo(a)pyrene	Max <5 FW
Pyrene	Max <5 FW
Winter flounder, *Pseudopleuronectes americanus*	
Edible portions	
New York Bight, 1980	
Total PAHs	315 FW
Benzo(a)pyrene	21 FW
Long Island Sound, 1980	
Total PAHs	103 FW
Benzo(a)pyrene	ND
Windowpane, *Scopthalmus aquosus*	
Edible portions	
New York Bight, 1980	
Total PAHs	536 FW
Benzo(a)pyrene	4 FW

TABLE 15 (CONTINUED)
PAH Concentrations in Selected Shellfish and Finfish from Estuarine and Coastal Marine Waters

Taxonomic Group, Compound, and Other Variables	Concentration[a]
Long Island Sound, 1980	
Total PAHs	86 FW
Benzo(a)pyrene	ND
Red hake, *Urophycus chuss*	
Edible portions	
New York Bight, 1980	
Total PAHs	412 FW
Benzo(a)pyrene	22 FW
Long Island Sound, 1980	
Total PAHs	124 FW
Benzo(a)pyrene	5 FW
Several species	
Edible portions	
Greenland	
Benzo(a)pyrene	65 FW
Italy	
Benzo(a)pyrene	5–8 DW[d]

[a] Concentrations in mg/kg.
[b] Fresh weight.
[c] Not detected.
[d] Dry weight.

From Eisler, R., Polycyclic Aromatic Hazards to Fish, Wildlife, and Invertebrates: A Synoptic Review, Biol. Rep. 85(1.11), U.S. Fish and Wildlife Service, Washington, D.C., 1987.

bivalve mollusks and fish from estuarine and coastal marine waters. In general, organisms with high lipid content, poor MFO systems, and activity patterns or distributions coinciding with the location of the PAH source display highest tissue concentrations.

Fish exposed to PAHs commonly develop lesions and tumors, and some investigators have reported a correlation between tissue levels of PAHs and neoplasia in mollusks.[68] However, more sampling and analysis is required.[69] Unmetabolized PAHs can be directly toxic to marine organisms. In addition, reactive metabolites of some PAHs (e.g., epoxides and dihydrodiols) have the ability to bind to cellular proteins and DNA, causing biochemical disruptions and cell damage that lead to mutations, developmental malformations, tumors, and cancer.[50] Correlations have been demonstrated between hepatic neoplasms in fish (e.g., hepatocellular carcinoma, hepatocellular adenoma, and cholangiocellular carcinoma), as well as other abnormalities, and PAH concentrations in sediments, particularly for bottom-dwelling species.[59,70]

Less data have been collected on the responsiveness of other animal groups (e.g., seabirds and mammals) to PAHs. While information on the effects of individual

PAH compounds on seabirds is rather sparse, laboratory experiments with mammals indicate that the toxic effects on various organ systems are coupled to alkylated PAHs and metabolites of unsubstituted PAHs.[50] Much data on PAH metabolism in birds, mammals, as well as fish, derive from experimental studies in the laboratory. The database of PAH metabolism by feral populations in areas chronically contaminated with multiple pollutants is not comprehensive. Nevertheless, accumulated evidence suggests that PAH metabolism poses a significant health hazard to these organisms.

Bacteria oxidize PAHs to carbon dioxide and water and in the process produce dihydrodiols and catechols. Fungi metabolize PAHs in a manner similar to that observed in mammals, utilizing the cytochrome P450-dependent MFO system. Algae assimilate PAHs; however, because the compounds tend to be poorly metabolized, they accumulate to very high levels in the plant tissues. At low concentrations (5–100 μg/l), PAH compounds may stimulate or inhibit growth and cell division in marine algae and bacteria. At higher concentrations (0.2–10.0 mg/l), the same compounds interfere with cell division in these organisms. Mortality also increases in some species exposed to the elevated PAH levels.[50,59]

Marine community changes caused by individual PAHs are more difficult to delineate than those generated by oil spills. This is so because PAHs affect organisms through toxic action (Table 16), whereas oil spills impact organisms through multiple pathways (e.g., physical contact, smothering, toxic action, and habitat modification). The concentrations of individual PAHs in marine environments, with the exception of heavily polluted sites, are usually much lower than the concentrations acutely toxic to marine organisms. Although acute lethal effects associated with elevated PAH levels (e.g., local mass fish kills) have been rarely observed in nature, sublethal effects manifested at much lower concentrations are a chronic problem in some regions. Sublethal chronic effects of PAH compounds may be reflected in impaired metabolic pathways, reduced growth, decreased fecundity, and reproductive failure of individuals. These effects may be of great significance to long-term survival and abundance of populations in marine communities. However, they often are difficult to quantify and distinguish from population changes produced by other environmental stressors or even natural variability. More data are needed to determine how marine populations vary in abundance in response to changing environmental conditions (e.g., temperature, salinity, food supply, predation, and competition) and pollutant stressors.

Sublethal effects of PAHs may take many forms in marine organisms — biochemical, behavioral, physiological, and pathological — and they may be more evident at a particular life history stage. For instance, juvenile or adult marine invertebrates typically are less sensitive to PAHs than are egg and larval stages.[50] Some responses to PAH contamination require months to develop in individual organisms, and population and community impacts may be detected years after initial PAH exposure. The degree of change in the structure and dynamics of marine communities depends on the severity and duration of lethal and sublethal impacts on populations of organisms in an area.

TABLE 16
Toxicities of Selected PAHs to Marine Organisms

PAH Compound, Organism, and Other Variables	Concentration in Medium[a]	Effect
Benzo(a)pyrene		
Sandworm, *Neanthes arenceodentata*	>1000	LC_{50} (96 h)
Chrysene		
Sandworm	>1000	LC_{50} (96 h)
Dibenz(a,h)anthracene		
Sandworm	>1000	LC_{50} (96 h)
Fluoranthene		
Sandworm	500	LC_{50} (96 h)
Fluorene		
Grass shrimp, *Palaemonetes pugio*	320	LC_{50} (96 h)
Amphipod, *Gammarus pseudoliminaeus*	600	LC_{50} (96 h)
Sandworm	1000	LC_{50} (96 h)
Sheepshead minnow, *Cyprinodon variegatus*	1680	LC_{50} (96 h)
Naphthalene		
Copepod, *Eurytemora affinis*	50	LC_{30} (10 d)
Pink salmon, *Oncorhynchus gorbuscha*, fry	920	LC_{50} (24 h)
Dungeness crab, *Cancer magister*	2000	LC_{50} (96 h)
Grass shrimp	2400	LC_{50} (96 h)
Sheepshead minnow	2400	LC_{50} (24 h)
Brown shrimp, *Penaeus aztecus*	2500	LC_{50} (24 h)
Amphipod, *Elasmopus pectenicrus*	2680	LC_{50} (96 h)
Coho salmon, *Oncorhyncus kisutch*, fry	3200	LC_{50} (96 h)
Sandworm	3800	LC_{50} (96 h)
Mosquitofish, *Gambusia affinis*	150,000	LC_{50} (96 h)
1-Methylnaphthalene		
Dungeness crab, *Cancer magister*	1900	LC_{50} (96 h)
Sheepshead minnow	3400	LC_{50} (24 h)
2-Methylnaphthalene		
Grass shrimp	1100	LC_{50} (96 h)
Dungeness crab	1300	LC_{50} (96 h)
Sheepshead minnow	2000	LC_{50} (24 h)
Trimethylnaphthalenes		
Copepod, *Eurytemora affinis*	320	LC_{50} (24 h)
Sandworm	2000	LC_{50} (96 h)
Phenanthrene		
Grass shrimp	370	LC_{50} (24 h)
Sandworm	600	LC_{50} (96 h)
1-Methylphenanthrene		
Sandworm	300	LC_{50} (96 h)

[a] Concentrations in μg/l.

Modified from Eisler, R., Polycyclic Aromatic Hazards to Fish, Wildlife, and Invertebrates: A Synoptic Review, Biol. Rep. 85(1.11), U.S. Fish and Wildlife Service. Washington, D.C., 1987.

5. Halogenated Hydrocarbons

a. Sources

More insidious than the effects of sewage sludge dumping or major oil spills are the impacts of halogenated hydrocarbons that slowly accumulate in the marine hydrosphere. These organic compounds rank among the most persistent, ubiquitous, and toxic pollutants in marine ecosystems. Of particular concern are the lipophilic, high-molecular-weight halogens that tend to biomagnificate through marine food chains and pose a significant health hazard to man. Organochlorine compounds, such as DDT, chlordane, and PCBs, provide examples (Figure 10). These chemical contaminants enter marine waters via multiple routes including urban and agricultural runoff, sewage waste disposal, industrial waste input, and atmospheric deposition.[71,72] Heaviest contaminant loads occur in estuarine and coastal marine waters near metropolitan centers and regions of intense industrial activity. However, the application of agrochemicals (i.e., pesticides, herbicides, etc.) in more remote areas also may contribute significant concentrations of the compounds.

Two other classes of halogenated hydrocarbons, the chlorinated dibenzo-p-dioxins (CDDs) and chlorinated dibenzofurans (CDFs), are considerably toxic to marine organisms. These aromatic heterocyclic compounds originate from numerous anthropogenic sources, most notably industrial discharges from pulp and paper mills, wood treatment plants, municipal wastewaters, industrial incinerators, and atmospheric fallout. They are widely distributed from industrialized and heavily populated areas.[73] Some of the compounds have been shown to be embryotoxic, teratogenic, and carcinogenic to animals.

Many halogenated hydrocarbons are broad-spectrum poisons that affect entire biotic communities. Being chemically stable in nature, hydrophobic, and highly mobile, they have become dispersed in marine environmental compartments worldwide. Similar to many other types of pollutants in marine ecosystems, halogenated hydrocarbons rapidly sorb to particulate matter and accumulate in bottom sediments, where they resist all types of degradation. The residues of the more recalcitrant compounds (e.g., PCBs) have the potential to exist relatively unchanged for decades or even centuries in marine ecosystems.

b. Chlorinated Hydrocarbons

The Joint Group of Experts on the Scientific Aspects of Marine Pollution[5] classifies chlorinated hydrocarbon compounds into five groups: (1) lower-molecular-weight compounds (up to three carbons); (2) aliphatic and aromatic herbicides (up to six carbons); (3) long-chain chlorinated paraffins; (4) chlorinated insecticides (e.g., mirex and camphenes); and (5) chlorinated aromatic industrial chemicals (e.g., PCBs). A diverse group of biocides (insecticides, herbicides, and fungicides) has been found in estuarine and nearshore oceanic waters. Among the well-known group of organochlorine insecticides are DDT and its metabolites (DDD and DDE), aldrin, chlordane, endrin, heptochlor, dieldrin, perthane, and toxaphene. Most of these compounds were banned from use in the United States in the 1970s because of their detrimental impacts on terrestrial and aquatic organisms, especially nontarget species. However, they are still utilized by other countries, particularly undeveloped

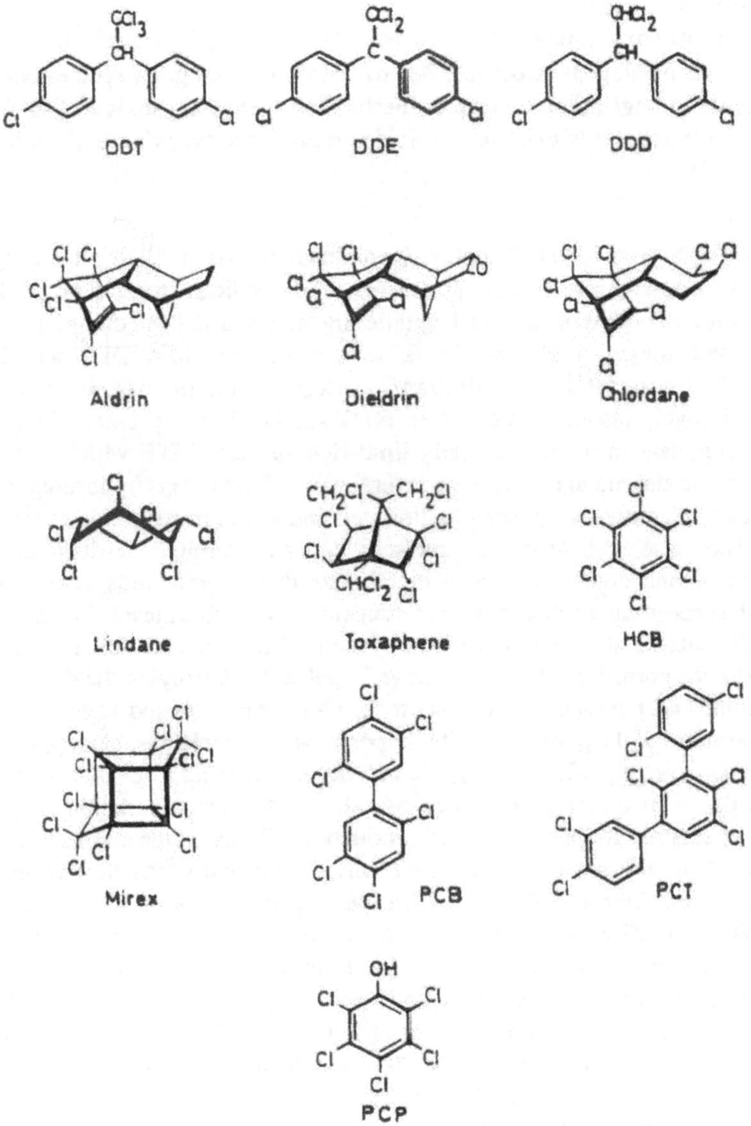

FIGURE 10 Molecular structure of selected organochlorine compounds. (From Reutergardh, L., in *Chemistry and Biogeochemistry of Estuaries*, Olausson, E. and Cato, I., Eds., John Wiley & Sons Ltd., Chichester, 1980, pp. 349–365. With permission.)

nations in tropical regions, where infectious diseases transmitted by insects and other pests remain rampant.

Fungicides of importance include chlorinated benzenes and phenols. For example, hexachlorobenzene (HCB) is an antifungal chemical once commonly employed in grain storage operations and as a component in wood preservatives.

Pentachlorophenol, a metabolite of HCB, is a frequently applied insecticide, fungicide, and slimicide.

Chlorophenoxy compounds, such as 2,4-D and 2,4,-T, are selective herbicides of note. These halogens effectively destroy broadleaf and grass species but also kill other types of vegetation. In general, herbicides are acutely toxic to fish. However, few documented cases exist of herbicide impacts on estuarine and marine biotic communities.

i.   DDT

The synthetic pesticide DDT and its breakdown derivative DDE pose a significant hazard to estuarine and marine organisms. The insidious impacts of DDT and its metabolites on both terrestrial and aquatic organisms have been chronicled since the 1950s, with the most serious effects seen in marine birds. DDT and DDE are extremely persistent and recalcitrant to degradation in marine environments. Although highly insoluble in seawater, DDT and DDE are extremely lipophilic, and they accumulate in biota, especially lipid-rich species. DDT, which is potentially carcinogenic and mutagenic to man, enters marine food chains by sorbing to external surfaces of plankton or by being assimilated and stored in the lipids of these organisms. They also sorb to detrital surfaces and pass through detritus food chains. Because marine organisms cannot metabolize these compounds, they biomagnify through succeeding trophic levels and concentrate to highest levels in top carnivores.

DDT attacks the central nervous system of insects and nontarget species. It precludes the normal conduction of nerve impulses by destroying the delicate balance of sodium and potassium within neurons. Organisms exposed to sufficiently high concentrations of the pesticide exhibit hyperactivity, convulsions, paralysis, and death. DDT also adversely affects estrogenic activity by inducing a breakdown of sex hormones that regulate the mobilization of calcium. It inhibits carbonic anhydrase, an enzyme necessary for proper eggshell production of birds. Some marine birds exposed to DDT (e.g., the brown pelican, *Pelecanus occidentalis*; the herring gull, *Larus argentatus*) experienced substantial decreases in population sizes in some areas during the 1960s and 1970s as a consequence of severe thinning of their egg shells.

As a universal contaminant, DDT has been detected in seawater, sediment, and biotic samples from most estuarine and coastal regions worldwide, as well as from many offshore and deep-sea locations. Total DDT (tDDT) concentrations in open ocean waters are generally between 0.005 and 0.06 ng/l and in coastal waters, usually less than 5 ng/l. In coastal and nearshore bottom sediments, tDDT concentrations vary widely from <0.01 to >1000 ng/g dry weight. Levels of tDDT residues in sentinel organisms (i.e., mussels) from coastal waters of the northeast Pacific and northwest Atlantic (United States) range from 2.8 to 1109 ng/g dry weight. Substantially higher residue levels have been recorded in higher-trophic-level organisms, most notably fish, dolphins, porpoises, whales, and seals.[74] Figure 11 depicts DDT concentrations in finfish samples collected along the east coast of the United States. Results of various U.S. monitoring programs conducted during the past 30 years (e.g., National Pesticide Monitoring Program, U.S. EPA Mussel Watch Project, NOAA National Status and Trends Program) indicate that DDT contamination in estuarine and marine organisms declined dramatically during the 1970s (Table 17).

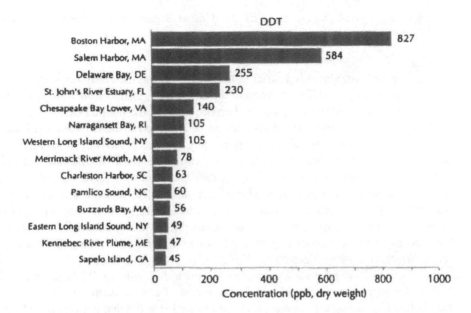

DDT

Boston Harbor, MA	827
Salem Harbor, MA	584
Delaware Bay, DE	255
St. John's River Estuary, FL	230
Chesapeake Bay Lower, VA	140
Narragansett Bay, RI	105
Western Long Island Sound, NY	105
Merrimack River Mouth, MA	78
Charleston Harbor, SC	63
Pamlico Sound, NC	60
Buzzards Bay, MA	56
Eastern Long Island Sound, NY	49
Kennebec River Plume, ME	47
Sapelo Island, GA	45

Concentration (ppb, dry weight)

FIGURE 11 Concentrations of DDT in Atlantic Coast fish liver tissue sampled from various U.S. estuaries by the NOAA Status and Trends Program. (From Larsen, P. F., *Rev. Aquat. Sci.*, 6, 67–86, 1992. With permission.)

TABLE 17
Median or Geometric Mean DDT Concentrations in Bivalves and Finfish for Several National U.S. Surveys

	tDDT or DDE (ppm wet weight) by Sampling Period			
Organism: Substrate	1965–1972	1972–1975	1976–1977	1984–1986
Bivalves	0.024[a]		0.01[b]/0.001[c]	0.003
Fish, whole juvenile		0.014[d]	ND	ND
Fish, muscle		0.110[e]	0.012[f]	
Fish, liver			0.220[f]	0.054[g]
Fish, whole F.W.	0.7–1.1	0.4–0.6	0.370	

Note: See original source for particular studies. ND = not determined; F.W. = fresh weight.

[a] Median of 8180-site means composited from 7839 samples.
[b] Median of 89-site means composited from 188 samples.
[c] Median of 80-site values or site means.
[d] Median of 144-site means composited from 1524 composites.
[e] Median of area or site means from samples.
[f] Median of 19-site means from samples.
[g] Median of 42-site medians from 126 composites.

From Mearns, A. J., Matta, M. B., Simecek-Beatty, D., Buchman, M. F., Shigenaka, G., and Wert, W. A., *PCB and Chlorinated Pesticide Contamination in U.S. Fish and Shellfish: A Historical Assessment Report*, NOAA Tech. Mem. NOS OMA 39, National Oceanic and Atmospheric Administration, Seattle, WA, 1988.

Mearns et al.[75] demonstrated that tDDT concentrations decreased 80- to 100-fold in estuarine and coastal marine biota of the United States between 1965–1972 and 1984–1986.

ii. Polychlorinated Biphenyls

Among the most important industrial contaminants in marine ecosystems are polychlorinated biphenyls (PCBs), a group of synthetic halogenated aromatic hydrocarbons that has been linked to a number of environmental and public health concerns. In the United States, commercial production of PCBs occurred between 1929 and 1977, and during this interval the contaminants became universally distributed in ocean waters and in nearly all marine plant and animal species.[2,28] PCBs enter marine waters most directly via waste discharges from manufacturing facilities and industries. Secondary sources include leaching from dumpsites, volatilization by vaporization from plastics, and inefficient burning in incinerators followed by adsorption onto particulates, transport, and eventual atmospheric fallout. Chlorobiphenyls in the atmosphere average about 1 ng/m^3 worldwide.[76] Atmospheric transport is a major reason for the global distribution of PCBs.

Because of their great stability, persistence, and lipophilicity, PCBs accumulate to high levels in environmental and biotic media. These contaminants are poorly metabolized by biological systems, and both sublethal and lethal impacts have been delineated in marine biota. However, precise toxicological effects of PCBs on these organisms are often difficult to characterize. When exposed to sufficiently high PCB levels, fish experience a greater incidence of epidermal lesions, blood anemia, altered immune responses, and fin erosion, and marine mammals (e.g., porpoises, seals, sea lions, whales), an array of reproductive abnormalities, such as pathological changes in reproductive organs and accompanying depression of reproductive potential. PCBs are also suspected of being a human carcinogen and a cause of various chronic diseases in man (e.g., skin lesions, liver maladies, and reproductive disorders).

PCB concentrations range from about 0.1 to 1000 ng/l in estuarine and coastal marine waters and 0.01 to 150 ng/l in the open ocean (Table 18). In these waters, PCB levels typically peak in the surface microlayer. They decrease to approximately 1.5 to 2.0 pg/l at a depth of 3500 to 4000 m in the deep sea.[2,28] PCBs have been recorded at depths greater than 5000 m.

As in the case of DDT, PCBs rapidly sorb to fine-grained sediments and other particulate matter and subsequently settle to the seafloor. Therefore, highest concentrations of the contaminants exist in bottom sediments of estuarine and coastal marine environments, particularly those in close proximity to industrialized centers (Figure 12). For example, peak levels of PCBs in U.S. coastal and nearshore bottom sediments occur in Escambia Bay, FL (<30 to 480,000 ng/g dry weight), New Bedford Bay, MA (8400 ng/g dry weight), Palos Verdes, CA (80 to 7420 ng/g dry weight), New York Bight, NJ (0.5 to 2200 ng/g dry weight), and Hudson-Raritan Bay, NY (286 to 1950 ng/g dry weight).

The concentrations of PCBs increase by a factor of 10 to 100 times when proceeding upward through marine food chains. For instance, PCB levels in marine zooplankton range from <0.003 to 1.0 ppm. Fish and mammals (i.e., seals) have PCB levels in their tissues between 0.03 and 212 ppm. Lethal levels of PCBs in fish

TABLE 18
Ranges in PCB Concentrations (ng/l) Reported for Open
Ocean, Coastal Waters, and Estuaries or Rivers

Area	Location	Range in PCB (ng/l)
Open oceans	North Atlantic	<1–150
		0.4–41.0
	Sargasso Sea	0.9–3.6
	North-South Atlantic	0.3–8.0
	Mediterranean Sea	0.2–8.6
Coastal waters	Southern California	2.3–36.0
	Northwest Mediterranean	1.5–38.0
	Atlantic coast, United States	10–700
	Baltic coasts	0.3–139.0
		0.1–28.0
	Dutch coast	0.7–8.0
Estuaries/rivers	Wisconsin rivers, United States	<10–380
	Rhine-Meuse system, Holland	10–200
	Tiber estuary, Italy	9–1000[a]
	Brisbane River, Australia	ND–50
	Hudson River, United States	<100–2.8 × 10[6]

Note: ND, not detectable (no limits quoted).

[a] As decachlorinated biphenyl equivalents.

From Phillips, D. J. H., in *PCBs in the Environment*, Vol. 1, Waid, J. S., Ed., CRC Press, Boca Raton, FL, 1986, 130. With permission.

based on laboratory experiments fall in the 10 to 300 ppm range.[77] Besides the lipid concentration of an organism, which is the most important factor governing PCB concentrations in tissues, the condition of the organism, its size, and the season of sampling all potentially influence PCB levels in a given marine species.

Results of NOAA and U.S. EPA biomonitoring programs conducted in U.S. waters during the past 25 years indicate no clear evidence of a largescale nationwide decrease of PCB concentrations in estuarine and coastal marine environments. The most dramatic reductions in PCB residues in shellfish and finfish have taken place since 1980 in areas nearby known industrial contaminant sources and other "hot spot" areas. These reductions are ascribed to more effective cleanup operations and tighter regulatory controls. Hence, PCBs will continue to be the subject of ongoing estuarine and marine monitoring programs in U.S. waters.

c. Chlorinated Dibenzo-p-dioxins and Dibenzofurans

Another group of halogenated hydrocarbons that has received considerable attention in recent years because of the multiple impacts of the contaminants on aquatic organisms includes the chlorinated dibenzo-*p*-dioxins (CDDs) and chlorinated dibenzofurans (CDFs). These aromatic heterocyclic compounds appear to be highly toxic

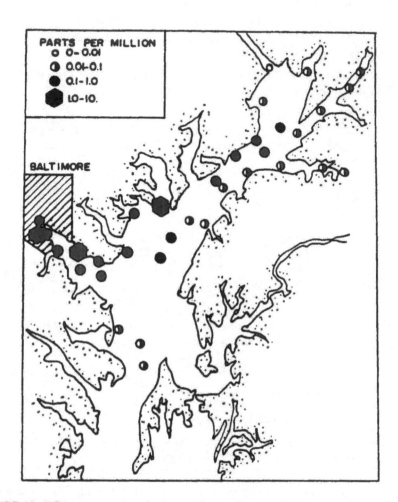

FIGURE 12 PCB concentrations in the surface sediments of the upper Chesapeake Bay. Note highest concentrations near Baltimore. (From Munson, T. O., *Upper Bay Survey*, Vol. 2, Final Report to the Maryland Department of Natural Resources, Westinghouse Electric Corporation, Ocean Sciences Division, Annapolis, MD, 1976. Permission granted by Northrop Grumman Corporation, successor in interest to the Westinghouse Electric Corporation.)

to marine organisms, and they can induce serious sublethal effects (e.g., cardiovascular changes, developmental abnormalities, impaired reproduction, liver disorders, immunosuppression, hormonal and histopathological alterations, and progressive weight loss). Occurring as trace contaminants in organic chemicals, such as chlorinated phenols, PCBs, and phenoxy herbicides, CDDs and CDFs enter estuarine and marine environments via wastewater discharges, groundwater inflow, sewer overflows, storm drains, and atmospheric deposition. In addition, combustion processes (i.e., municipal incineration, chemical waste combustion, and wood burning) represent important pathways of CDD and CDF delivery to marine waters.

CDDs and CDFs are widely distributed in seawater and sediments at low parts-per-trillion concentrations. As in the case of most halogenated hydrocarbon compounds, they attain highest concentrations in densely populated and heavily industrialized regions. Bottom sediments act as a sink for the contaminants.

Being environmentally persistent and highly lipophilic compounds, CCDs and CDFs tend to bioaccumulate in marine organisms by food chain transfer rather than by direct uptake from seawater, suspended matter, or bottom sediments. Cooper[78] noted that aquatic invertebrates, vertebrates, and fish-eating birds appear to be at greatest risk from polychlorinated dibenzo-*p*-dioxins and polychlorinated dibenzo-furans contamination due to the persistence, bioaccumulation, and potency of the compounds. Most investigations have focused on the toxicity of CCDs and CDFs to fish and mammals, with few studies conducted on invertebrates. Overall, the effects of CDDs and CDFs on estuarine and marine organisms are far from well understood.

d. Effects on Biotic Communities

Halogenated hydrocarbons pose a potentially serious threat to the structure of estuarine and marine communities because some species, particularly those occupying upper trophic levels, accumulate the contaminants to very high levels and, hence, are susceptible to chronic chemical exposure. Through sublethal exposure and biomagnification, effects of the chemicals are most pronounced in fish and mammals which commonly experience reproductive disorders and reduced population sizes subsequent to contaminant uptake. The high mortalities of harbor seals, together with diminished hatching success of marine birds exposed to DDT and PCBs, provide examples. Population changes such as these can adversely affect entire biotic communities in an area by altering feeding relationships and energy flow, thereby contributing to significant alterations of community structure. Acute (i.e., lethal) effects of DDT, PCBs, and other halogenated hydrocarbons on these larger organisms are rare, however, since a single large dose of the contaminants — amounting to several g/kg of body weight — is required to cause death. Such high concentrations do not occur in marine environments.

Additional toxicological testing of halogenated hydrocarbons on estuarine and marine organisms is necessary to properly assess the factors that influence the bioconcentration, toxicokinetics, and metabolism of these compounds. This laboratory experimentation must be supplemented with observations of the effects of halogenated hydrocarbons on marine communities in nature. Field investigations will continue to be extremely difficult to conduct because a single sample collected at an impacted site may contain as many as 100–150 different halogenated hydrocarbon compounds. It will not only be problematical and costly to identify the compounds, but also most difficult to ascertain which of the compounds is responsible for the observed biological effects in the community.

6. Heavy Metals

a. Composition and Sources

One of the less desirable by-products of an industrialized society is the increase of heavy metal accumulation in marine environments. During the 20th Century, heavy metal pollution has become commonplace in marine systems worldwide. Because of their extreme persistence, high toxicity, and tendency to bioaccumulate, heavy metals are the focus of numerous ecotoxicological investigations. They remain especially troublesome in harbors, embayments, and estuaries where human activities are responsible for large inputs of the contaminants.

Heavy metals may be grouped into two categories: transitional metals and metalloids. Transitional metals (e.g., copper, cobalt, iron, and manganese) include those elements essential for metabolic function of organisms at low concentrations but may be toxic at high concentrations. In contrast, metalloids (e.g., arsenic, cadmium, lead, mercury, selenium, and tin) are generally not required for metabolic activity but may be toxic at low concentrations.[1,16] Aside from these heavy metals, organometal pollutants (e.g., tributyl tin, alkylated lead, and methylmercury) are of primary concern, because they are particularly toxic to marine organisms and potentially deleterious to man.

Estuarine and oceanic waters receive heavy metals from both natural and anthropogenic sources. Natural processes (e.g., weathering and erosion of rocks, leaching of soils, eruption of volcanoes, and emissions of deep-sea hydrothermal vents) release significant concentrations of heavy metals to marine systems. However, anthropogenic inputs are usually much greater in coastal waters near urban or industrialized centers (e.g., Boston Harbor, Newark Bay, and Commencement Bay). Anthropogenic sources of heavy metals in the sea are numerous and diverse. Mining, smelting, refining, electroplating, and other industrial operations add substantial quantities of heavy metals to estuarine and coastal waters. Municipal wastes, landfill leachates, dredged-spoil dumping, ash disposal, fossil-fuel combustion, as well as boating and shipping activities also contribute large amounts of heavy metals to these waters.

Riverine inflow, atmospheric deposition (Table 19), and anthropogenic point sources are the principal routes of entry of heavy metals into the sea. Since heavy metals are generally particle-reactive and rapidly sorb to suspended sediments and other particulate matter, the seafloor serves as the ultimate repository for a large fraction of the contaminants. Estuaries, in particular, tend to trap these pollutants, owing to their unique circulation patterns, and only small amounts generally escape to oceanic waters.[8,28]

b. Metal Toxicity

Above a threshold availability, heavy metals are toxic to marine organisms. The approximate order of increasing toxicity of heavy metals, according to Abel,[79] is as follows: Co, Al, Cr, Pb, Ni, Zn, Cu, Cd, and Hg. However, the toxicity of a given metal varies in estuarine and marine organisms for several reasons. The capacity of the organisms to take up, store, remove, or detoxify heavy metals differs considerably. Many intrinsic and extrinsic factors influence heavy metal uptake: (1) intra- and interspecifically variable intrinsic factors, such as surface impermeability, nutritional state, stage of molt cycle, and throughput of water by osmotic flux; and (2) extrinsic physico-chemical factors, such as dissolved metal concentration, temperature, salinity, presence or absence of chelating agents, and presence or absence of other metals. Some of these factors may also influence the accessibility of metals

TABLE 19
Atmospheric Deposition to the Ocean of Primarily Anthropogenic Trace Metals (10^9 g/atm)

	Cd	Cu	Ni	Zn	Aa	Pb	Hg
North Pacific	0.47–0.81	3.2–10.6	4.5–5.7	9.1–46.5	1.6–0.7	17.9	0.58
South Pacific	0.06–0.10	0.4–1.3	0.6–0.7	1.1–5.7	0.2–0.09	2.2	0.33
North Atlantic	1.47–2.54	10.2–33.3	14.1–18.1	28.6–147.0	5.1–2.3	56.5	0.42
South Atlantic	0.11–0.19	0.8–2.5	1.1–1.3	2.1–10.9	0.4–0.17	4.2	0.09
North Indian	0.12–0.22	0.9–2.8	1.2–1.5	2.4–12.5	0.4–0.2	4.8	0.07
South Indian	0.06–0.11	0.4–1.4	0.6–0.8	1.2–6.2	0.2–0.1	2.4	0.2
Global total	2.3–4.0	16–52	22–28	44–228	7.9–3.6	88.0	1.7

From The United Nations Joint Group of Experts on the Scientific Aspects of Marine Pollution, Atmospheric Input of Trace Species to the World Ocean. Rep. GESAMP (38), Rome, 1989.

to organisms (i.e., metal bioavailability).[80] Unfortunately, the processes controlling bioavailability of trace metals are poorly constrained.

Once assimilated by marine organisms, heavy metals may be sequestered by metallothioneins and lysosomes, thus faciliting detoxification processes. Metal binding activity of metallothioneins and lysosomes is also variable in marine organisms. While some species have the capability of regulating the body concentration of trace metals to approximately constant levels over a wide range of ambient dissolved metal availabilities, others do not. Therefore, heavy metals may be either maintained in a metabolically available form or may be detoxified subsequent to their accumulation by the organisms.

The toxicity of trace metals has been coupled to the free metal ionic activity regardless of the total metal concentrations. It is the bioavailability of a metal rather than its mere presence in the environment that must be considered in toxicological studies, and the bioavailability depends strongly on its chemical speciation. In marine environments, heavy metals occur in dissolved form (as free ions, complexed ions, etc.) or in the solid state (as colloids, sorbed onto particle surfaces, in mineral matrices, etc.). They may be present as inorganic and organic species. Consequently, the identification of the biologically active chemical and physical forms of heavy metals is critical to assessing heavy metal pollution and biotic impacts in marine environments.

At toxic levels, heavy metals act as enzyme inhibitors in marine organisms. They also adversely affect cell membranes. Organometallic species often damage reproductive and central nervous systems.

Estuarine and marine organisms exhibit a range of pathological responses to heavy metal pollution. Some individuals experience neoplasm formation and genetic derangement, and others, tissue inflammation and degeneration. Other disruptions include changes in physiology and development. Feeding behavior, digestive efficiency, and respiratory metabolism commonly are adversely affected; in addition, growth abnormalities may be severe as exemplified by growth inhibition in crustaceans, mollusks, echinoderms, hydroids, protozoans, and algae.[28] Together, these toxic effects can lead to serious insidious impacts on biotic communities.

c. Metal Concentrations

This section reviews the concentrations of selected heavy metals in seawater, sediments, and biotic samples from estuarine and marine environments (Tables 20–22; Figure 13). The elements selected for discussion are those considered to be environmentally and toxicologically significant in these environments. The heavy metal concentrations reported are largely derived from the work of Kennish,[1,2] Clark,[16] Fowler,[74] Bruland and Franks,[81] and Bryan and Langston.[82]

i. Copper

The levels of copper in estuarine waters (~0.2 to >100 μg/l) are usually much greater than those in oceanic waters (~0.1 μg/l). Because dissolved copper sorbs to particulate matter, the concentrations in bottom sediments can be substantial. In estuarine bottom sediments, for example, copper concentrations range from approximately 1–2000 μg/g dry wt (Table 20). A positive correlation exists between the concentrations of copper in sediments and those observed in many benthic species. In contaminated estuaries, copper concentrations in bivalves often exceed 100 μg/g dry wt and occasionally 1000 μg/g dry wt (Table 21). Based on laboratory studies, dissolved copper concentrations of 1–10 μg/l can significantly impact many estuarine and coastal marine organisms.[2,82]

ii. Cadmium

The estimated input of cadmium to the world's oceans amounts to more than 7000 mt/yr.[16] The concentrations of cadmium in coastal waters removed from industrialized centers range from 1–100 ng/l. Where inputs are elevated because of industrial or municipal waste disposal, pronounced onshore-offshore gradients in cadmium concentrations may be evident. In open ocean waters, the levels of cadmium decline to about 0.1–60.0 ng/l, with lowest values found in gyres.

Cadmium accumulates in bottom sediments at a rate of more than 2600 mt/yr, most rapidly along the continental shelf. In estuarine sediments, cadmium concentrations generally range from 0.1–10.0 μg/g dry wt, while in deep-sea sediments, cadmium levels typically are less than 0.5 μg/g dry wt. Concentrations of the metal in contaminated coastal sediments may exceed 200 μg/g dry wt.

Mussels (Mytilus spp.) used as sentinel organisms in coastal regions have accumulated cadmium to mean levels of 1–5 μg/g dry wt. The gastropods Nucella lapillus and Patella vulgata inhabiting the Bristol Channel-Severn estuary contain much higher cadmium values, amounting to 144 and 277 μg/g dry wt, respectively. Concentrations in fish muscle of samples collected from coastal waters range from approximately 0.02–0.7 mg/kg. Reported concentrations of cadmium in euphausiids from oceanic waters of the northeast Pacific average 5.5 μg/g dry wt. However, some oceanic species (e.g., the amphipod Thermisto gaudichaudii) have been recovered with cadmium concentrations in excess of 50 μg/g dry wt.[82]

TABLE 20
Heavy Metals in Sediments from Selected U.S. Estuaries (μg/g dry wt)

Estuary	Chromium	Copper	Lead	Zinc	Cadmium	Silver	Mercury
Casco Bay, ME	92.10	16.97	29.13	76.27	0.15	0.09	0.12
Merrimack River, MA	41.15	6.47	23.25	35.75	0.07	0.05	0.08
Salem Harbor, MA	2296.67	95.07	186.33	238.00	5.87	0.88	1.19
Boston Harbor, MA	223.67	148.00	123.97	291.67	1.61	2.64	1.05
Buzzards' Bay, MA	73.66	25.02	30.72	97.72	0.23	0.37	0.12
Narragansett Bay, RI	93.60	78.95	60.25	144.43	0.35	0.56	0.00
East Long Island Sound, NY	37.63	11.26	22.13	58.83	0.11	0.15	0.09
West Long Island Sound, NY	131.50	111.00	69.75	243.00	0.73	0.68	0.48
Raritan Bay, NJ	181.00	181.00	181.00	433.75	2.74	2.06	2.34
Delaware Bay, DE	27.76	8.34	15.04	49.66	0.24	0.11	0.09
Lower Chesapeake Bay, VA	58.50	11.32	15.70	66.23	0.38	0.08	0.10
Pamlico Sound, NC	79.67	14.13	30.67	102.67	0.33	0.09	0.11
Sapelo Sound, GA	51.80	5.93	16.00	38.33	0.09	0.02	0.03
St. Johns River, FL	37.67	9.77	26.00	67.67	0.18	0.11	0.07
Charlotte Harbor, FL	26.47	1.17	4.33	7.20	0.08	0.01	0.02
Tampa Bay, FL	23.70	4.97	4.67	9.10	0.15	0.08	0.03
Apalachicola Bay, FL	69.17	16.93	30.67	111.67	0.05	0.06	0.06
Mobile Bay, AL	93.00	17.40	29.67	161.00	0.11	0.11	0.12
Mississippi River Delta, LA	72.27	19.40	22.67	90.00	0.47	0.17	0.06
Barataria Bay, LA	52.07	10.50	18.33	59.33	0.19	0.09	0.05
Galveston Bay, TX	41.13	8.03	18.33	33.97	0.05	0.09	0.03
San Antonio Bay, TX	39.43	5.57	11.33	32.00	0.07	0.09	0.02
Corpus Christi Bay, TX	31.43	6.63	13.00	56.00	0.19	0.07	0.04
Lower Laguna Madre, TX	24.53	5.83	11.33	36.00	0.09	0.07	0.03
San Diego Harbor, CA	178.00	218.67	50.97	327.67	0.99	0.76	1.04
San Diego Bay, CA	49.70	7.67	11.61	58.67	0.04	0.76	0.04
Dana Point, CA	39.80	10.03	18.80	53.67	0.22	0.80	0.13
Seal Beach, CA	108.33	26.00	27.37	125.00	0.17	1.27	0.59
San Pedro Canyon, CA	106.50	31.33	17.33	118.33	1.17	1.20	0.32
Santa Monica Bay, CA	53.53	10.53	33.37	46.67	0.18	0.51	0.01
San Francisco Bay, CA	1466.67	160.71	67.39	501.66	0.51	0.37	0.25
Bodega Bay, CA	246.33	0.06	2.17	38.33	0.18	1.74	0.14
Coos Bay, OR	110.30	1.47	4.65	32.00	0.62	0.31	0.11
Columbia River Mouth, OR/WA	29.53	17.00	15.90	107.67	0.86	2.14	0.25
Nisqually Reach, WA	118.07	13.33	24.57	105.33	0.68	2.62	0.32
Commencement Bay, WA	69.50	51.33	34.63	101.00	0.77	5.90	0.01
Elliott Bay, WA	114.37	96.00	20.23	166.00	0.84	1.18	0.11
Lutak Inlet, AK	58.27	26.67	15.90	180.33	0.96	0.09	0.24
Nahku Bay, AK	23.27	9.80	43.30	191.33	1.09	4.37	0.23
Charleston, SC	86.33	16.03	27.33	72.67	—	—	—

From Young, D. and Means, J., in *National Status and Trends Program for Marine Environmental Quality. Progress Report on Preliminary Assessment of Findings of the Benthic Surveillance Project, 1984*, National Oceanic and Atmospheric Administration, Rockville, MD, 1987; U.S. Geological Survey, National Water Summary, 1986.

TABLE 21
Heavy Metal Concentrations (μg/g) in Contaminated Estuarine and Marine Sediments[a]

Metal	Baltic Sea[b] (Various Sources)	Bristol Channel/Severn Estuary, United Kingdom[c] (Industry Sewage)	Mersey Estuary, United Kingdom[d] (Sewage, Industry Including Chlor-Alkali)	Los Angeles Outfall, California[e] (Sewage)	Derwent Estuary, Tasmania[f] (Refinery, Chlor-Alkali)	Restronguet Creek, United Kingdom[c] (Mining)	Port Pirie, Australia[g] (Smelter)
As		8.0	71.0			2520 (13)	151 (1.0)
Cd	8.1 (<0.01)	1.1	3.9	66 (0.3)	862	1.2 (0.3)	267 (0.5)
Cu	283 (1.0)	54.0	144.0	940 (8.3)	>400	2540 (19)	122 (3.0)
Hg	9 (0.01)	0.48	6.2	5.4 (0.04)	1130	0.22 (0.12)	8.0
Ni	920 (1.0)	33.0	44.0	130 (9.7)	42	32 (28)	19.4 (12)
Pb	400 (2)	88.0	205.0	580 (6.1)	>1000	400 (2)	5270 (2)
Zn	2090 (6)	255.0	255.0	2900 (43)	>10,000	2090 (6)	16,667 (11)

[a] Maximum concentrations shown together with local background values (in parentheses), where given.
[b] Data from Brugman, L., Mar. Pollut. Bull., 12, 214, 1981.
[c] Data from Bryan, G. W. et al., J. Mar. Biol. Assoc. U.K., Pub. 4, 1985.
[d] Data from Langston, W. J., Estuarine Coastal Shelf Sci., 23, 239, 1986.
[e] Data from Mason, A. Z. and Simkiss, K., Exp. Cell Res., 139, 383, 1982.
[f] Data from Kojoma, Y. and Kagi, J. H. R., Trends Biochem. Sci., 3, 403, 1978.
[g] Data from Ward, T. J. et al., in Environmental Impacts of Smelters, Nriagu, J. O., Ed., John Wiley & Sons, New York, 1984, 1.

From Langston, W. J., in Heavy Metals in the Marine Environment, Furness, R. W. and Rainbow, P. S., Eds., CRC Press, Boca Raton, FL, 1990, 115. With permission.

iii. Lead

The concentrations of lead in the open ocean range from 0.001–0.014 μg/l. Somewhat higher values have been documented in enclosed seas (e.g., Baltic and Mediterranean Seas). Lead levels in estuaries and coastal marine waters exceed those in the open ocean by a factor of 10 or more.

In nearshore sediments, lead concentrations range from 10–100 μg/g dry wt. Much lower quantities of the metal (8–80 μg/g dry wt) occur in deep-sea sediments. Highest concentrations in bottom sediments have been recorded at contaminated coastal sites, such as the Southern California Bight (540 μg/g dry wt) and New York Bight (270 μg/g dry wt). At a waste dumpsite in the Firth of Clyde, Scotland, lead concentrations are as high as 320 μg/g dry wt, and in the Gannel estuary, they approach 2000 μg/g dry wt. In contaminated sediments near a coastal landfill in the Sea of Japan, lead reaches 530 μg/g dry wt.[83]

Organolead compounds are generally more toxic to marine organisms than inorganic forms. The mean concentrations of lead in mussels from coastal regions around the world range from 1–16 μg/g dry wt. However, concentrations up to about 3000 μg/g dry wt have been observed in mussels from contaminated waters. The amount of lead in the clam *Scrobicularia plana* from contaminated sediments of the Gannel Estuary, England, is nearly 1000 μg/g dry wt. In the same estuary, the polychaete *Nereis diversicolor* has concentrations of approximately 700 μg/g dry wt.[82] Plaice (*Pleuronectes platessa*) and flounder (*Platichthys flesus*) along the Danish marine coast have lead levels of 0.05–0.6 μg/g dry wt. The amount of lead in marine mammals (i.e., dolphins, porpoises, seals, and whales) from waters around the British Isles is 0.05–7.0 μg/g wet wt.[2]

iv. Mercury

Important chemical forms of mercury in marine environments include elemental mercury, divalent mercury ions (Hg^{2+}), and methylmercury ((CH_3)$_2$Hg). Mercury concentrations in the open ocean typically range from 0.001–0.004 μg/l, which are an order of magnitude less than those generally observed in coastal waters. Estuarine waters have highly variable levels of the metal. Hot spots of mercury contamination have been chronicled in Minamata Bay, Japan, (50–70 ng/l) and areas of the New York Bight (10–90 ng/l).

Mercury readily sorbs to particulate matter and tends to accumulate in bottom sediments. Concentrations in coastal seafloor sediments vary from <0.1 μg/g dry wt at untainted sites to >1000 μg/g dry wt at contaminated locations (Tables 20 and 21). The levels of mercury in sediments of the Ems and Mersey estuaries are 1–5 μg/g and 0.4–6.2 μg/g, respectively.[82] Deep-sea sediments in the North Atlantic have mercury levels of 0.008–0.6 μg/g dry wt.

Most of the mercury accumulating in shellfish (40–90%) and finfish (>90%) is methylmercury. The concentrations of mercury in mussels (*Mytilus* spp.) from coastal regions generally vary between 0.1 and 0.4 μg/g dry wt. In oceanic waters, mercury levels in euphausiids are 0.026–0.497 μg/g dry wt. While the amount of mercury in the muscle of oceanic finfish averages about 150 μg/kg, much higher levels have been recorded in samples from contaminated regions.[16] Large pelagics (i.e., marlin — *Makaira indica*, tuna — *Thunnus* spp., and swordfish — *Xiphias*

TABLE 22
Heavy Metal Concentrations in Marine Fauna (µg/g dry wt)

Metal	Geometric Mean and Location	Phytoplankton	Algae	Mussels	Oysters	Gastropods	Crustaceans	Fish	Seals, Mammals
Arsenic	Geometric mean	—	20.0	15.0	10.0	20.0	30.0	10.0	
	Newfoundland	—	9.8–17 (b)	1.6–5	—	4.0–11.5	3.8–7.6	0.4–0.8	
	England	—	26.0–54 (b)	1.8–15	2.6–10.0	8.1–38.0	16.0	1.7–8.7	
	Greenland	—	36.0 (b)	14–17	—	—	63–80	14.7–307	
Cadmium	Geometric mean	2.0	0.5	2.0	10.0	6.0	1.0	0.2	
	Spain	—	0.8–4 (b)	0.5–8	2.9–3.5	1.1–9.0	0.7–32	<0.4–4.3	
	England	—	0.2–53 (b)	3.7–65	6–54	3.5–1,120	2.8–33	0.06–3.96+	2.2–11.6+
	Australia	—	—	4.2–83	9–174	2.8–30.0	—	0.05–0.4+	—
	Norway	—	1.0–13.0 (b)	1.9–140	—	0–51	1.9–7.0	<0.01–0.03+	
Copper	Geometric mean	7.0	15.0	10.0	100	60.0	70.0	3.0	
	Spain	—	5–26 (g)	6–14	120–435	5–50	110–435	<0.6–10.0	
	England	—	4–141 (b)	7–15	20–6480	0–1750	6–64+	0.5–14.6+	
	California	—	—	7–77	10–2100	3–177	(4–150)	(16–29.3)	14.5–386.0 (m)
	Norway	—	9–170 (g)	3–120	—	17–190	2–90	—	
Lead	Geometric mean	4.0	4.0	5.0	3.0	5.0	1.0	3.0	
	Spain	—	4–20 (g)	2–15	4–11	10–27	<1.2–11.0	<1.2–2.2	
	England	—	16–66 (g)	7–19	5–17	0.2–0.8	8.0	14–28	0.4+
	California	—	—	0.3–42.0	—	0.6–21.0	—	<0.001–5.3	0.3–34.2 (s)
	Norway	—	3–1200 (g)	2–3100	—	0–39	8.3	—	
Mercury	Geometric mean	0.17	0.15	0.4	0.4	0.2	0.4	0.4	0.6–103.0 (m)
	Hawaii	—	—	—	—	0–0.03	0.03–0.12	0.02–23.0	0.1–700.0 (s)
	California	—	—	—	—	<0.01–0.07+	0.02–0.04+	0.02–0.2	
	Atlantic	0.2–5.3	<0.01–0.07+	<0.01–0.13+	0.02–180.0+	—	<0.05–0.6+	0.1–9.0	—
	Mediterranean	—	<0.5–0.7+	0.25–0.4	—	0.1–3.5+	0.3–4.5+	0.1–29.8+	—

Nickel	Australia	—	—	0.05–0.23+	1.5–8.2	0.32–0.65	—	0.3–16.5	0.1–106.0 (m)
	Norway	0.5–25.2	—	0.24–0.84	—	0.61	0.31–0.39	0.14–7.3	0.4–225.0 (s)
	England	—	<0.01–25.5 (g)	0.64–1.86	0.56–1.2	0.02–1.84	0.98	0.02–1.8	
	Geometric mean	3.0	3.0	3.0	1.0	2.0	1.0	1.0	
	England	—	4–33 (g,b)	5–12	2–174	8.8–12.3	1.1–12.3	0.5–10.6	
	California	—	—	3.3–20.0	—	1.8–18.5	—	—	
Silver	Geometric mean	0.2	0.2	0.3	—	1.0	0.4	0.1	
	California	—	—	0.7–46.0	—	0.4–10.7	—	0.1–1.2 (m)	
Zinc	Geometric mean	38.0	90.0	100	1700	200.0	80.0	80.0	
	South Africa	0.6–710.0	5.6+ (g)	73–113	400–886	12.0+	17.0+	3.2–7.2+	
	California	0.1–725.0	46–244	70–8430	1.7–288.0	—	78–875	78–875	
	Spain	—	63–345 (g)	190–370	310–920	60–120	79–330	21–220	
	Australia	—	—	170–1350	3740–38,700	56–1050	—	4–375	
	England	—	28–1240 (g)	12–779	1830–99,200	9.7–1500.0	36–82	2–342	
	Norway	—	20–2310 (g)	105–2370	—	87–2900	12–32	—	

Note: +, ppm wet weight; all other values in ppm dry weight; b, brown algae; g, green algae; m, mammals; s, seals.

From Förstner, U., in *Chemistry and Biogeochemistry of Estuaries*, Olausson, E. and Cato, I., Eds., John Wiley & Sons, Chichester, U.K., 1980, 307. With permission.

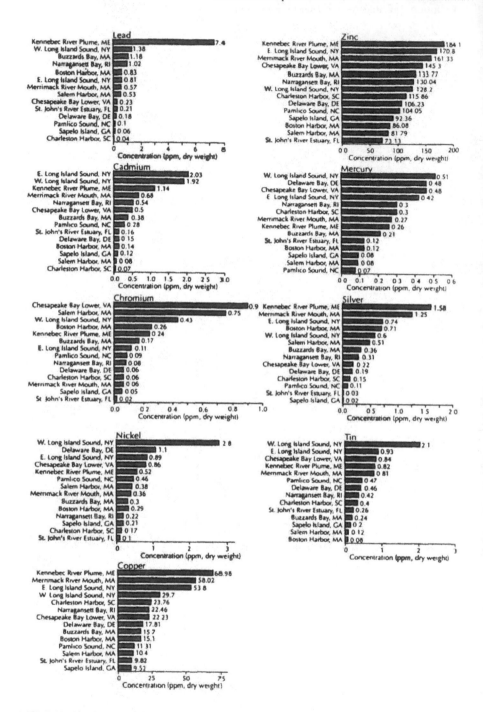

FIGURE 13 Concentrations of heavy metals in Atlantic Coast fish liver tissue sampled from various U.S. estuaries by the NOAA Status and Trends Program. (From Larsen, P. F., *Rev. Aquat. Sci.*, 6, 67–86, 1992. With permission.)

gladias) exhibit the highest mercury values (1000-5000 µg/kg) in muscle. Total mercury concentrations in the liver of seabirds commonly range from 4.9–306.0 µg/g dry wt. The levels of mercury in marine mammals (i.e., dolphins, porpoises, and seals) generally are between 0.3 and 430.0 µg/g wet wt.[2]

v. Tin

Inorganic tin concentrations in open ocean waters vary from 0.003–0.008 µg/l. Localized elevated levels of dissolved tin, approaching 50 µg/l, have been documented in coastal waters receiving industrial discharges (e.g., Poole Harbor, England).[82] Of greater toxic concern than the inorganic forms in marine waters are organotin compounds, notably tributyl tin oxide and tributyl tin fluoride (TBT). TBT levels vary widely in estuarine waters from less than 1 ng/l in open areas to 600 ng/l near marinas. TBT is extremely toxic, and at concentrations as low as 0.01 µg/l and 1.0 µg/l cause reduced growth and increased mortality, respectively, in various marine organisms.

Concentrations of total tin in some estuarine sediments (e.g., Hayle and Helford estuaries, England) are greater than 1000 µg/g dry wt. Levels of TBT in estuarine bottom sediments typically range from 0.005–0.5 µg/g dry wt but can be much higher. For example, values of <3–3935 ng/g dry wt have been detected in bottom sediments of east coast estuaries in England.[84]

Marine invertebrates can accumulate significant levels of tin. In Poole Harbor, England, maximum concentrations of total tin in *Nereis diversicolor*, *Littorina littorea*, *Scrobicularia plana*, and *Mya arenaria* equal 7.03, 8.81, 11.53, and 21.4 µg/g dry wt, respectively. Maximum total organotin in these organisms amounts to 6.49, 4.34, 10.4, and 21.4 µg/g dry wt, respectively.[82] Uptake and assimilation of organotin compounds from marine waters appear to be rapid. Over a 41-wk culture period, Pacific oysters (*Crassostrea gigas*) were shown to accumulate tin (1.41 mg/kg) and TBT (0.87 mg/kg) from antifouling paints but readily lost this burden during depuration.[2]

d. Effects on Biotic Communities

There is frequent reference in the literature to heavy metal impacts on estuarine and marine organisms. For instance, dissolved copper concentrations of 1–10 µg/l have been shown in the laboratory to increase mortality of young bay scallops (*Argopecten irradians*) and surf clams (*Spisula solidissima*), as well as isopods (*Idotea balthica*).[85,86] Rygg[87] concluded that the most sensitive species in benthic communities of Norwegian fjords were absent at sites where the sediment concentrations of copper exceeded 200 µg/g dry wt. In addition, species diversity of benthic faunal communities in Norwegian fjords showed a strong negative correlation with copper concentrations in bottom sediments. Bryan and Langston[82] also reported that cockles (*Cerastoderma edule*), clams (*Macoma balthica*), and mussels (*Mytilus edulis*) were missing from areas of the Fal estuary in England where bottom sediments contained more than 2000 µg/g dry wt of copper.

Even at extremely low levels, cadmium may adversely affect benthic organisms. Reproduction in the polychaete *Neanthes arenaceodentata* was inhibited at a seawater concentration of only 1 μg/l.[88] Luoma et al.[89] demonstrated that the condition of clams (*Corbicula* sp.) decreased as the concentration of cadmium in sediments increased from 0.1–0.4 μg/g dry wt. Fabiano et al.[90] found a negative correlation between bacterial biomass and cadmium concentrations in the northern Tyrrhenian Sea, with high cadmium concentrations possibly arresting bacterial development.

Plankton populations are also very sensitive to cadmium. Kayser and Sperling[91] discerned growth inhibition in the phytoplanker *Prorocentrum micans* at cadmium concentrations of about 1 μg/l. Paffenhöfer and Knowles[92] observed reduced reproduction in the copepod *Psuedodiaptomus coronatus* at cadmium levels of 5 μg/l. At this concentration, growth of juvenile plaice (*Pleuronectes platessa*) and abundance of isopods (*Idotea balthica*) also decreased.[86,93]

The toxicity of inorganic lead to marine organisms is less than that of many other heavy metals. Under experimental conditions, marine organisms display adverse effects of inorganic lead only when concentrations are about 100 times greater than those in coastal waters (i.e., 0.01–0.1 μg/l). High levels of inorganic lead in bottom sediments also do not seem to be especially toxic to benthic species. For example, lead concentrations of nearly 1000 μg/g dry wt in the clam *Scrobicularia plana* from the Gannel estuary did not result in any observable effects, as did lead levels of about 100 μg/g dry wt in the limpet *Acmaea digitalis* from the California coast.[2,82] Mussels from contaminated waters of the Sorfjord in Norway have accumulated lead to levels of approximately 3000 μg/g dry wt without any evident impacts.[16]

Organolead compounds are more toxic to marine organisms than inorganic lead. For instance, Marshall and Jarvie[94] revealed that trimethyl and triethyl lead concentrations of 70–100 μg/l were lethal to *S. plana*. In 1979, some 2400 shorebirds died in the Mersey estuary, owing to the accumulation of trialkyl lead compounds (10 μg/g wet wt) in liver tissue.

A major fraction of mercury accumulating in the tissues of marine organisms is methylmercury, although inorganic mercury may be the dominant form in seawater and bottom sediments. Mercury, when methylated, undergoes biological magnification and, hence, poses a threat to man, as demonstrated by the death of more than 100 people in Minamata, Japan, during the 1950s and 1960s due to the mobilization of methylmercury through the food chain of Minamata Bay. High mortality of invertebrate and fish populations in the bay was also ascribed to this mercury pollution. Organic forms of mercury are more toxic to marine mammals and man than inorganic forms because they are poorly excreted and, thus, can accumulate to very high levels.

Apart from the effects of food chain biomagnification of mercury compounds, there are surprisingly few reports of major impacts of mercury pollution on estuarine and marine communities. However, inferences have been made. For example, where levels of mercury in bottom sediments are high, such as in the Derwent (1100 μg/g dry wt), Ems (1–5 μg/g dry wt), and Mersey (0.4–6.2 μg/g dry wt) estuaries, intertidal communities are absent.[95-97] Nevertheless, unequivocal demonstration of the toxic nature of mercury in these sediments is lacking, and caution must be exercised in extrapolating toxic implications to observed changes in benthic communities.

The presence of organotin compounds in marine waters has assumed increasing significance in recent years in regard to impacts on biotic communities. Many investigations of organotins conducted to date have focused on sublethal effects on invertebrates, notably molluscs (i.e., clams, oysters, mussels, and scallops). When exposed to tributyl tin (TBT) concentrations as low as 0.01 μg/l, marine invertebrates often exhibit changes in growth and reproduction. Oyster shell abnormalities, reduced growth rates in mussels, imposex in stenoglossan gastropods, failure of recruitment in clams, and breakdown of sexual differentiation and failure of settlement in oysters have all been linked to TBT exposure (Table 23).[2] These alterations can significantly impact invertebrate communities by eliminating susceptible populations in heavily polluted coastal regions.[98]

Organotins can be magnified to relatively high proportions in invertebrates (commonly >60%), even though they constitute a small fraction of the total tin in some media (e.g., estuarine bottom sediments). Food chain magnification of TBT does not appear to occur in higher organisms (i.e., crustaceans, fish, and mammals), however, because they possess the necessary enzymes to break down the contaminant relatively rapidly.[82] While TBT is extremely toxic to many marine organisms, it is not very stable and degrades to less toxic compounds in a few weeks in marine environments. The use of mussels as sentinel organisms to monitor TBT promises to provide an effective measure of long-term changes of contaminant inputs to coastal systems.[99]

TABLE 23
Effects of Tributyltin (TBT) on Marine and Estuarine Organisms

Species	Concentration of TBT in water (ng/l as Sn)	Effect
Nucella lapillus, dogwhelk	1–10	Imposex, impaired reproduction
Crassostrea gigas, oyster[a]	8	Shell thickening
	20	Reduced growth, viability
Mytilus edulis, mussel[a]	40	Reduced viability
Venerupis decussata, clam[a]	40	Reduced growth, viability
Gammarus oceanicus, amphipod[a]	120	Reduced growth
Homarus americanus, lobster[a]	400	Reduced viability
Pavlova lutheri *Skeletonema costatum* } microalgae *Dunaliella tertiolecta*	40–400	Reduced growth

[a] Larvae.

From Langston, W. J., in *Heavy Metals in the Marine Environment*, Furness, R. W. and Rainbow, P. S., Eds., CRC Press, Boca Raton, FL, 1990, 117. With permission.

7. Radioactivity

a. Sources

i. Natural Sources

Radionuclides in estuarine and marine environments derive from both natural and anthropogenic sources (Tables 24–26). Natural background radiation is attributable to cosmogenic and primordial radionuclides. Cosmogenic radionuclides originate from the interaction of primary cosmic rays with matter in the atmosphere and on the surface of the earth. The collision of high energy particles (cosmic radiation) with nitrogen, oxygen, argon, and other atoms in the upper atmosphere produces secondary particles, primarily neutrons and protons, together with pions, kaons, and

---

TABLE 24
Comparison of Annual Doses Received by Marine Organisms from Natural Sources of Radiation (mGy/yr)

	Marine Organisms				
	Phytoplankton	Zooplankton	Mollusca	Crustacea	Fish
Cosmic	4.4×10^{-2}	4.4×10^{-2}	4.4×10^{-2}	4.4×10^{-2}	4.4×10^{-2}
Water	3.5×10^{-2}	1.8×10^{-2}	9×10^{-3}	9×10^{-3}	9×10^{-3}
Sediments $\beta + \gamma$	0	0	0.27–3.2	0.27–3.2	0–3.2
Internal	0.17–0.64	0.23–1.4	0.65–1.3	0.69–1.9	0.24–0.37
Total	0.27–0.72	0.29–1.7	0.97–4.6	1.0–5.2	0.29–3.7

From International Atomic Energy Agency, *Effects of Ionizing Radiation on Aquatic Organisms and Ecosystems*, Tech. Rep. Ser. No. 172, International Atomic Energy Agency, Vienna, 1976. With permission.

TABLE 25
Activity at 12 Radioactive Solid Waste Dumpsites in the North Atlantic Ocean, 1949–1979

	Dumped Activity (Bq)	
	α	β-γ
Average per site[a]	3.7×10^{13}	2.6×10^{15}
Range per site[b]	$(1.1–2.8) \times 10^{17}$	$(1.9–3.7) \times 10^{17}$
Maximum Bq (per site per year)	5.2×10^{13}	3.5×10^{15}
Total for all 12 sites	4.8×10^{14}	3.1×10^{16}

[a] The sites used for more than 4 yr.
[b] Range of years the sites were used was 1 to 14.

From Preston, A., in *Wastes in the Ocean*, Vol. 3, *Radioactive Wastes and the Ocean*, Park, P. K., Kester, D. R., Duedall, I. W., and Ketchum, B. H., Eds., John Wiley & Sons, New York, 1983, 107. With permission.

TABLE 26
Inventory of Radioactive Waste Input to the Ocean during the Period
1957–1979

	^{239,240}Pu	^{137}Cs	^{90}Sr	^{14}C	^{3}H
Total global fallout (by early 1970s)	1.2×10^{16}	6.2×10^{17}	4.3×10^{17}	2.2×10^{17}	1.1×10^{20}
North Atlantic Ocean (early 1970s)	2.3×10^{15}	1.2×10^{17}	8.5×10^{16}	—	2.4×10^{19}
Windscale discharge (1957–1978)	5.2×10^{14}	3.1×10^{16}	4.8×10^{15}	—	1.4×10^{16}
	Total α-emitters	Total β- and α-emitters (other than ^{3}H)			
NEADS (1967–1979)	3.1×10^{15}	—	9.5×10^{15}	—	9.7×10^{15}

ᵃ Values in Bq.

From Needler, G. T. and Templeton, W. L., *Oceanus*, 24, 60–67, 1981. With permission.

electromagnetic radiation.[100] Secondary cosmic radiation accounts for nearly all cosmic rays impinging on the sea surface.

Primordial radionuclides are those that were generated at the time of formation of the earth. Derived from internal radioactive sources, primordial radionuclides principally occur in the earth's lithosphere. Primordial radionuclides of significance in the earth's upper crust include ^{40}K (~3 ppm), ^{232}Th (10–15 ppm in granitic rocks), and ^{238}U (together with ^{232}U and ^{234}U averages 3–4 ppm in granite).[101] All have a half life greater than 10^9 years. ^{40}K is the major, naturally occurring source of internal radiation in estuarine and marine organisms.

During the decay of a radioactive substance, a parent radionuclide undergoes spontaneous disintegration of its nucleus with the emission of one or more radiations and the formation of a daughter nuclide.[102] Radioactive particles release ionizing radiation — α, β, and neutron particles, as well as γ-rays — that can damage biological tissue and, hence, pose a threat to living organisms. The degree of radiation damage to living tissue depends on both the type and dose of radiation incurred by an organism.

Background radioactivity is measured in the International System of Units (i.e., Becquerel [Bq], Gray [Gy], and Sievert [Sv]). Cosmic rays supply a dose rate of about 4×10^{-8} Sv/hr at the sea surface, which decreases to about 5×10^{-9} Sv/hr at 20 cm depth and nearly negligible rates below 100 m. The total background radioactivity in surface seawater (~12.6 Bq/l) is less than that in marine sands (185–370 Bq/kg) and muds (740–1110 Bq/kg). Marine organisms receive radiation doses generally less than 5 mGy/yr from natural sources (Table 24).[28,103]

ii. Anthropogenic Sources

Since 1944, artificial radioactivity has entered estuarine and marine environments from several major anthropogenic sources, namely nuclear weapons testing, the nuclear fuel cycle (e.g., wastes from nuclear power plants and nuclear fuel reprocessing facilities), accidents involving nuclear material, military related projects (e.g., disposal of military hardware), and direct dumping of low level radioactive waste in the deep sea. The relative importance of these anthropogenic sources of radionuclides has changed over the past 50 years. Between 1945 and 1980, most artificial radionuclides entering the sea originated from nuclear weapons testing. More than 1200 nuclear weapons tests were conducted throughout the world during these 35 years, with the major bomb testing programs conducted between 1954 and 1962. After the Partial Nuclear Test Ban Treaty of 1963, underground nuclear weapons testing became most important, and atmospheric weapons fallout to the sea diminished substantially. Atmospheric fallout of radionuclides from nuclear weapons testing has been about four times greater in the northern hemisphere than in the southern hemisphere. A number of fission products of nuclear detonations in the atmosphere are biologically significant in the sea, notably carbon (^{14}C), cesium (^{137}Cs), strontium (^{89}Sr, ^{90}Sr), and iodine (^{131}I) isotopes. Atmospheric nuclear explosions have produced large quantities of radioactivity. For example, the 14 kiloton nuclear explosion at Hiroshima generated approximately 8×10^{24} Bq of activity.[101] Nuclear weapons testing has produced about 5×10^{15} Bq/yr of ^{14}C since 1945, compared to approximately 1×10^{15} Bq/yr of ^{14}C generated naturally by cosmic ray interactions.

Since 1980, other anthropogenic sources of radioactive materials in the sea have become important, particularly those associated with the nuclear fuel cycle. Radioactive waste is produced at various stages of the nuclear fuel cycle (i.e., mining, milling, conversion, isotopic enrichment, fuel element fabrication, reactor operation, and fuel reprocessing) (Figure 14). In addition, the use of radioisotopes in agriculture, industry, and medicine accelerated significantly after 1975 and have contributed to the pool of artificial radionuclides. The number of nuclear power plants on line worldwide grew from 66 in 1970 to 430 in 1996, with another 5 units under construction. Low level radioactive wastes are discharged from coastal nuclear power plants and fuel reprocessing plants into the sea. Nuclear-powered submarines and ships also release some radioactivity, albeit in smaller amounts than the aforementioned facilities. The total activity discharged from nuclear power plants generally ranges from about 1–8 TBq/yr of non-tritium isotopes and 1–350 TBq/yr of tritium.

Major accidents at nuclear power plants, such as at Three Mile Island in Pennsylvania and Chernobyl in the Ukraine, release considerable activity that may be detected in seawater samples. The Three Mile Island nuclear power plant accident in 1979 released approximately 10^{17} Bq of activity, and the Chernobyl nuclear power plant accident in 1986, about 1.85×10^{18} Bq.[101] The release of radionuclides to the environment during these accidents resulted in contamination of marine waters. In the case of Chernobyl, radiocesium levels in muscle of fish from the southern Baltic Sea increased three- to four-fold, and ^{134}Cs, ^{137}Cs, and ^{106}Ru in fish in the Danube River increased five-fold.[104] Apart from power plant accidents, the loss of nuclear-powered submarines at sea (~10) contributed additional artificial radioactivity to marine waters.

The disposal of radioactive materials in the sea has been restricted to low level wastes released via discharges from land-based sources or direct dumping of packaged wastes. While direct discharges of low level radioactive wastes still occur worldwide, packaged dumping at sea ceased during the early 1980s. Licensed under

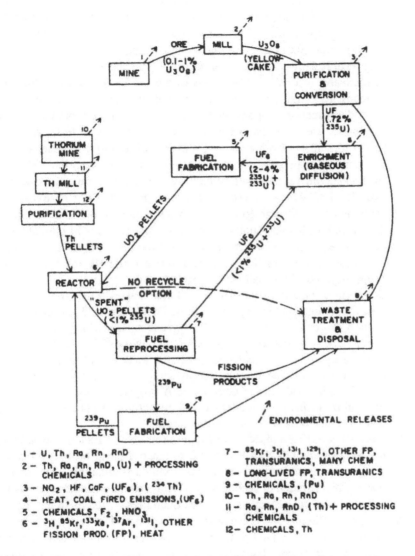

1 – U, Th, Ra, Rn, RnD
2 – Th, Ra, Rn, RnD, (U) + PROCESSING CHEMICALS
3 – NO_2, HF, CaF, (UF_6), (^{234}Th)
4 – HEAT, COAL FIRED EMISSIONS, (UF_6)
5 – CHEMICALS, F_2, HNO_3
6 – ^{3}H, ^{85}Kr, ^{133}Xe, ^{37}Ar, ^{131}I, OTHER FISSION PROD. (FP), HEAT

7 – ^{85}Kr, ^{3}H, ^{131}I, ^{129}I, OTHER FP, TRANSURANICS, MANY CHEM
8 – LONG-LIVED FP, TRANSURANICS
9 – CHEMICALS, (Pu)
10 – Th, Ra, Rn, RnD
11 – Ra, Rn, RnD, (Th) + PROCESSING CHEMICALS
12 – CHEMICALS, Th

FIGURE 14 Components of the nuclear fuel cycle, illustrating radioactive releases to the environment. (From Whicker, F. W. and Schultz, V., *Radioecology: Nuclear Energy and the Environment*, Vol. 1, CRC Press, Boca Raton, FL, 1982, 257 pp. With permission.)

the London Convention, the dumping of packaged low level radioactive wastes at sea by Euopean countries and the United States amounted to about 2000 mt/yr prior to the time it was halted.

b. Types of Radioactive Waste

Six categories of radioactive waste have been defined.[105,106] These include:

1. **High level wastes** — These wastes are either fuel assemblies that are discarded after having served their useful life in a nuclear reactor (spent

fuel) or the portion of the wastes generated in the reprocessing that contain virtually all of the fission products and most of the actinides not separated out during reprocessing. The wastes are being considered for disposal in geologic repositories or by other technical options designed to provide long-term isolation of the wastes from the biosphere.

2. **Transuranic wastes** — These wastes are produced primarily from the reprocessing of defense spent reactor fuels, the fabrication of plutonium to produce nuclear weapons, and, if it should occur, plutonium fuel fabrication for use in nuclear power reactors. Transuranic wastes contain low levels of radioactivity, but varying amounts of long-lived elements above uranium in the periodic table of elements, mainly plutonium. They are currently defined as materials containing >370 Bq/g of transuranic activity.

3. **Low level wastes** — These wastes contain <370 Bq/g of transuranic contaminants. Although low level wastes require little or no shielding, they have low, but potentially hazardous concentrations of radionuclides, and, consequently, require management. Low level wastes are generated in almost all activities involving radioactive materials and are presently being disposed of by shallow land burial.

4. **Uranium mine and mill tailings** — These wastes are the residues from uranium mining and milling operations. They are hazardous because they contain low concentrations of radioactive materials which, although naturally occurring, contain long-lived radionuclides. The tailings, with a consistency similar to sand, are generated in large volumes, about 10^{10} kg/yr in the United States, and are presently stored in waste piles at the site of mining and milling operations. A program is underway either to immobilize or bury uranium mine and mill tailings to prevent them from being dispersed by wind or water erosion.

5. **Decontamination and decommissioning wastes** — As defense and civilian reactors and other nuclear facilities reach the end of their productive lifetimes, parts of them will have to be handled as either high- or low-level wastes, and disposed of accordingly. Decontamination and decommissioning activities will generate significant quantities of wastes in the future.

6. **Gaseous effluents** — These wastes are produced in many defense and commercial nuclear facilities, such as reactors, fuel fabrication facilities, uranium enrichment plants, and weapons manufacturing facilities. They are released into the atmosphere in a controlled manner after passing through successive stages of filtration and mixing with air, where they are diluted and dispersed.

i.   Concentrations

Typical maximum activity concentrations in low level radioactive waste are 4 GBq/t (α) and 12 GBq/t (β, γ). Maxmium specific activities of intermediate level waste amount to about 2×10^{12} Bq/m^3 (α) and 2×10^{-14} Bq/m^3 (β, γ). By comparison, typical maximum activities of high level waste are 4×10^{14} Bq/m^3 (α) and 8×10^{16} Bq/m^3 (β, γ).[101]

High level radioactive wastes, which cannot be dumped at sea, are defined as those containing, per ton of material: more than 37,000 TBq tritium; 37 TBq β- and γ-emitters, 3.7 TBq ^{90}Sr and ^{137}Cs; or 0.037 TBq α-emitters with half-lives of more than 50 years.[16] Provisions of the London Convention of 1972, now ratified by more than 90 states, prohibit the dumping of high level radioactive waste in the sea. In addition to this international control, the U.S. Marine Protection, Research and Sanctuaries Act of 1972, as well as the Low Level Radioactive Waste Policy Act of 1980 and it amendments (1985), regulate the disposal of low level radioactive wastes at sea by the United States.

Initial dumping of radioactive solid wastes at sea occurred in 1946, with low and intermediate level radioactive wastes being dumped there until 1982.[16] The United States alone dumped about 107,000 containers of low level radioactive wastes ($\sim$4.3 × 10^{15} Bq) between 1946 and 1970, nearly all (95–98%) at four sites: (1) the Farallon Island site in the Pacific Ocean west of San Francisco; (2) the 2800-m site in the northwest Atlantic Ocean off the mid-Alantic states; (3) the 3800-m site in the northwest Atlantic off the mid-Atlantic states; and (4) the Massachusetts Bay site. The most frequently used dumpsite was the 2800-m site in the northwest Atlantic Ocean. Over this same 25-yr period, eight European countries (i.e., Belgium, England, France, Germany, Italy, The Netherlands, Sweden, and Switzerland) dumped low and intermediate level radioactive wastes at 10 sites in the northeast Atlantic. After 1971, dumping was centered at a single site, the Northeast Atlantic Dumpsite (NEADS), a rectangular area bounded by 45°59′N to 46°10′N by 16° to 17°30′W at 4.4-km depth. Use of this site was suspended in 1983. Between 1949 and 1980, European countries dumped 1 × 10^5 mt of radioactive waste at sea (3 × 10^{16} Bq) nearly all in the northeast Atlantic (Table 25).

The ocean received considerable amounts of radioactive waste from several sources during the 1950s, 1960s, and 1970s (Table 26). Today, industrial, scientific, medical, and military applications of radioactivity produce a large volume of low level radioactive waste, including contaminated solutions and solids, laboratory ware and chemicals, cleaning and decontamination material, and machinery components and other equipment.[101] Perhaps of more direct concern to estuarine and coastal marine environments are the low level radioactive liquid wastes operationally discharged from nuclear power plants and fuel reprocessing facilities. Nuclear power plants contribute only a very small amount of the total low level radioactive waste introduced into marine waters. In shallow coastal marine waters, fuel reprocessing facilities may account for greater volumes of low level radioactive waste. For example, the annual discharge of α-emitters in liquid effluents from nuclear fuel reprocessing facilities at Windscale and Downreay in England and La Hague in France from 1972 to 1977 ranged from 1.1 × 10^{11} to 1.7 × 10^{14} Bq, and that of β- and γ-emitters, approximately 2.0 × 10^{13} to 7.7 × 10^{15} Bq. The North Sea and North Atlantic have served as the receiving water bodies for these effluents.

c. Effects on Biotic Communities

The exposure of marine organisms to radiation may lead to no observable effects, or it can elicit a wide range of responses — such as alterations of growth and development, physiological changes (e.g., on hemopoietic and reproductive systems), genetic changes, cancer, decreased longevity, and death — depending on total dose, dose rate, type of radiation, and exposure period.[103] Both acute and chronic effects of radiation exposure have been repeatedly demonstrated in the laboratory. As shown by Whicker and Schultz,[102] major taxonomic groups of organism exhibit large differences in sensitivity to radiation. For instance, acute lethal doses (LD_{90} values) for blue-green algae range from about 4000–12,000+ Gy, and for other algae 30–1200 Gy. Among animal groups, mollusks are relatively insensitive, with acute lethal doses (LD_{50} values) of approximately 200–1090 Gy. Crustacea ($LD_{50} = 56$–566 Gy) and fish ($LD_{50} = 11$–56 Gy) appear to be much more sensitive to radiation exposure than mollusks but less sensitive than mammals ($LD_{50} = 2$–13 Gy). In general, less advanced and complicated organisms tend to be more tolerant to ionizing radiation than are higher organisms, as reflected by the following order of radiation sensitivity exhibited by major taxonomic groups in the laboratory: bacteria < algae < crustaceans < fish < birds < mammals (Figure 15). Adults are less sensitive than gametes and larval stages to radiation damage.[16]

In nature, effects of radiation on marine populations are likely to be sublethal due to chronic exposures of individuals to relatively low radiation doses typically encountered in marine environments. Under such conditions, dose rates are not uniform, and the total dose depends on both internal irradiation (i.e., resulting from absorption and ingestion) and external irradiation. Chronic exposures of <10 mGy would not be expected to cause measurable deleterious changes in estuarine or marine communities. Direct mortality is not significant for any taxonomic group until a dose of ~2 Gy is reached. An entire community may be disrupted at acute lethal exposures of 10 Gy or more, which would be incurred only during catastrophic nuclear events.[103]

Only one case of radiation mortality of marine organisms has been documented in nature: the exposure of carnivorous fish in the Marshall Islands to ^{131}I from fallout during nuclear weapons testing.[107] Existing radiation conditions in the sea have not resulted in any measurable environmental impact on estuarine and marine communities.[16] This does not infer that marine organisms have not experienced any tissue damage from chronic radiation exposure. Tissue repair processes may keep pace with damage caused by low level, chronic radiation exposure. However, most studies of chronic radiation exposure of marine communities have not uncovered major detrimental effects either to populations or communities.

Besides radiation-induced somatic effects, genetic disturbances are the inevitable consequence of natural or artificial radiation exposure. Genetic research has attempted to ascertain mutations and chromosomal aberrations in marine organisms subsequent to acute and chronic exposures to radiation. Heritable mutations arising from radiation exposure are a serious concern. Any dose of ionizing radiation, no matter how small, can induce mutation, although the degree of impact increases proportionally to the radiation dose. Genetic changes have been measured in aquatic biota from areas contaminated with radiation, and DNA changes have been documented in field-collected waterfowl. However, it has been difficult to establish a connection between low level, chronic exposure and abnormalities or defects developing from irradiation of the gonads. For instance, the overall fecundity of aquatic

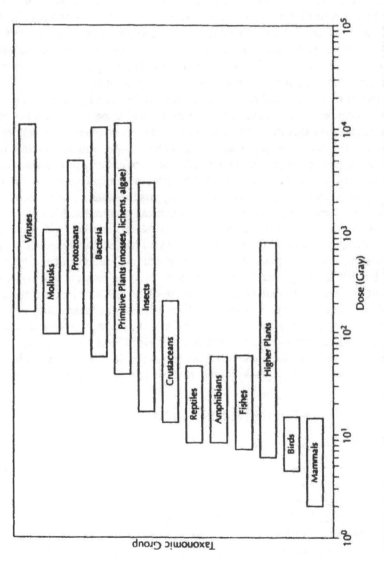

FIGURE 15 Ranges of acute lethal radiation doses to various taxonomic groups of organisms. (From Whicker, F. W. and Schultz, V., *Radioecology: Nuclear Energy and the Environment*, Vol. 1, CRC Press, Boca Raton, FL, 1982, 257 pp. With permission.)

populations inhabiting areas contaminated with radiation does not appear to be adversely impacted.[103]

Radiation exposure of marine organisms occurs internally due to absorption or ingestion and externally as a consequence of radiation present in seawater and sediments, as well as radionuclides sorbed to exterior body surfaces. Biotic accumulation of radionuclides is greatest near nuclear fuel processing plants, nuclear power plants, and facilities producing nuclear explosives. Little evidence exists, however, for biomagnification at higher trophic levels, although a few radionuclides preferentially accumulate in higher trophic level organisms (e.g., ^{137}Cs in fish). Phytoplankton and zooplankton, characterized by large surface area-to-volume ratios, accumulate radionuclides relatively quickly. For larger organisms (e.g., macroalgae, macroinvertebrates, fish, and mammals), the degree of uptake is low.

The transfer of radionuclides through marine food chains is a significant human health concern because radiation damages reproductive and somatic cells leading to chromosome aberrations that may result in human cancer or other malfunctions. Since radioactive substances readily sorb to particulate matter and accumulate in bottom sediments, benthic organisms and benthic feeders coming in contact with contaminated sediments may be expected to have higher radionuclide concentrations than pelagic organisms and pelagic feeders inhabiting pristine waters. Hence, benthic organisms of direct dietary importance to man, which can accumulate high concentrations of radionuclides (e.g., clams, mussels, and oysters), are the focal point of many biomonitoring programs. For example, the NOAA's National Status and Trends Program conducted a study of the concentrations of artificial radionuclides (^{241}Am, $^{239+240}$Pu, ^{238}Pu, ^{137}Cs, ^{110}Ag, ^{90}Sr, ^{65}Zn, ^{58}Co, and ^{60}Co) in mussels and oysters collected along the coastal United States during the early 1990s.[108] Results of this study indicate that over the last 15 years, $^{239+240}$Pu and ^{137}Cs concentrations in the bivalves decreased significantly and that in many cases the ^{241}Am activities are also lower in 1990 samples than those collected in the mid–1970s by other programs. These reductions in radionuclide concentrations reflect the decrease in atmospheric fallout of the contaminants subsequent to the Partial Nuclear Test Ban Treaty of 1963.

IX. OTHER ANTHROPOGENIC IMPACTS

A. Electric Generating Stations

The siting of electric generating stations in the coastal zone, particularly large (> 500 MW) units, can affect biotic communities in several ways, most notably by: (1) the calefaction of receiving waters due to waste heat discharges; (2) the release of biocides and other chemicals; (3) impingement of larger organisms on intake screens; and (4) entrainment of smaller life forms in cooling water systems. Heated effluent discharged from power stations with open-cycle, once-through cooling systems averages about 12°C above ambient for 100 MWe oil or coal-fired facilities and approximately 15°C above ambient for 1000 MWe nuclear-powered units. Calefaction or thermal loading commonly alters physiological processes in organisms (e.g., enzyme activity, respiration, and reproduction) and can lead to a significant reduction in growth as well as increased mortality and changes in community structure (Table 27). Episodes of heat

TABLE 27
Thermal Effects on Major Groups of Marine Organisms

Organism	Temperature		
	Critical Temp (°C)	ΔT (°C)	Effect
Algae	34	7–10	Major shift in community composition
Seagrass		4–5	Destruction of bed
Mangroves	37–38	5–10	Diminished photosynthesis; recruitment failure
Copepods	35–40		Mass mortality
Corals		3–4	High mortality
Fish	34.0–37.5		Incipient death

From GESAMP, *Thermal Discharges in the Marine Environment*, Report and Studies No. 24, FAO, Rome, 1984.

shock and cold shock mortality are manifested most conspicuously in near-field regions closest to power plants. However, avoidance and attraction reactions and other behavioral responses of biota may be evident even in far-field regions at greater distances from the plant sites, where temperature increases are no more than 1 or 2°C.

While thermal shock mortality of up to one million or more finfish during a single event has been documented in the literature,[109] the trend is for much lower mortality in recent years because of improved methods of plant operation and structural upgrading of intake and discharge systems. Sublethal effects of thermal discharges, such as reduced hatching success of eggs and developmental inhibition of larvae, may be even more detrimental to the long-term viability of impacted populations. In extreme cases, thermal loading can cause a shift in the structure of biotic communities inhabiting estuaries or shallow marine waters, with changes in species composition, species diversity, and population density occasionally being observed in near-field regions.[38]

Increases in outfall temperatures from 10–38°C favor the proliferation of less desirable, more heat-tolerant blue-green algae and the demise of diatoms, green algae, and other more favorable forms. Noxious phytoplankton blooms can develop near the discharge zone, leading to large biochemical oxygen demands. The resultant degradation in water quality often impacts the benthic community as well.

Elevated temperatures may greatly influence the properties of receiving waters in near-field regions. As the temperature rises, the density, viscosity, surface tension, and nitrogen solubility of seawater decrease. In addition, higher temperatures, along with increased organic loading and bacterial respiration in summer, can promote anoxia or hypoxia.

Various chemicals released in liquid effluents of power plants pose a potential danger to estuarine and coastal marine organisms because of their toxicity. Chief among these toxic chemicals are biocides used to control biofouling on heat exchanger surfaces (Figure 16). Chlorine — in liquid, gaseous, or hypochlorite form — provides the most effective biofouling control; however, it can easily kill

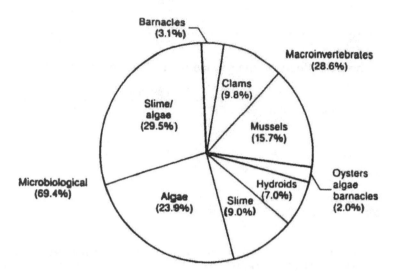

FIGURE 16 Biofouling organisms in electric generating station cooling water systems based on a survey of 365 units. (From Chow, W., in *Condenser Biofouling Control: the State-of-the Art*, Chow, W. and Massalli, Y. G., Eds., Technical Report, Electric Power Research Institute, Palo Alto, CA, 1985. With permission.)

nontarget organisms entrained in the cooling water systems as well as organisms in receiving waters exposed to residual concentrations. Chlorine has the capacity to form toxic residual organic compounds (e.g., chloramines). As a result, organisms in areas considerably removed from the cooling water systems of power plants may also be potentially threatened. Because plankton populations are especially sensitive to chlorine toxicity, acute reductions in phytoplankton productivity are generally observed in near-field regions of large power plants. Benthic invertebrates, finfish, and other organisms are not immune to the effects of chlorine toxicity either, and they too may be significantly impacted by the biocide.

Power plant chlorination is strictly controlled and tightly regulated because of the potential adverse effects of chlorine on nontarget organisms. Hence, at each power plant unit permitted to use chlorine in the United States, the total residual chlorine concentration cannot exceed 0.2 mg/l, and chlorine discharges are permitted for no more than 2 hr a day. These restrictions are designed to ensure the viability of balanced indigenous communities in receiving waters.

The greatest mortality of estuarine and marine organisms at coastal power plants is attributable to impingement and entrainment (Table 28). Annual losses of macro-invertebrates and fish alone on intake water screening devices commonly exceed a million individuals at larger stations.[2] Impingement rates depend on a number of factors, most importantly the configuration of the intake structure, intake flow velocities, type of intake screens, and the behavior and physiology of the organisms themselves. Modifications of intake structures and improvements in intake screening technology have substantially lowered impingement mortality.

Entrainment mortality of plankton, microinvertebrates, and small juvenile fish passively drawn into cooling water systems remains a significant problem. At larger

TABLE 28
Average Yearly Entrainment and Impingement
Mortality at the Brunswick Electric Generating
Station[a]

Species	Entrainment[b]	Impingement[c]	Impingement[d]
Spot	186,000	724	350
Croaker	123,000	356	235
Shrimp	171,000	675	760
Flounder	7200	21	12
Mullet	5200	34	18
Trout	38,500	169	205
Menhaden	32,500	9744	4000
Anchovy	913,000	2748	1600

[a] Number of organisms × 10^3.
[b] Computed from weekly average plant flows from September 1976 through August 1978 and 5-year average entrainment densities for the same weekly period. These flows are close to the full plant flow.
[c] Two-year averages of measured impingement losses from September 1976 through August 1978.
[d] Five-year averages of measured impingement losses from January 1974 through August 1978.

From Lawler, J. P., Hogarth, W. T., Copeland, B. J., Weinstein, M. P., Hodson, H. G., and Chen, H. Y., in *Issues Associated With Impact Assessment, Proc. 5th Natl. Workshop on Entrainment and Impingement*, Jensen, L. D., Eds., EA Communications, Sparks, MD, 1981, 159. With permission.

power plant units, annual entrainment losses of eggs, larvae, and juveniles of some species may exceed a billion individuals. Mortality results from mechanical damage and hydraulic shocks during in-plant passage, thermal stresses due to elevated temperatures, or chemical toxicity owing to biocide releases. Secondary entrainment in outfall waters further increases mortality of organisms surviving primary entrainment passage. Critical factors affecting entrainment include the seasonal occurrence and density of entrainable organisms in intake waters, as well as the sizes, life stages, and susceptibility of entrained organisms to injury during in-plant passage. Although entrainment mortality at coastal power plants is unequivocally large, the projected impact of these losses on biotic communities in receiving waters has been difficult to determine. At present, there are no documented cases of long-term, system-wide biological problems in coastal waters attributable to a single power plant unit despite the large absolute numbers of organisms lost on-site.

B. DREDGING AND DREDGED-SPOIL DISPOSAL

Rivers, estuaries, harbors, and ports are regularly dredged by mechanical and hydraulic devices to maintain navigable waterways (Table 29; Figure 17). This practice

TABLE 29
Types of Dredging Devices and Their Relationship to Sediment
Type and Disposal Method

Dredge Type	Sediment Type	Disposal Conveyance
Mechanical devices		
Dipper dredge	Blasted rock	Vessel
Bucket dredge	Coarse-grain size	Vessel
Ladder dredge	Fine-grain size	Vessel
Agitation dredge	Mud, clay	Prevailing current
Hydraulic devices		
Agitation dredge	Mud, clay	Prevailing current
Hopper dredge	Fine-grain size	Vessel
Suction dredge	Soft mud, clay	Pipeline
Cutterhead dredge	Consolidated, coarse-grain size	Pipeline
Dustpan dredge	Sand	Pipeline
Sidecasting dredge	Fine-grain size	Short Pipe

From Kester, D. R., Ketchum, B. H., Duedall, I. W., and Park, P. K., in *Wastes in the Ocean*, Vol. 2, *Dredged Material Disposal in the Ocean*, Kester, D. R., Ketchum, B. H., Duedall, I. W., and Park, P. K., Eds., John Wiley & Sons, New York, 1983, 3. With permission.

directly impacts benthic habitats and communities. The disposal of dredged spoils likewise causes environmental alteration. Land-based spoil disposal alternatives are often excessively costly and environmentally less attractive than marine disposal alternatives.[110] Hence, the dumping of dredged spoils most often occurs at sea. In the United States, the volume of dredged spoils dumped in marine waters exceeds that dumped on land by a factor of three.

The most severe impacts of dredging and dredged-spoil disposal on biotic communities result from physical effects. The removal of bottom sediments and entrained organisms during dredging operations effectively destroys the benthic habitat and causes mass mortality of bottom-dwelling organisms. Mortality of the benthos occurs from mechanical damage by the dredge itself, as well as by smothering of sediment when the organisms are picked up or deposited. Recovery of the benthic community varies considerably at both dredged and dredged-spoil disposal sites, being contingent upon the time of dredging or dredged-spoil disposal relative to the reproductive periods of benthic populations endemic to these areas. The dispersal capabilities of larval stages also play an important role.[1,2]

Repopulation of the impacted sites may take place soon after the termination of dredging or dredged-spoil disposal, with recovery of the communities typically requiring months or years to complete. Initial colonizers are usually opportunistic pioneering fauna that are later supplanted by equilibrium assemblages in a successional sequence. Species originally inhabiting the impacted sites usually return to recolonize the disturbed areas. However, when dredged spoils have sediment characteristics much different than those of seafloor sediments at the dumpsite, significant

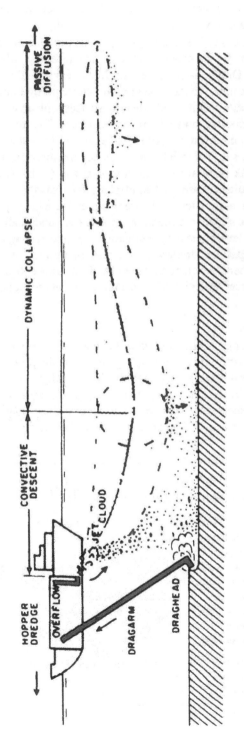

FIGURE 17   Schematic diagram of hydraulic (hopper) dredging of Chesapeake Bay sediments. Note upper turbidity plume produced by overflow discharge and a near-bottom turbidity plume produced by draghead agitation and settling of particulates from the upper plume. Three transport phases for hopper overflow discharge are depicted: convective descent, dynamic collapse, and passive diffusion. (From Nichols, M., Diaz, R. J., and Schaffner, L. C., *Environ. Geol. Water Sci.*, 15, 31–43, 1990. With permission.)

differences in species composition, abundance, and diversity may arise.[28] Thus, it is essential to conduct a comprehensive assessment of various physical, sedimentary, and biological criteria prior to selecting a waste disposal site (Table 30).

Changes in water quality by dredging and dredged-spoil disposal also can adversely affect plankton and nekton by increasing nutrients and turbidity, while decreasing dissolved oxygen levels. Dredged-spoils derived from heavily industrialized estuaries and harbors are commonly contaminated with petroleum hydrocarbons, organochlorine compounds, heavy metals, and other substances which can be remobilized when dumped and transported away from disposal sites (Figure 18). Chemically contaminated dredged spoils must be thoroughly analyzed for potential environmental impacts prior to disposal. Three laboratory procedures used in this regard are: (1) elutriate tests; (2) bulk or total sediment analysis; and (3) bioassay tests (liquid-phase, suspended particulate phase, and solid-phase bioassays).

There are some positive effects associated with dredging and dredged-spoil disposal. For example, dredging can improve circulation in estuaries and shallow embayments. By increasing nutrient levels, primary production may increase in the system. When environmentally compatible, dredged spoils may be utilized in salt marsh creation, spoil-island development, beach nourishment, and substrate enhancement. As such, dredged-spoils have great potential value in coastal habitat restoration.

C. LITTER

Litter, or marine debris, is an escalating global problem in the marine environment. While many individuals consider marine litter principally an aesthetic problem, GESAMP as well as other national and international organizations allude to ecological detriments of this material in the oceans. Environmentally persistent debris, particularly plastics, has been responsible for devastating impacts on fish, marine birds, sea turtles, cetaceans, and pinnipeds.[111] Discarded synthetic fishing nets and

---

TABLE 30
Site Selection Considerations for Waste Disposal in the Ocean

Physical	Sedimentary	Biological
Ocean flow	Physical and chemical properties	Fishing grounds
Surface waves	of wastes and sediment	Aquaculture sites
Wind-driven surface currents	Sorption capacity	Breeding and nursery grounds
Interior circulation	Distribution coefficients	Migration routes
Turbulent diffusion	Sedimentation	Productivity
Shear diffusion	Sedimentation dispersion	Recreational areas
Vertical mixing	Bioturbation	
Modeling advection and diffusion	Sediment stability	

From GESAMP (IMCO/FAO/UNESCO/WMO/WHO/IAEA/UN/UNEP Joint Group of Experts on Scientific Aspects of Marine Pollution), *Scientific Criteria for the Selection of Waste Disposal Sites at Sea*. Reports and Studies No. 16, Inter-Governmental Maritime Consultative Organization, London, 1982.

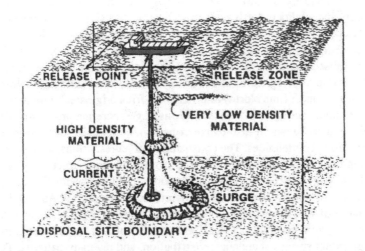

FIGURE 18 Sediment transport processes operating during open-water, dredged-spoil disposal. (From Pequegnat, W. E., Pequegnat, L. H., James, B. M., Kennedy, E. A., Fay, R. R., and Fredericks, A. D., *Procedural Guide for Designation Surveys of Ocean Dredged-Material Disposal Sites*, Technical Report EL-81-1, U.S. Army Corps of Engineers, Washington, D.C., 1981.)

rope, monofiliment fishing line, and plastic packing bands can entangle, trap, and kill these larger animals. In addition, plastic pellets, bags, containers, and packaging material are frequently life-threatening when swallowed. Waldichuk[112] notes that some potentially affected species (e.g., the Hawaiian monk seal, green sea turtle, and British Columbia sea otter) are considered to be endangered, and certain human-induced mortalities associated with ghost fishing and ingestion of plastics can further jeopardize them.

Negative economic impacts are also coupled to marine litter in estuarine and marine environments. Fouled fishing gear, entangled propellers on boats, gear and vessel repair costs, cancelled fishing trips, and shoreline cleanups all contribute to considerable financial losses for fishermen as well as the general public.[113] Enforcement of national environmental regulations and international treaties on pollution should help to mitigate marine litter problems in the sea. Annex V of the MARPOL 1973/78 Convention, which prohibits dumping at sea of all plastics and regulates the distance from shore that all other materials may be dumped, is a major step forward in this process. It entered into force in 1988, and to date, 79 countries have signed the treaty. It promises to vastly improve global pollution — especially plastics — in marine waters.

D. COASTAL DEVELOPMENT IMPACTS

Development impacts on coastal ecosystems are numerous and varied. Construction of ports and harbors, commercial and industrial installations, and domestic housing and tourist facilities cause radical physical alterations of coastal habitats that exacerbate pollution effects. For instance, the removal of inland forests and understory, dredging of channels, and modification of river catchment areas often greatly affect the quantity and quality of freshwater reaching the coastal zone. As a result, increased sediment and chemical loads are common in nearby estuaries. However, in some cases the sediment input decreases and the salinity of normally brackish waters increases subsequent to the inland construction of dams and the diversion of freshwater from rivers for agricultural irrigation. Due to dam-building, at least 20% of the freshwater runoff from North America and Africa originates from impoundments.

Other human activities in coastal regions (e.g., shoreline protection, drainage of wetlands, and reclamation) also cause considerable physical disruption of habitats that harm biotic communities. The construction of jetties, groins, wharfs, weirs, bank protection (e.g., bulkhead, revetment, and riprap), moored structures (e.g., pilings), moored floating vessels, marinas, boat ramps, and recreational docks, likewise destroys or substantially degrades many hectares of habitat.[10,114] As noted above, the maintenance of waterways and harbors directly impacts estuarine and coastal marine communities, with constant physical perturbation of bottom sediments modulating to a large extent species abundance, distribution, and diversity patterns. The successional pattern of benthic communities depends on the frequency and magnitude of the disturbances as well. By mapping successional mosaics, it is possible to document long-term changes in benthic community structure ascribable to human activity. When anthropogenic disturbances are intense, the communities are likely to be dominated by opportunistic pioneering species rather than equilibrium forms characteristic of more stable systems.[8] Therefore, careful biomonitoring of these communities should be a high priority in assessing coastal development impacts.

X. MARINE POLLUTION MONITORING AND ASSESSMENT

A need exists for sensitive, robust, and repeatable methods for detecting and assessing marine pollutants. The most comprehensive and effective approach for assessing marine pollution involves the integration of chemical, ecological, and toxicity data. While chemical surveys are important for detecting hot spots of contamination in the environment, they alone cannot provide sufficient information for complete assessment of impacts. Similarly, ecological surveys of populations or communities in areas likely to be affected by contaminant inputs mainly yield indirect or circumstantial evidence of contaminant impacts. Conclusive proof of adverse contaminant effects requires the implementation of both chemical and ecological surveys coupled with bioassay (toxicity) tests. Direct evidence of contaminant impacts on biota can only be properly obtained by exposing organisms to contaminated waters or sediments from or in the study site and carefully documenting their responses.[115]

A. BIOASSAY TESTS

Toxicity testing has become an important element in marine pollution studies. Bioassays, using indigenous species (if available) or "standardized" sensitive forms usually not indigenous to a study site, typically measure acute toxicity, sublethal effects, and genotoxicity of contaminants in the laboratory. Specific information that can be culled from laboratory tests on field-collected samples includes changes in physiological processes, behavioral changes, impairment of reproductive success, enzymatic responses, mutagenicity, histopathological conditions, mortality, and other responses.

The toxicity of a chemical contaminant to an organism is a function of its concentration and the duration of exposure. The toxic action of a chemical contaminant may be manifested in the death of the organism (i.e., lethal toxicity), in effects on the organism other than its death (i.e., sublethal toxicity), in effects that manifest themselves quickly — by convention, within a period of a few days (i.e., acute toxicity) — and in effects that manifest themselves over a longer period measurable in weeks or months rather than days (i.e., chronic toxicity). The most common endpoint used is mortality. Experimental conditions in toxicity tests typically expose a sample of test organisms to a certain concentration of toxin, and observations are made of the time of death of the organisms. Exposure conditions and durations have become standardized.[116] The median lethal concentration, sometimes referred to as the median tolerance limit (TL), defines the concentration of a toxin which causes 50% of the organisms to die within a specific period of time. The time is usually 24, 48, 72, and 96 h, with the median lethal concentration recorded as 24 h LC_{50}, 48 h LC_{50}, 72 h LC_{50}, or 96 h LC_{50}.

Some test endpoints other than death (e.g., growth and reproduction) are still widely used. However, a few, such as biochemical, behavioral, or pathological endpoints, have never been accepted in any broad or consistent way. Little effort has been expended on identifying biologically relevant endpoints, especially for chronic data. In addition to endpoints, the choice of species, exposure duration, relevance of exposure concentrations used, water quality characteristics, and methods of data evaluation are important considerations when utilizing the single species data in the natural environment.[116]

B. Ecological Surveys

Aside from ecological and toxicological assessment, monitoring programs are most useful in investigations of biological impacts of waste disposal in the sea. The goals of most marine monitoring programs are: (1) to determine that human health is not threatened; (2) to ensure that waste disposal has not harmed marine ecosystems or marine resources; and (3) to supply managers with data that enables them to make informed decisions regarding continued, reduced, or expanded use of the ocean for waste disposal and other activities. It is necessary to monitor marine ecosystems to establish the magnitude, spatial distribution, and temporal distribution of anthropogenic impacts in the receiving environment.[33]

Because they are very sensitive to habitat disturbance related to waste disposal and provide significant quantitative site-specific information, benthic infauna have great utility in marine monitoring programs. More specifically, changes in benthic community structure have been an important element for years in monitoring biological effects of pollutants. Some investigators have successfully quantified the degree of change in benthic community structure by applying univariate, multivariate, and graphical methods of data analysis.[117] Nevertheless, some workers have advocated phasing out benthic infauna from these programs in favor of laboratory sediment bioassay tests (e.g., acute toxicity and genotoxicity tests) which typically are less labor intensive, time consuming, and expensive. However, the monitoring of benthic infauna confers several important advantages over the use of laboratory bioassays alone. First, resident benthic infauna yield data on impacts to the ecosystem of concern, whereas laboratory bioassays often use species not indigenous to the test site. Second, *in situ* benthic infauna may reflect conditions in the field better than laboratory bioassay species. Third, benthic species at field sites may be responding to chronic contaminant stresses, but effective long-term chronic bioassay tests in marine monitoring programs are generally lacking, owing in part to their prohibitive costs.

In situ sampling of marine benthic communities will continue to be conducted in future marine pollution monitoring and assessment programs. A recent focus of these programs has been to develop an index that both integrates parameters of macrobenthic community structure and distinguishes between polluted and unpolluted areas. For example, Engle et al.[118] have formulated an index for estuarine macrobenthos in the Gulf of Mexico that discriminates between areas with degraded environmental conditions and reference sites with ungraded conditions. This type of index holds great promise in evaluating the spatial patterns of degraded benthic resources over extensive areas.

C. NATIONAL POLLUTANT MONITORING AND ASSESSMENT PROGRAMS

In the United States, two major programs are underway to monitor and assess the environmental quality of estuarine and coastal marine waters nationwide. These programs are the NOAA's National Status and Trends (NS&T) Program and the U.S. EPA Environmental Monitoring and Assessment Program (EMAP). Both are multi-year programs which provide data on chemical contamination in the nation's estuarine and coastal marine environments.

1. National Status and Trends Program

Since 1984, NOAA has monitored the concentration of toxic organic compounds and trace metals in bottom-feeding fish, shellfish, and sediments through its NS&T Program. The objective of this program is to determine the status and long-term trends of toxic contamination in estuarine and coastal marine environments throughout the United States.[34] The NS&T Program consists of two main projects: (1) the National Benthic Surveillance Project, which was initiated in 1984, analyzes contaminants in bottom fish and sediments collected from 149 sampling sites; and (2) the Mussel Watch Project, which commenced in 1986, analyzes contaminants from more than 250 sampling sites (Figure 19). The NS&T Program also initiated Biological Effects Surveys and Research in 1986 to further investigate those regions where laboratory analyses of samples indicated a potential for substantial environmental degradation and biological impacts of the contaminants. Historical Trends Assessment studies were likewise added to the program to more closely examine

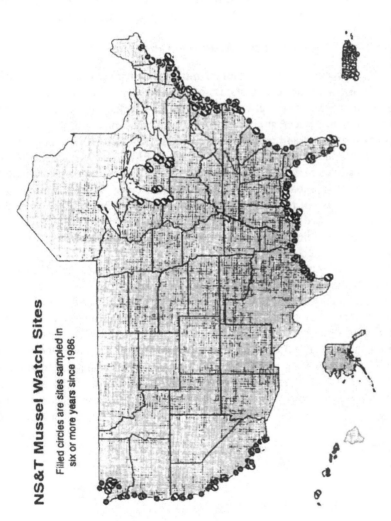

NS&T Mussel Watch Sites

Filled circles are sites sampled in six or more years since 1986.

FIGURE 19 Map showing monitoring and assessment sites of the National Status and Trends Mussel Watch Project in U.S. coastal waters. (From O'Connor, T. P. and Beliaeff, B., *Recent Trends in Coastal Environmental Quality: Results from the Mussel Watch Project*, NOAA Technical Report, Department of Commerce, Rockville, MD, 1995, 40 pp.)

the environmental conditions in different regions of the country. The NS&T Program assesses the levels of more than 70 chemical contaminants and certain associated effects in biota and sediments from the sampling sites (Table 4).

Results of the NS&T Program to date reveal that the highest chemical contamination occurs in estuarine and coastal sediments near urban centers (i.e., Boston, New York, San Diego, Los Angeles, and Seattle). Temporal trends in contaminant concentrations in mollusk samples over the period from 1986 to 1993 show generally declining values in coastal U.S. waters, most probably due to environmental regulations that have banned or curtailed the use of many toxic substances.[119]

2. Environmental Monitoring and Assessment Program

EMAP is a comprehensive environmental monitoring network of the U.S. EPA designed to: (1) estimate the current status and trends in the condition of the nation's ecological resources on a regional basis, with known confidence; (2) seek associations between human-induced stress and ecological condition; and (3) provide periodic statistical summaries and interpretive reports on ecological status and trends to resource managers and the public. Several features of EMAP are particularly advantageous. First, estimates of status and trends can be made with quantifiable confidence since EMAP is probability based. Second, monitoring and assessments focus on biological indicators of response to natural and human-induced stress; indicators of pollutant exposure and habitat conditon are sampled simultaneously to provide a context for interpreting biological indicators. Third, EMAP monitoring is regional and national in scale, rather than local.[120]

The near-coastal component of EMAP, termed EMAP-Estuaries (EMAP-E), gives a quantitative assessment of the regional extent of coastal environmental problems by measuring status and change in selected indicators of ecological condition. Environmental problems investigated by EMAP-E in near-coastal systems include: eutrophication, hypoxia, sediment contamination, and habitat loss. In 1990, the U.S. EPA initiated a multiyear EMAP-E demonstration project in the estuaries of the Virginian Province, an area extending from Cape Cod south to the mouth of Chesapeake Bay.[121] The ultimate goal of this project is to produce an assessment report that will yield estimates of the ecological condition of the estuarine resources in the aforementioned northeastern region. Other regional studies are planned. Together with the NOAA's NS&T Program, EMAP-E should serve as a basis for accurately identifying emerging environmental problems in the nation's near-coastal ecosystems.

XI. STATUS OF MARINE POLLUTION

Because 80% of global marine pollution stems from land-based sources, estuarine and nearshore environments exhibit the greatest anthropogenic impacts, while the open ocean appears to be relatively clean.[6] Estuaries are most susceptible to pollution effects (Table 31). Excessive development and the wide diversity of human activities occurring in coastal zones throughout the world have often compromised marine communities and habitats. In the open sea, however, the atmosphere and shipping are the main pollution sources.[7] When oil slicks, litter, and other evidence of human activity are observed in the open ocean, they tend to be concentrated along the sea lanes of the world, with minimal biotic effects being reported.[4] This has led to the "dirty coast/clean seas" hypothesis accepted by many members of the marine science community. However, a paucity of data exists on the concentrations and biological effects of numerous contaminants in the open ocean.[3] The general inadequacy of the available database, therefore, warrants more detailed attention to open ocean contamination.

Many estuarine and coastal marine ecosystems are heavily degraded by land-based sources of pollution. Considering the possible irreversible degradation of these shallow water systems and the unknown vulnerability of the open ocean to various anthropogenic impacts, international cooperative action must be pursued to avert potentially critical environmental consequences. Marine pollution problems can no longer be strictly addressed by coastal nations with their own legislation programs and agenda. It is now clear that to effectively resolve such complex and pervasive problems, regional and global cooperation is required. The Governing Council of the United Nations Environment Program (UNEP) has strongly endorsed a regional approach to the control of marine pollution, implementing the Regional Seas Program in 1974 which includes 12 regions and approximately 140 participating coastal states and territories. This program incorporates a comprehensive management strategy to combat marine pollution by focusing on both the causes and consequences of environmental degradation in the sea.[122] Scientists from national institutes of participating states conduct research, monitoring, assessment, and other studies for the program to achieve its long-term mission.

On the global scale, various initiatives developed over the past 25 years have engendered international cooperation to control marine pollution problems. GESAMP, established in 1969, has played a leading role in the assessment of pollution problems and during the past 15 years has prepared several global reports on the state of the marine environment. In forthcoming years, GESAMP will be concentrating on hazard assessments of various toxic substances in the sea (e.g., halogenated hydrocarbon compounds, oil, and oil spill disperants), harmful substances carried by ships, environmental impacts of coastal aquaculture, effects of anthropogenically mobilized sediments, evaluation of waste disposal in the oceans, and global change and the air/sea exchange of chemicals.[122]

The Global Investigation of Pollution in the Marine Environment (GIPME) is another important ocean monitoring program. Initiated in 1974, GIPME ensures maximum coordination of the marine pollution programs of its sponsors, the Intergovernmental Oceanographic Commission and the UNEP. GIPME has been involved in numerous cooperative ventures with the Regional Seas Program on marine pollution monitoring and research initiatives.

International treaties have been periodically developed and signed by member states to control marine pollution both regionally and globally. The Convention on the Prevention of Marine Pollution by Dumping of Wastes and Other Matter of 1972 (i.e., the London Convention) is the most notable example. It was brought into force on August 30, 1975, and since that time control of waste dumping at sea has been accomplished mainly through national legislation and regulations of states contracting

TABLE 31
Summary of Pollutant Concentration Susceptibility in Estuaries

Most Susceptible Systems	Least Susceptible Systems
General population	General population
Brazos River**	Willapa Bay**
Ten Thousand Islands	St. Catherines/Sapelo Sound**
San Pedro Bay**	Penobscot Bay**
North-South Santee Rivers**	Humboldt Bay**
Galveston Bay**	Broad River**
Suisun Bay**	Hood Canal**
Sabine Lake**	Coos Bay**
St. Johns River*	Casco Bay**
Apalachicola Bay	Grays Harbor
San Antonio Bay**	Chincoteague Bay*
Connecticut River*	Bogue Sound**
Great South Bay*	St. Andrew/St. Simons Sound*
Merrimack River	Sheepscot Bay*
Atchafalaya/Vermillion Bays*	Apalachee Bay
Matagorda Bay*	Rappahannock River
Heavy industry	St. Helena Sound
Brazos River**	Puget Sound*
North/South Santee Rivers**	Heavy industry
Galveston Bay**	St. Catherines/Sapelo Sound**
Sabine Lake**	Hood Canal**
San Pedro Bay**	Penobscot Bay**
Connecticut River*	Casco Bay**
Calcasieu Lake	Humboldt Bay**
Hudson River/Raritan Bay	Buzzards Bay
Charleston Harbor	Boston Bay
Perdido Bay	Coos Bay**
Potomac River	Broad River**
San Antonio Bay**	Willapa Bay**
Mobile Bay	Bogue Sound**
Suisun Bay**	Puget Sound*
Great South Bay*	Narragansett Bay
Baffin Bay	Santa Monica Bay
Chesapeake Bay	Saco Bay
Agricultural activities	St. Andrew/St. Simons Sound*
Brazos River**	Agricultural activities
Suisun Bay**	Humboldt Bay**
North/South Santee Rivers**	Hood Canal**
St. Johns River*	Penobscot Bay**
Matagorda Bay*	Coos Bay**
Atchafalaya/Vermillion Bays*	St. Catherines/Sapelo Sound**
San Pedro Bay**	Chincoteague Bay*
Sabine Lake**	Bogue Sound**
Corpus Christi Bay	Long Island Sound

TABLE 31 (CONTINUED)
Summary of Pollutant Concentration Susceptibility
in Estuaries

Most Susceptible Systems	Least Susceptible Systems
Galveston Bay**	Casco Bay**
San Antonio Bay**	Willapa Bay**
Winyah Bay	Broad River**
Albermarle Sound	Sheepscot Bay*
Neuse River	Klamath River
Laguna Madre	

*,** Systems that are present in all three categories are marked with two asterisks; systems present in two categories are marked by one asterisk.

From Biggs, R. B., DeMoss, T. B., Carter, M. M., and Beasley, E. L., *Rev. Aquat. Sci.*, 1, 203, 1989. With permission.

to the convention (Table 2). Additional controls on waste dumping at sea are afforded by provisions of the International Convention for the Prevention of Pollution from Ships of 1973, which were later amended by a protocol in 1978 (MARPOL 1973/78). Annex V of this global treaty places restrictions on the dumping of litter and bans on the dumping of all plastics into the ocean.

Global marine pollution problems are clearly ongoing. While there is evidence for a decrease in the concentration of certain constituents of four classes of critical contaminants in the global marine environment (i.e., petroleum hydrocarbons, halogenated hydrocarbons, heavy metals, and radionuclides), other types of pollution (e.g., plastics and other forms of litter) appear to be increasing.[112] Many countries must formulate new legislation and more stringent regulations to control local and national marine pollution problems. In doing so, they will also help to mollify marine pollution both regionally and globally.

REFERENCES

1. Kennish, M. J., Ed., *Practical Handbook of Marine Science*, 2nd ed., CRC Press, Boca Raton, FL, 1994.
2. Kennish, M. J., Ed., *Practical Handbook of Estuarine and Marine Pollution*, CRC Press, Boca Raton, FL, 1996.
3. Davis, W. J., Contamination of coastal versus open ocean surface waters: a brief meta-analysis, *Mar. Pollut. Bull.*, 26, 128, 1993.
4. McIntyre, A. D., The state of the marine environment, *Mar. Pollut. Bull.*, 21, 403, 1990.
5. GESAMP, State of the Marine Environment, Reports and Studies No. 39, UNEP, Nairobi, 1990, 111 pp.
6. Karau, J., The control of land-based sources of marine pollution: recent international initiatives and prospects, *Mar. Pollut. Bull.*, 25, 80, 1992.
7. McIntyre, A. D., The current state of the oceans, *Mar. Pollut. Bull.*, 25, 28, 1992.

8. Kennish, M. J., *Ecology of Estuaries*, Vol. 2, *Biological Aspects*, CRC Press, Boca Raton, FL, 1990.
9. Nordstrom, K. L., Developed coasts, in *Coastal Evolution*, Carter, R. W. G. and Woodroffe, C. D., Eds., Cambridge University Press, Cambridge, 1994, 477.
10. Barcena, A., An overview of the oceans in Agenda 21 of the 1992 United Nations Conference on environment and development, *Mar. Pollut. Bull.*, 25, 107, 1992.
11. Goldberg, E. D., Competitors for coastal ocean space, *Oceanus*, 36, 12, 1993.
12. GESAMP, Scientific Criteria for the Selection of Waste Disposal Sites at Sea. Reports and Studies No. 16, Inter-Governmental Maritime Consultative Organization, London, 1982.
13. GESAMP, Report of the twenty-third session, Reports and Studies No. 51, FAO, London, 1993.
14. GESAMP, Report of the twenty-fifth session, Reports and Studies No. 56, Rome, 1995.
15. Bacci, E., *Ecotoxicology of Organic Contaminants*, Lewis Publishers, Ann Arbor, MI, 1994.
16. Clark, R. B., *Marine Pollution*, 3rd ed., Clarendon Press, Oxford, 1992.
17. Grassle, F., Sludge reaching bottom at the 106 site, not dispersing as plan predicted, *Oceanus*, 33, 61, 1990.
18. Swanson, R. L. and Mayer, G. F., Ocean dumping of municipal and industrial wastes in the United States, in *Oceanic Processes in Marine Pollution*, Vol. 3, *Marine Waste Management: Science and Policy*, Camp, M. A. and Park, P. K., Eds., Robert E. Krieger Publishing Company, Malaabar, FL, 1989, 35.
19. Gage, J. D. and Tyler, P. A., *Deep-Sea Biology: A Natural History of Organisms at the Deep-Sea Floor*, Cambridge University Press, Cambridge, 1991.
20. Chandler, G. T., Coull, B. C., and Davis, J. C., Sediment- and aqueous-phase fenvalerate effects on meiobenthos: implications for sediment quality criteria development, *Mar. Environ. Res.*, 37, 313, 1994.
21. Haebler, R. and Moeller, R. B., Jr., Pathobiology of selected marine mammal diseases, in *Pathobiology of Marine and Estuarine Organisms*, Couch, J. A. and Fournie, J. W., Eds., CRC Press, Boca Raton, FL, 1993, 217.
22. Office of Technology Assessment, Pollutant discharges to surface waters in coastal regions, U.S. Congress, Office of Technology Assessment, Washington, D.C., 1986.
23. Ott, M. N., *Environmental Statistics and Data Analysis*, Lewis Publishers, Ann Arbor, MI, 1995.
24. Blumberg, A. F., Signell, R. P., and Jenter, H. L., Modeling transport processes in the coastal ocean, in *Environmental Science in the Coastal Zone: Issues for Further Research*, National Academy Press, Washington, D. C., 1994, 20.
25. Kjerfve, B., *Hydrodynamics of Estuaries*, Vol. 1, CRC Press, Boca Raton, FL, 1988.
26. Rutherford, J. C., *River Mixing*, John Wiley & Sons, New York, 1995.
27. Capuzzo, J. M., Burt, W. V., Duedall, I. W., Park, P. K., and Kester, D. R., The impact of waste disposal in nearshore environments, in *Wastes in the Ocean*, Vol. 6, *Nearshore Waste Disposal*, Ketchum, B. H., Capuzzo, J. M., Burt, W. V., Duedall, I. W., Park, P. K., and Kester, D. R., Eds., John Wiley & Sons, New York, 1985, 3.
28. Kennish, M. J., *Ecology of Estuaries: Anthropogenic Effects*, CRC Press, Boca Raton, FL, 1992.
29. McDowell, J. E., How marine animals respond to toxic chemicals in coastal ecosystems, *Oceanus*, 36, 56, 1993.

30. Howells, G., Calamari, D., Gray, J., and Wells, P. G., An analytical approach to assessment of long-term effects of low levels of contaminants in the marine environment, *Mar. Pollut. Bull.*, 21, 371, 1990.
31. Agard, J. B. R., Gobin, J., and Warwick, R. M., Analysis of marine macrobenthic community structure in relation to oil pollution, natural oil seepage and seasonal disturbance in a tropical environment (Trinidad, West Indies), *Mar. Ecol. Prog. Ser.*, 92, 233, 1993.
32. Austen, M. C., McEvoy, A. J., and Warwick, R. M., The specificity of meiobenthic community responses to different pollutants: results from microcosm experiments, *Mar. Pollut. Bull.*, 28, 557, 1994.
33. Bilyard, G. R., The value of benthic infauna in marine pollution monitoring studies, *Mar. Pollut. Bull.*, 18, 581, 1987.
34. O'Connor, T. P., The National Oceanic and Atmospheric Administration (NOAA) National Status and Trends Mussel Watch Program: national monitoring of chemical contamination in the coastal United States, in *Environmental Statistics, Assessment, and Forecasting*, Cothern, C. R. and Ross, N. P., Eds., CRC Press, Boca Raton, FL, 1994, 331.
35. Department of the Environment (U.K.), Inputs of dangerous substances to water: proposals for a unified system of control. The government's consultative proposals for tighter controls over the most dangerous substances entering the aquatic environment, *The Red List*, July, 1988.
36. Valiela, I., *Marine Ecological Processes*, 2nd ed., Springer-Verlag, New York, 1995.
37. National Research Council, Managing Wastewater in Coastal Urban Areas, National Academy Press, Washington, D.C., 1993, 477.
38. Kennish, M. J., Pollution in estuaries and coastal marine waters, *J. Coastal Res.*, Spec. Iss. 12: Coastal Hazards, pp. 27, 1994.
39. D'Avanzo, C. and Kremer, J. N., Diel oxygen dynamics and anoxic events in an eutrophic estuary of Waquoit Bay, Massachusetts, Estuaries 17, 131, 1994.
40. Shear, N. M., Schmidt, C. W., Huntley, S. L., Crawford, D. W., and Finley, B. L., Evaluation of the factors relating combined sewer overflows with sediment contamination of the lower Passaic River, *Mar. Pollut. Bull.*, 32, 288, 1996.
41. Costello, M. J. and Read, P., Toxicity of sewage sludge to marine organisms: a review, *Mar. Environ. Res.*, 37, 23, 1994.
42. European Environmental Statistics Handbook, Gale Research International Ltd., Andover, England, 1993.
43. Lester, J. N., Sewage and Sewage sludge treatment, in *Pollution: Causes, Effects, and Control*, 2nd ed., Harrison, R. M., Ed., Royal Society of Chemistry, Cambridge, 1990, 33.
44. Aubrey, D. G. and Connor, M. S., Boston Harbor: fallout over the outfall, *Oceanus*, 36, 61, 1993.
45. Bascom, W. N., Disposal of sewage sludge via ocean outfalls, in *Oceanic Processes in Marine Pollution*, Vol. 3, *Marine Waste Management: Science and Policy*, Camp, M. A. and Park, P. K., Eds., Robert E. Krieger Publishing Co., Malabar, FL, 1989, 27.
46. Moore, D. C. and Rodger, G. K., Recovery of a sewage sludge dumping ground. II. Macrobenthic community, *Mar. Ecol. Prog. Ser.*, 75, 301, 1991.
47. Hardy, F. G., Evans, S. M., and Tremayne, M. A., Long- term changes in the marine macroalgae of three polluted estuaries in northeast England, *J. Exp. Mar. Biol. Ecol.*, 172, 81, 1993.
48. Weston, D. P., Quantitative examination of macrobenthic community changes along an organic enrichment gradient, *Mar. Ecol. Prog. Ser.*, 61, 233, 1990.

49. Spies, R., Benthic-pelagic coupling in sewage-affected marine ecosystems, *Mar. Environ. Res.*, 13, 195, 1984.

50. Albers, P. H., Petroleum and individual polycyclic aromatic hydrocarbons, in *Handbook of Ecotoxicology*, Hoffman, D. J., Rattner, B. A., Burton, G. A., Jr., and Cairns, J., Jr., Eds., Lewis Publishers, Boca Raton, FL, 1995, 330.

51. Ambrose, P., Oil pollution on decrease, *Mar. Pollut. Bull.*, 22, 262, 1991.

52. Bishop, P. L., *Marine Pollution and its Control*, McGraw-Hill, New York, 1983.

53. Bishop, J. M., *Applied Oceanography*, John Wiley & Sons, New York, 1984.

54. Doerffer, J. W., *Oil Spill Response in the Marine Environment*, Pergamon Press, Oxford, 1992.

55. Teal, J. M. and Howarth, R. W., Oil spill studies: a review of ecological effects, *Environ. Mgmt.*, 8, 27, 1984.

56. Thorhaug, A., Oil spills in the tropics and subtropics, in *Pollution in Tropical Aquatic Systems*, Connell, D. W. and Hawker, D. W., Eds., CRC Press, Boca Raton, FL, 1992, 99.

57. May, R. F., Marine conservation reserves, petroleum exploration and development, and oil spills in coastal waters of western Australia, *Mar. Pollut. Bull.*, 25, 147, 1992.

58. McElroy, A. E., Farrington, J. W., and Teal, J. M., Bioavailability of polycyclic aromatic hydrocarbons in the aquatic environment, in *Metabolism of Polycyclic Aromatic Hydrocarbons in the Aquatic Environment*, Varanasi, U., Ed., CRC Press, Boca Raton, FL, 1989, 1.

59. Eisler, R., Polycyclic Aromatic Hydrocarbon Hazards to Fish, Wildlife, and Invertebrates: A Synoptic Review, Biol. Rep. 85(1.11), U.S. Fish and Wildlife Service, Washington, D.C., 1987.

60. Guerin, W. F. and Jones, G. E., Estuarine ecology of phenanthrene-degrading bacteria, *Est. Coastal Shelf Sci.*, 29, 115, 1989.

61. Wright, L. D., *Morphodynamics of the Inner Continental Shelf*, CRC Press, Boca Raton, FL, 1994.

62. Guzzella, L. and De Paolis, A., Polycyclic aromatic hydrocarbons in sediments of the Adriatic Sea, *Mar. Pollut. Bull.*, 28, 159, 1994.

63. Huggett, R. J., de Fur, P. O., and Bieri, R. H., Organic compounds in Chesapeake Bay sediments, *Mar. Pollut. Bull.*, 19, 454, 1988.

64. McLeese, D. W., Ray, S., and Burridge, L. E., Accumulation of polynuclear aromatic hydrocarbons by the clam *Mya arenaria*, in *Wastes in the Ocean*, Vol. 6, *Nearshore Waste Disposal*, Ketchum, B. H., Capuzzo, J. M., Burt, W. V., Duedall, I. W., Park, P. K., and Kester, D. R., Eds., John Wiley & Sons, New York, 1985, 81.

65. O'Connor, J. M., Klotz, J. B., and Kneip, T. J., Sources, sinks, and distribution of organic contaminants in the New York Bight ecosystem, in *Ecological Stress and the New York Bight: Science and Management*, Mayer, G. F., Ed., Estuarine Research Federation, Columbia, SC, 1982, 631.

66. Barrick, R. C. and Prahl, F. G., Hydrocarbon geochemistry of the Puget Sound region. III. Polycyclic aromatic hydrocarbons in sediments, *Est. Coastal Shelf Sci.*, 25, 175, 1987.

67. Readman, J. W., Mantoura, R. F. C., Rhead, M. M., and Brown, L., Aquatic distribution and heterotrophic degradation of polycyclic aromatic hydrocarbons (PAH) in the Tamar estuary, *Est. Coastal Shelf Sci.*, 14, 369, 1982.

68. Sindermann, C. J., *Ocean Pollution: Effects on Living Resources and Humans*, CRC Press, Boca Raton, FL, 1996.

69. Law, R. J. and Biscaya, J. L., Polycyclic aromatic hydrocarbons (PAH) — problems and progress in sampling, analysis, and interpretation, *Mar. Pollut. Bull.*, 29, 235, 1994.

70. Macias-Zamora, J. V., Distribution of hydrocarbons in recent marine sediments off the coast of Baja, California, *Environ. Pollut.*, 92, 45, 1996.

71. Miskiewicz, A. G. and Gibbs, P. J., Organochlorine pesticides and hexachlorobenzene in tissues of fish and invertebrates caught near a sewage outfall, *Environ. Pollut.*, 84, 269, 1994.

72. Harrad, S. J., Sewart, A. P., Alcock, R., Boumphrey, R., Burnett, V., Duarte-Davidson, R., Halsall, C., Sanders, G., Waterhouse, K., Wild, S. R., and Jones, K. C., Polychlorinated biphenyls (PCBs) in the British environment: sinks, sources, and temporal trends, *Environ. Pollut.*, 85, 131, 1994.

73. Wenning, R. J., Paustenback, D. J., Harris, M. A., and Bedbury, H., Principal components analysis of potential sources of polychlorinated dibenzo-*p*-dioxin and dibenzofuran residues in surficial sediments from Newark Bay, New Jersey, *Arch. Environ. Contam. Toxicol.*, 24, 271, 1993.

74. Fowler, S. W., Critical review of selected heavy metal and chlorinated hydrocarbon concentrations in the marine environment, *Mar. Environ. Res.*, 29, 1, 1990.

75. Mearns, A. J., Matta, M. B., Simececk-Beatty, D., Buchman, M. F., Shigenaka, G., and Wert, W. A., PCB and Chlroinated Pesticide Contamination in U.S. Fish and Shellfish: A Historical Assessment Report, NOAA Tech. Mem. NOS OMA 39, National Oceanic and Atmospheric Administration, Seattle, WA, 1988.

76. Oliver, B. G., Baxter, R. M., and Lee, H.-B., Polychlorinated biphenyls, in *Analysis of Trace Organics in the Aquatic Environment*, Afghan, B. K. and Chau, A.S.Y., Eds., CRC Press, Boca Raton, FL, 1989, 31.

77. Rice, C. P. and O'Keefe, P., Sources, pathways, and effects of PCBs, dioxins, and dibenzofurans, in *Handbook of Ecotoxicology*, Hoffman, D. J., Rattner, B. A., Burton, G. A., Jr., and Cairns, J., Jr., Eds., Lewis Publishers, Boca Raton, FL, 1995, 424.

78. Cooper, K. R., Effects of polychlorinated dibenzo-*p*-dioxins and polychlorinated dibenzofurans on aquatic organisms, *Rev. Aquat. Sci.*, 1, 227, 1989.

79. Abel, P. D., Water Pollution Biology, Ellis Horwood, Chichester, 1989.

80. Rainbow, P. S., The significance of trace metal concentrations in marine invertebrates, in *Ecotoxicology of Metals in Invertebrates*, Dallingerr, R. and Rainbow, P. S., Eds., Lewis Publishers, Ann Arbor, MI, 1993, 3.

81. Bruland, K. W. and Franks, R. P., Mn, Ni, Cu, Zn, and Cd in the western North Atlantic, in *Trace Metals in Sea Water*, Wong, C. S., Boyle, E., Bruland, K. W., Burton, J. D., and Goldberg, E. D., Eds., Plenum Press, New York, 1983, 395.

82. Bryan, G. W. and Langston, W. J., Bioavailability, accumulation and effects of heavy metals in sediments with special reference to United Kingdom estuaries: a review, *Environ. Pollut.*, 76, 89, 1992.

83. Kim, E. Y., Murakami, T., Saeki, K., and Tatsukawa, R., Mercury levels and its chemical form in tissues and organs of seabirds, *Arch. Environ. Contam. Toxicol.*, 30, 259, 1996.

84. Dowson, P. H., Bubb, J. M., and Lester, J. N., Organotin distribution in sediments and waters of selected east coast estuaries in the UK, *Mar. Pollut. Bull.*, 24, 492, 1992.

85. Nelson, D. A., Miller, J. E., and Calabrese, A., Effect of heavy metals on bay scallops, surf clams, and blue mussels in acute and long-term exposures, *Arch. Environ. Contam. Toxicol.*, 17, 596, 1988.

86. Giudici, M. de N. and Guarino, S. M., Effects of chronic exposure to cadmium or copper on *Idotea balthica* (Crustacea, Isopoda), *Mar. Pollut. Bull.*, 20, 69, 1989.

87. Rygg, B., Effects of sediment copper on benthic fauna, *Mar. Ecol. Prog. Ser.*, 25, 83, 1985.

88. Jenkins, K. D. and Mason, A. Z., Relationships between subcellular distribution of cadmium and perturbations in reproduction in the polychaete *Neanthes arenaceodentata, Aquat. Toxicol.*, 12, 229, 1988.
89. Luoma, S. N., Dagovitz, R., and Axtmann, E., Temporally intensive study of trace metals in sediments and bivalves from a large river-esturarine system: Suisun Bay Delta in San Francisco Bay, *Sci. Tot. Environ.*, 97/98, 685, 1990.
90. Fabiano, M., Danovaro, R., Magi, E., and Mazzucotelli, A., Effects of heavy metals on benthic bacteria in coastal marine sediments: a field result, *Mar. Pollut.*, 28, 18, 1994.
91. Kayser, H. and Sperling, K.-R., Cadmium effects and accumulation in cultures of *Prorocentrum micans* (Dinophyta), *Helgolander Meeresunters*, 33, 89, 1980.
92. Paffenhöfer, G.-A. and Knowles, S. C., Laboratory experiments on feeding, growth, and fecundity of and effects of cadmium on *Pseudodiaptomus coronatus, Bull. Mar. Sci.*, 28, 574, 1978.
93. von Westernhagen, H., Dethlefsen, V., and Rosenthal, H., Correlation between cadmium concentration in the water and tissue residue levels in dab, *Limanda limanda* L., and plaice, *Pleuronectes platessa* L., *J. Mar. Biol. Assoc. UK*, 60, 45, 1978.
94. Marshall, S. J. and Jarvie, A. W. P., The toxicity of alkyl lead compounds to *Scrobicularia plana* (da Costa), *Appl. Organometal Chemistry*, 2, 143, 1988.
95. Bloom, H. and Ayling, G. M., Heavy metals in the Derwent estuary, *Environ. Geol.*, 2, 3, 1977.
96. Essink, K., Mercury pollution in the Ems estuary, *Helgoländer Meeresunters.*, 33, 111, 1980.
97. Langston, W. J., Metals in sediments and benthic organisms in the Mersey estuary, *Est. Coastal Shelf Sci.*, 23, 239, 1986.
98. Langston, W. J., Bryan, G. W., Burt, G. R., and Gibbs, P. E., Assessing the impact of tin and TBT in estuarine and coastal regions, *Func. Ecol.*, 4, 433, 1990.
99. Page, D. S., A six-year monitoring study of tributyltin and dibutyltin in mussel tissues from the Lynher River, Tamar estuary, UK, *Mar. Pollut. Bull.*, 30, 746, 1995.
100. Eisenbud, M., *Environmental Radioactivity from Natural, Industrial, and Military Scources*, 3rd ed., Academic Press, New York, 1987.
101. Hewitt, C. N., Radioactivity in the environment, in *Pollution: Causes, Effects, and Control*, 2nd ed., Harrison, R. M., Ed., Royal Society of Chemistry, Cambridge, 1990, 343.
102. Whicker, F. W. and Schultz, V., *Radioecology: Nuclear Energy and the Environment*, Vol. 2, CRC Press, Boca Raton, FL, 1982.
103. Talmage, S. S. and Meyers-Schone, L., Nuclear and thermal, in *Handbook of Ecotoxicology*, Hoffman, D. J., Rattner, B. A., Burton, G. A., Jr., and Cairns, J., Jr., Eds., Lewis Publishers, Ann Arbor, MI, 1995, 469.
104. Eisler, R., Ecological and toxicological aspects of the partial meltdown of the Chernobyl Nuclear Power Plant reactor, in *Handbook of Ecotoxicology*, Hoffman, D. J., Rattner, B. A., Burton, G. A., Jr., and Cairns, J., Jr., Eds., Lewis Publishers, Ann Arbor, MI, 1995, 549.
105. U.S. Interagency Review Group on Nuclear Waste Management, Report to the President by IRG, TID-29442, Technical Report, U.S. Department of Commerce, Washington, D.C., 1979.
106. U.S. Interagency Review Group on Nuclear Waste Management, Subgroup Report on Alternate Technology Strategies for the Isolation of Nuclear Waste, TID-28318, Technical Report, U.S. Department of Commerce, Washington, D.C., 1979.

107. Langford, T. E., *Electricity Generation and the Ecology of Natural Waters*, Liverpool University Press, Liverpool, England, 1983.
108. Valette-Silver, N. J. and Lauenstein, G. G., Radionuclide concentrations in bivalves collected along the coastal United States, *Mar. Pollut. Bull.*, 30, 320, 1995.
109. Kennish, M. J., Roche, M. B., and Tatham, T. R., Anthropogenic effects on aquatic communities, in *Ecology of Barnegat Bay, New Jersey*, Kennish, M. J. and Lutz, R. A., Eds., Springer-Verlag, New York, 1984, 318.
110. Engler, R. M. and Mathis, D. B., Dredged-material disposal strategies, in *Oceanic Processes in Marine Pollution*, Vol. 3, *Marine Waste Management: Science and Policy*, Camp, M. A. and Park P. K., Eds., Robert E. Krieger Publishing Co., Malabar, FL, 1989, 53.
111. Beck, C. A. and Barros, N. B., The impact of debris on the Florida manatee, *Mar. Pollut. Bull.*, 22, 508, 1991.
112. Waldichuk, M., The state of pollution in the marine environment, *Mar. Pollut. Bull.*, 20, 598, 1989.
113. Nash, A. D., Impacts of marine debris on subsistence fishermen: an exploratory study, *Mar. Pollut. Bull.*, 24, 150, 1992.
114. Nordstrom, K. F., Beaches and dunes of human-altered coasts, *Prog. Phys. Geogr.*, 18, 497, 1994.
115. Chapman, P. M. and Long, E. R., The use of bioassays as part of a comprehensive approach to marine pollution assessment, *Mar. Pollut. Bull.*, 14, 81, 1983.
116. Mount, D. I., Development and current use of single species aquatic toxicity tests, in *Ecological Toxicity Testing: Scale, Complexity, and Relevance*, Cairns, J., Jr. and Niederlehner, B. R., Eds., Lewis Publishers, Ann Arbor, MI, 1995, 97.
117. Wilson, J. G. and Jeffrey, D. W., Benthic biological pollution indices in estuaries, in *Biomonitoring of Coastal Waters and Estuaries*, Kramer, K. J. M., Ed., CRC Press, Boca Raton, FL, 1994, 311.
118. Engle, V. D., Summers, J. K., and Gaston, G. R., A benthic index of environmental condition of Gulf of Mexico estuaries, *Estuaries*, 17, 372, 1994.
119. NOAA, Recent Trends in Coastal Environmental Quality: Results from the Mussel Watch Project, Technical Report, Department of Commerce, Rockville, MD, 1995, 40.
120. USEPA, Virginian Province Demonstration Report: EMAP-Estuaries — 1990, Technical Report EPA/620/R-93/006, U.S. Environmental Protection Agency, Washington, D.C., 1993.
121. USEPA, Statistical Summary EMAP-Estuaries: Virginian Province — 1990 to 1993, Technical Report EPA/620/R-94/026, U.S. Environmental Protection Agency, Washington, D.C., 1995.
122. Gerges, M. A., Marine pollution monitoring, assessment, and control: UNEP's approach and strategy, *Mar. Pollut. Bull.*, 28, 199, 1994.

2 Case Study 1: New York Bight Apex

I. INTRODUCTION

A number of coastal ocean regions have been severely degraded as a result of long-term disposal of human wastes. Among the most heavily impacted systems worldwide is the New York Bight Apex, which has long received large volumes of organic carbon and toxic contaminant inputs from the Hudson-Raritan estuary (since the 1800s), from the dumping of dredged material (1914–1995), sewage sludge (1924–1987), and acid wastes (1949–1983), and from an array of smaller and more poorly documented sources. Originally viewed as a relatively inexpensive disposal site, the New York Bight Apex has changed considerably during the 20th Century due to multiple waste disposal activities, which have caused significant alteration of marine communities and benthic habitats. Fisheries resources, including both finfish and shellfish populations, have also been adversely affected. However, with tighter pollution control legislation and enforcement during the past 25 years culminating in the closure of waste dumpsites in the 1980s and 1990s, environmental conditions in the Hudson-Raritan estuary, New York Harbor, and the entire New York Bight Apex have gradually improved. Upgraded conditions promise to return the ecosystem to a more healthy state — favorable perhaps to the eventual proliferation of balanced indigenous biotic communities.

II. AREA DESCRIPTION AND POLLUTANT SOURCES

The New York Bight Apex is the region along the inner edge of the Atlantic continental shelf extending from the New Jersey and Long Island coasts to approximately 6–7 km offshore (74°00'W to 73°36'W longitude; 40°15'N to 40°35'N latitude) (Figure 1). Covering an area more than 1000 km², the apex encompasses that portion of the Atlantic continental shelf of North America most greatly affected by human use, due to long-term pollutant inputs from various point and nonpoint sources, including direct dumping of municipal wastes (sewage sludge), industrial wastes, and dredged material, as well as sewage treatment plant effluents, riverine inflow (e.g., Hudson River), atmospheric deposition, and urban runoff.[1-15] These anthropogenic wastes have resulted in elevated levels of toxic pollutants (e.g., organochlorine compounds, heavy metals, polycyclic aromatic hydrocarbons, and radioactive substances) in the apex.[4,10,11,16] The accumulation of large volumes of organic wastes, toxic chemicals, and pathogens in seafloor sediments at designated dumpsites has been responsible for lasting alterations in the benthic environment. Despite the long history of waste disposal over extensive areas of the New York Bight Apex,

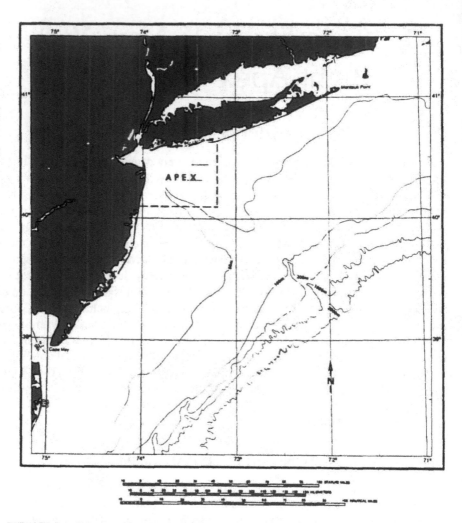

FIGURE 1   Map showing the New York Bight Apex.

the most dramatic changes in benthic community structure are evident over an area of only 15–200 km².[17]

The Hudson-Raritan system exerts strong influence on the New York Bight Apex. The Hudson River plume and dredged spoils have been the largest sources of pollutants in the apex over the years. While wastewater discharges and runoff have historically contributed most of the organic carbon, nutrient (nitrogen and phosphorus), and microbial loads to the apex, dredged material has accounted for the bulk of the heavy metal input (cadmium, chromium, copper, iron, lead, and zinc).

Sewage and industrial effluents released from the Hudson and Raritan Rivers may stimulate episodic blooms (i.e., 1×10^5 cells/ml) of "red tide" flagellates in the apex, and in mid-summer, periods of severe dissolved oxygen depression occasionally develop in the lower Hudson-Raritan estuary and the apex, owing to the decomposition of organic matter following this excess plant growth. Anoxia in the benthos may arise from seasonal and annual variations in productivity and stratification of the water column.[18] Additionally, water quality over the sewage sludge dumpsite may be compromised by the release of nitrogen and phosphorus which can exacerbate these conditions. A major concern at the sewage sludge dumpsite is that excessive nutrient concentrations may overwhelm the capacity of the area to assimilate them, thereby periodically culminating in red tides, brown tides, or eutrophic conditions that often alter the trophic structure of the system through the loss of large numbers of heterotrophs.

Natural processes interacting with anthropogenic wastes may hasten habitat destruction in the apex. For example, strong currents associated with storm surges, hurricanes, and other meteorological events (particularly during the winter months) can roil bottom sediments and disperse pollutants from impacted sites to originally unaffected areas. Subsequently, lipophilic organic compounds, xenobiotics, and pathogens tend to settle and accumulate once again in bottom sediments, where they can adversely affect food webs and the functioning of the ecosystem.

An understanding of both natural and anthropogenic processes affecting benthic biota and habitats in the apex requires interactive studies of bottom sediments, bottom topography, sediment transport, biogeochemistry, and population ecology, all of which must be characterized within the broader context of physical oceanographic processes (e.g., sediment transport, water mass patterns, upwelling) that act as major controls on the benthic regime.

III. HISTORICAL DEVELOPMENTS

A. POINT SOURCE POLLUTION

The New York Bight Apex served as a major repository of anthropogenic wastes for more than 75 years, receiving substantial volumes of sewage sludge, dredged material, acid-iron waste, and construction and demolition debris (i.e., cellar dirt) at designated dumpsites along the upper Hudson Shelf Valley. During the 1960s, 1970s, and 1980s, more than 10 million metric tons (mt) of solid wastes were dumped annually in this area. The dredged-material dumpsite was first opened in 1914, and the volume of spoils dumped there increased to 4.3×10^6 m³/yr by the 1970s.[2] Between 1924 and 1987, sludge from some 200 sewage treatment plants in the New York metropolitan area was dumped at the 12 Mile Dumpsite located about 22 km southeast of the Hudson-Raritan estuary mouth immediately adjacent to the Christiaensen Basin, a bathymetric depression at the head of the Hudson Shelf Valley. By 1983, sludge dumping at this site reached a maximum volume of 8.3×10^6 mt (wet),[6] which was greater than at any other sludge dumpsite in the world.[19] The U.S. Environmental Protection Agency (U.S. EPA) officially closed the sewage sludge dumpsite at the end of 1987. Today, the apex continues to receive significant inputs of pollutants from the Hudson-Raritan estuary.

Historically, dredged spoils constitute the single largest point source of pollution in the apex, since they contain contaminants originally transported by influent systems, such as the Hudson River, as well as substances in runoff, domestic wastes, and industrial wastes discharged or spilled into the New York Harbor that accumulate in bottom sediments.[5,20] During dredging of navigable waterways in the harbor or disposal of these sediments at the dredged material dumpsite, contaminants have been released to surrounding waters. Contaminated sediment removed from Newark Bay has also been dumped at the dredged material dumpsite. Hence, the disposal of large quantities of dredged spoils and associated toxic halogenated compounds, heavy metals, and other contaminants has long been an issue of potential environmental impact. Over the past 200 years, the dredged spoil disposal sites have moved several times, with the most recent site located west of the 12-mile sewage sludge dumpsite (Figure 2).

In October 1996, Governor Whitman of New Jersey announced plans to dump more than 5 million (mt) of moderately contaminated dredged spoils at the dredged material dumpsite in 1997. These dredged spoils will consist of sediments removed from New York Harbor. Disposal of the contaminated sediments will take place prior to final closure of the dumpsite which was scheduled for September 1, 1997.

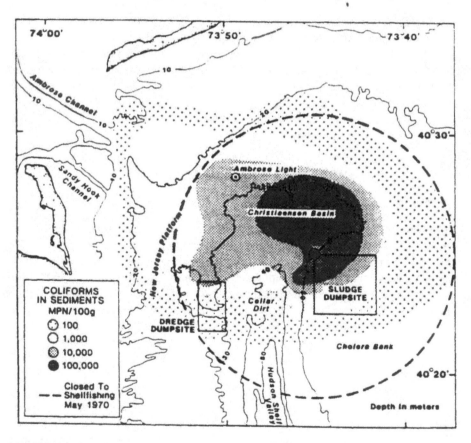

FIGURE 2   Location of the 12-mile dumpsite, dredged material dumpsite, and the apex area closed to commercial shellfishing in the New York Bight. MPN = most probable number. (From Studholme, A. L. and O'Reilly, J. E., in *Effects of the Cessation of Sewage Sludge Dumping at the 12-Mile Site*, Studholme, A. L., O'Reilly, J. E., and Ingham, M. C., Eds., NOAA Tech. Rept. NMFS 124, NOAA Woods Hole, MA, 1995.)

The topographic depression in the Christiaensen Basin (~33 m deep) lies between the sewage sludge and dredged material dumpsites. Consequently, particulates eroded from the dumpsites tend to accumulate in the basin, and the fine silts and muds found there appear black and oily, presumably from contaminants. The Christiaensen Basin, therefore, is also the site of high pollutant concentrations. Organic enrichment and changes in water stratification periodically cause localized oxygen depletion in the basin.[5] These conditions create severe stresses for the benthos.

Allied Chemical Company and NL industries both used the acid-waste dumpsite for chemical waste disposal. NL industries commenced dumping of acid-iron wastes in 1949 and terminated dumping in 1982. These wastes originated from the production of titanium dioxide. Allied Chemical Company continued to dump material through 1983, disposing of 3.14×10^4 mt of chemical wastes that year. The total volume of waste dumped at the site decreased by nearly two thirds between 1973 and 1983.[17] As in the case of sewage sludge, the dumping of acid-iron wastes in the New York Bight Apex has also been discontinued. When active, the industrial waste and sewage sludge dumpsites comprised major sources of pollutant inputs to the apex.

The dumping of cellar dirt in the New York Bight Apex dates back to the early 1800s.[21] Cellar dirt has received less attention than the sewage sludge, dredged material, and acid wastes because it consists of relatively nontoxic solid materials. In addition, smaller volumes of cellar dirt have been dumped in the apex compared to the other types of waste.

Waste dumping in the New York Bight Apex peaked in the 1970s and 1980s. At that time, the estimated volume of wastes reaching the New York Bight amounted to: sewage sludge, $3.0-4.3 \times 10^6$ m³/yr; dredged spoils, $>7 \times 10^6$ m³/yr; acid wastes, $>2 \times 10^6$ m³/yr; construction debris and cellar dirt, 4.5×10^5 m³/yr; atmospheric fallout of Cd, Cr, Cu, Fe, Pb, and Zn, 3940 to 32,000 mt/yr; plus suspended solids, 49,000 to 500,000 mt/yr; and nitrogen, 24,000 mt/yr, municipal and industrial wastewater containing oil and grease, 72,000 mt/yr; and total nitrogen, 79,000 mt/yr; and runoff and groundwater influx containing oil and grease, 124,000 mt/yr; and total nitrogen, 55,000 mt/yr.[1] During the 1990s, ocean dumping of waste materials in the New York Bight Apex has rapidly fallen in disfavor among both the marine science community and the general public. The disposal of all waste in the apex will be prohibited prior to the close of this decade.

Numerous scientific studies have been conducted in the New York Bight Apex during the past 50 years on the aforementioned anthropogenic wastes. Despite these studies, many natural physical and biological processes occurring in the apex are poorly understood. The dumpsite areas continue to be the target of assessment programs, and little is presently known about their recovery. Some important goals of these assessment programs are to determine: (1) the distribution and abundance of resource species in the vicinity of the dumpsites; (2) the resuspension and transport of contaminated sediment; and (3) the recovery of habitats subsequent to the removal of major waste loading.[4,9,10]

B. Nonpoint Source Pollution

The Hudson River plume represents the most important nonpoint source of contaminants in the New York Bight Apex. This riverine input, in turn, includes a myriad of pollutant sources, such as stormwater runoff, sewage treatment plant effluent, influx from shipping and boating activities, as well as numerous diffuse sources. Stormwater runoff and combined sewer overflows are potentially predominant sources of pollutants during periods of heavy precipitation. Direct discharges from coastal communities and atmospheric deposition deliver relatively minor amounts of the total pollutants.[5] While wastewater and runoff contributed the most nutrient (nitrogen and phosphorus), microbial, and organic carbon loads during the peak dumping years, dredged material accounted for the bulk of the heavy metal (cadmium, chromium, copper, iron, lead, and zinc) and organochlorine contaminant (PCBs) loads to the apex.[22] Nonpoint sources are now responsible for the overwhelming proportion of the total contaminant load in the apex, since the designated waste dumpsites have been closed.

IV. BIOTIC STUDIES

Effects of waste disposal on marine communities and fishery resource populations in the New York Bight Apex have been investigated since the late 1940s.[10,12,13,22-25] Results of these investigations indicate that the greatest impacts have occurred in benthic habitats, where communities of organisms have been altered significantly relative to other areas of the New York Bight. During periods of active waste dumping, for example, a marked reduction in benthic species diversity was recorded at impacted sites in the apex.[1,2] The direct effect of waste disposal on abundance, distribution, and health of local fishery resources is less conclusive, although bacterial and chemical contamination of some resource populations has been demonstrated in the area.[4,26,27] In addition, altered feeding responses have been detected in commercially important finfish — specifically winter flounder — affected by sewage sludge.[13] Among the most significant impacts chronicled on commercially and recreationally important finfish and shellfish in the apex have been increased incidences of fin rot disease in fish (e.g., winter flounder, *Pseudopleuonectes americanus*; summer flounder, *Paralichthys dentatus*) and exoskeleton deterioration and pathologies of gills in lobsters (*Homarus americanus*) and rock crabs (*Cancer irroratus*).[1] Winter flounder, in particular, have repeatedly exhibited deformed fin rays and associated skeletal abnormalities.

A. Pathogens and Disease

The Marine Ecosystems Analysis (MESA) Project initiated by The National Oceanic and Atmospheric Administration (NOAA) in 1973 identified several major contaminant impacts on marine resources in the New York Bight Apex due to human activities. In 1970, a circular area of the sewage sludge dumpsite with a radius of 11 km was closed to shellfishing by the U.S. Food and Drug Administration (FDA) because of high fecal coliform bacteria concentrations in sediments (3.3×10^4 per 100 g at the center of the dumpsite) and shellfish (Figure 2).[5] Within this circular area, bacterially polluted dredged spoils also covered 65 km^2. Apart from the sewage sludge and dredged spoils, ocean outfalls and the Hudson-Raritan estuary plume also delivered significant numbers of fecal coliforms to bottom sediments and shellfish of the inner bight.[22] As a result of this additional bacterial contamination, the FDA expanded the shellfish closure area in 1974. Pathogenic pollution, therefore, is derived from multiple sources, with a substantial fraction of microbes entering the apex from non-dumping pathways.

A detailed study of lobsters and rock crabs in the apex during the mid-1970s found exoskeletal "shell disease" primarily in specimens collected on or near the sewage sludge and dredged spoil dumpsites.[28] Skeletal erosion developed on appendages where contaminated sediments accumulated, predominantly on the tips of the walking legs, the ventral sides of chelipeds, exoskeletal spines, gill lamellae, and at points of skeletal articulation. Gill fouling, erosion, and necrosis were likewise documented in diseased animals. The gills were usually clogged with detritus and contained localized thickenings. The chitinous covering of the gills was often eroded, and the underlying tissues severely diseased or irreversibly damaged. Erosion and destruction of gill membranes may increase mortality rates by reducing oxygen uptake capability, which can be especially threatening to populations exposed to recurrent hypoxic conditions and occasional sulfide generation, such as observed in the Christiaensen Basin during the summer.[1,15,29]

High prevalences (up to 38%) of fin rot disease were recorded in samples of trawled marine fishes collected from the New York Bight in the 1970s and 1980s, with 22 species being affected.[15] Fin rot disease is grossly characterized as a progressive necrosis of the anal and dorsal fins and, less frequently, of the caudal fins.[1] During the sampling periods, it was commonly observed in flatfishes, notably winter flounder (*P. americanus*), summer flounder (*P. dentatus*), yellowtail flounder (*Limanda ferruginea*), fourspot flounder (*Paralichthys oblongus*), and windowpane (*Scopthalmus aquosus*). High prevalences of fin rot disease may signal an infectious etiology. While histopathological examination of diseased fish taken from the New York Bight Apex revealed the occurrence of bacteria of the genera *Aeromonas*, *Pseudomonas*, and *Vibrio*, a definitive bacterial cause of the finfish abnormalities has not been established.[15] It is possible that a noninfectious etiology is responsible, such as toxic chemicals (e.g., heavy metals).

B. BIOLOGICAL PRODUCTION

Nutrient and organic enrichment, together with chemical contaminant inputs, has led to elevated primary productivity and sediment organic concentrations, diseased fish and shellfish, and local changes in the structure of benthic communities in the New York Bight Apex. During periods of substantial waste disposal in the 1970s and 1980s, an area of sediments greater than 240 km^2 in the apex was enriched in carbon, petroleum hydrocarbons, halogenated compounds, and heavy metals due to direct waste dumping, discharges of ocean outfalls, and the Hudson-Raritan estuary plume. Heavily contaminated sediments typically contained depauperate benthic communities dominated by a few species with high standing crops. Changes in species composition and population structure at polluted sites were ascertained in both benthic macrofaunal and microfaunal species assemblages.[10,12,17,23,30,31] Despite these changes, secondary production remained relatively high in areas.

Two major symposia have been held to assess biological aspects of contaminants introduced into the New York Bight Apex.[4,32] A major goal of these symposia was to determine changes over time in fates and biological effects of the contaminants. The symposia considered water quality issues, primary productivity, phytoplankton dynamics, benthic communities, fisheries, and other subject areas. The following discussion draws heavily from these symposia and recent studies of the apex.

The biomass of primary producers, herbivores, carnivores, and decomposers averages about 2 g C/m^2, 2 g C/m^2, 6 g C/m^2, and 4 g C/m^2, respectively, over the entire New York Bight. Annual phytoplankton productivity in the outer bight ranges from 200–300 g $C/m^2/yr$.[33] In the apex, which is enriched in nutrients and organic carbon from several sources, phytoplankton productivity appears to substantially exceed that in the outer bight. Malone[34] reported phytoplankton productivity values in the apex between about 0.1 g $C/m^2/d$ (in December) and 6.4 g $C/m^2/d$ (in June). Benthic biomass in the bight decreases by a factor of nearly 20 when proceeding from silty sand bottoms (1800 g/m^2) to sand-gravel bottoms (94 g/m^2).[31] The mean macrofaunal biomass in the deeper, most severely contaminated areas of the apex amounted to 344 g/m^2 during the 1980–1982 period. Secondary production in the apex at this time was relatively high (383 $Kcal/m^2/yr$) and comparable to that of uncontaminated coastal areas.[24]

One of the most striking features of the New York Bight Apex under stratified conditions is the stimulation of phytoplankton productivity by nutrient enrichment from the Hudson-Raritan estuary.[35,36] Only a small quantity of nitrogen entering the Hudson-Raritan estuary complex (~10%) fuels primary production within the system.[37-40] Most of the nitrogen flushes into surface waters of the New York Bight Apex.[18] Up to 80% of the available nitrogen in the apex is used by phytoplankton within 20 km of the mouth of the Hudson-Raritan estuary.[41,42]

Hypoxia or anoxia along the seafloor of the inner bight may arise from seasonal pulses of phytoplankton production stimulated by nutrient enrichment from various sources, which create high oxygen demand. Duedall et al.[38] observed that sewage effluent discharged into waters surrounding the New York metropolitan region represented the main source of ammonium, nitrite, and phosphate in the apex, whereas nitrate was principally derived from the Hudson River. As noted by O'Connor,[36] nitrogenous loadings from sewage treatment plant outfalls along the New Jersey coast may have played an integral role in the genesis of nearshore oxygen depletion events chronicled in 1968, 1971, 1974, and 1977. However, with the passage of a series of environmental laws in the 1970s and the closure of waste dumpsites during the past decade, the levels of nearly all major pollutants have declined in New York Harbor and the New York Bight Apex. Water quality has generally improved across the region.

C. Benthic Communities

As a result of long-term organic carbon and chemical contaminant loading in the New York Bight Apex, extensive bottom habitats are heavily degraded and commonly contain depauperate benthic communities typified by reduced species diversity and very high standing crops of a few species, as noted previously. Studies of benthic macrofauna in the apex have been conducted for more than 30 years to assess impacts from waste disposal and responses of the fauna to abatement of waste dumping. Benthic fauna have been used as indicators of environmental change in the bight because of their relative immobility and close association with contaminant-accumulating sediments. Nearly 20 major sediment-benthos sampling projects have been conducted in the bight since 1966. Three of the most comprehensive include NOAA's Marine Ecosystem (MESA) — New York Bight Project undertaken from 1973–1976, NOAA's sampling of benthic macrofauna at 44 to 48 stations during the 1980–1985 period, and NOAA's 1986–1989 study of responses of the benthos to the phaseout of sewage sludge dumping in the apex.[43]

Results of the aforementioned benthic studies revealed conspicuous changes of benthic community structure and function areas of the inner New York Bight that were impacted by waste disposal. The greatest alterations occurred in benthos inhabiting areas at or in close vicinity of the sewage sludge and dredged material dumpsites. In an organically enriched, highly polluted zone of the dumpsites covering ~15 km², pollution-tolerant benthic species dominated, most notably the small deposit-feeding polychaetes *Capitella* spp. which include a complex of several species used worldwide as indicators of organic carbon enrichment. Some species characteristic of unstressed environments in the bight (e.g., *Ampelisca* spp., *Unicola irrorata*, and *Erichthonius rubricornis*) were almost completely eliminated from this zone.[5] In a less contaminated region within 4–8 km of the dumpsites, dense populations of *Cerianthiopris*, *Nephthys*, *Nucula*, and *Pherusa* occupied an area of ~200 km². More diverse assemblages of benthic organisms existed at increasing distances from the dumpsites. Ampelisid amphipods, considered pollution-intolerant taxa compared to the polychaetes which dominate much of the inner New York Bight, became more abundant in unstressed habitats away from the apex.[10,17] The following discussion provides details of benthic community changes in the apex during the last three decades, owing to multiple-waste disposal activities.

1. Early Studies (1966–1974)

Comprehensive investigations of benthic communities in the New York Bight during the 1960s and early 1970s by Pearce[44] and Pearce et al.[45] at the Sandy Hook Laboratory, National Marine Fisheries Service in Highlands, New Jersey, showed considerable impacts of waste disposal on benthos in the apex. Pearce[44] asserted that two areas in the apex covering a total of approximately 50 km² had impoverished benthic assemblages relative to unstressed areas in the bight. Pearce et al.[45] reported a reduction in the number of species and individuals in these areas in 1974.

Bottom sediments contained high concentrations of organic carbon and heavy metals (i.e., copper, chromium, lead, nickel, and zinc).[46] While a range of sedimentary facies occurs throughout the bight composed of silts to poorly sorted sand and gravel, fine- to medium-grained sands predominate.[47,48] At the sewage sludge and dredged spoil dumpsites, as well as the Christiaensen Basin, organic material blankets the natural sediments.[46] Total carbon concentrations during periods of active waste disposal peaked near the dredged spoil dumpsite.[47] Thomas et al.[29] documented high oxygen consumption in the Christiaensen Basin and upper Hudson Shelf Valley. In August 1974, dissolved oxygen concentrations dropped to 1 ppm in near-bottom waters, well below the tolerance of many benthic marine organisms.

2. MESA — New York Bight Project (1973–1976)

The largest benthic sampling program in the New York Bight entailed a 4-yr effort (1973–1976) by NOAA termed the Marine Ecosystems Analysis (MESA) — New York Bight Project. This project uncovered little contamination in the mid-shelf region of the bight relative to the apex. As a consequence of waste impacts delineated in the apex by this project, several recommendations were proposed for additional monitoring of sediments and benthos in the bight. Subsequently, a sampling plan was designed and annual sampling conducted from 1980 through 1985.[43,49] Critically important benthic data collected during this 6-yr period included macrobenthic species richness, numbers and biomass of dominant species, numbers of amphipods, and similarities of community structure over space and time.

3. Benthic Surveys (1979–1989)

Reid et al.[43] reported on benthic macrofaunal surveys in the New York Bight over the 1979–1989 period. Although their work focused on sampling conducted in the New York Bight Apex from 1980 through 1985, they also examined relevant data from NOAA's Northeast Monitoring Program (NEMP)[50] and data from a 1986–1989 study of the responses of the benthos to the phaseout of sewage sludge dumping in the apex.[10] Data collected over this 10-yr interval provided baselines against which responses to the phaseout of waste dumping could be measured.

Cluster analysis of benthic macrofaunal data obtained during 1980–1985 indicated the occurrence of five distinct assemblages: (1) sewage sludge dumpsite; (2) sewage sludge accumulation area (Christiaensen Basin); (3) inner Hudson Shelf Valley; (4) outer Hudson Shelf Valley; and (5) outer shelf. These assemblages experienced little change over the survey period. The numbers of species and amphipods per sample, both indicators of environmental stress, exhibited consistent spatial patterns, with lowest values recorded in the sludge accumulation area of the Christiaensen Basin and other inshore areas, and highest values in the outermost shelf and the outer Hudson Shelf Valley. A small area in the Christiaensen Basin contained a highly altered assemblage dominated by the pollution-tolerant polychaetes *Capitella* spp. Over most of the remaining basin, an enriched faunal assemblage was found. However, no consistently defaunated areas were observed during the sampling period.

The concentration of organic carbon in bottom sediments strongly influenced the species composition of benthic macrofauna. For example, large populations of a few small opportunistic species (e.g., *Capitella* spp.) often dominated areas of high organic loading within the Christiaensen Basin. For most of the basin with lower organic loading values, as well as areas immediately outside of the basin, high abundances of several benthic species were apparent. Apart from spatial variations in benthic assemblages, important temporal variations were also noted. In particular, statistically significant decreases in the numbers of species and amphipods were noted at most stations from 1980 to 1985. The most significant decreases in these parameters were registered in the sewage sludge accumulation area of the Christiaensen Basin and at a site 52 km southeast of that area in the Hudson Shelf Valley. The causes of these temporal decreases have not been unequivocally established, although increasing effects of sewage sludge or other waste inputs, natural factors, and sampling artifacts may all have played a role.

4. Sewage Sludge Dumping Phaseout (1986–1989)

The cessation of sewage sludge dumping in the New York Bight Apex at the 12-Mile Site took place on December 31, 1987. To study the responses of biota and habitats to abatement of sewage sludge dumping, the Environmental Processes Division of the Northeast Fisheries Center, National Marine Fisheries Service (NOAA) sampled the benthic environment in the apex using a sample design based on the technique of replication in time, also known as the Before/After, Control/Impact design. Sampling was initiated in June 1986 and extended through September 1989, focusing on three stations on the slope of the Hudson Shelf Valley: NY6 — heavily degraded; R2 — enriched; and NY11 — a reference area assumed to be least affected by sludge dumping (Figure 3).[10] Broadscale sampling was also conducted throughout a 350 km^2 area of the apex (Figure 4).

Similar water depths, hydrographic conditions, and sediment types characterize stations NY6, R2, and NY11 (Table 1). Station NY6, most strongly affected by sludge dumping, contained fine sandy sediments, high concentrations of total organic carbon (TOC), and elevated levels of chemical contaminants (e.g., >100 ppm chromium, >150 ppb polychlorinated biphenyls, and >7000 ppt polycyclic aromatic hydrocarbons). The macrobenthic community at this site was dominated by stress-tolerant taxa, low in diversity, and low in abundance. The mean numbers of species at station NY6 were consistently less than those at stations R2 and NY1 (Figure 5). When sewage sludge was actively dumped, seasonal hypoxia and detectable concentrations of sulfide occurred in bottom waters.

Located 3.4 km north-northwest of the sewage sludge dumpsite, station R2 also contained fine sands. However, it had lower TOC and chemical contaminant concentrations than station NY6. Benthic macrofauna at this station exhibited consistently high biomass (>200 g/m^2) compared to staions NY6 and NY11 (Figure 6).

Station NY11 was located approximately 10 km south of the sewage sludge dumpsite. Consisting of fine sands, station NY11 had relatively low concentrations of TOC and chemical contaminants. Benthic macrofauna were often abundant. Biomass values in nearby areas, which amounted to about 200 g/m^2 prior to the study, were somewhat lower during the study.[13]

The volume of sewage sludge dumped in the apex increased 89% between 1974 and 1983. Sludge dumping remained high until the phaseout of sludge disposal commenced in July 1986 (Figure 7). With the phaseout of sewage sludge disposal completed in December 1987, post-abatement changes recorded in benthic habitats at the study sites included reduced TOC and increased dissolved oxygen concentrations in

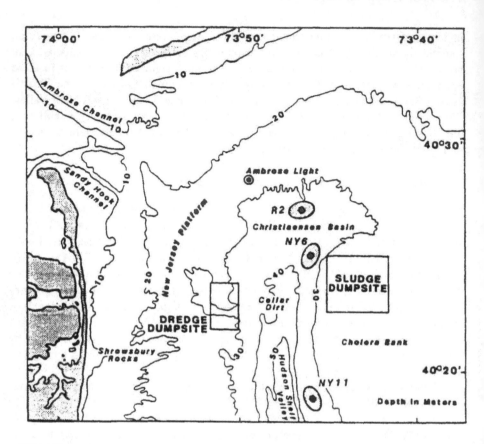

FIGURE 3 Location of the 12-mile sewage sludge and dredged material dumpsites and the replicate stations (NY6, R2, NY11) in the New York Bight Apex. (From Packer, D. et al., in *Effects of the Cessation of Sewage Sludge Dumping at the 12-Mile Site*, Studholme, A. L., O'Reilly, J. E., and Ingham, M. C., Eds., NOAA Tech. Rept. NMFS 124, NOAA Woods Hole, MA, 1995, 155.)

bottom waters during summer. Heavy metal contamination also decreased. Abundances of some benthic infaunal species likewise changed.[13]

Over the 39-month study period, the numbers of crustacean, molluscan, and total species increased most dramatically at stations previously influenced by sewage sludge (i.e., NY6 and R2), suggesting possible recovery of the impacted areas. The pollution-tolerant polychaetes *Capitella* spp., while abundant during sewage sludge dumping, declined substantially in 1988 and 1989 (Figure 8). The decline in abundance of these polychaetes may also have signified a positive response to the cessation of sewage sludge dumping.

Four species accounted for >90% of the megainvertebrate biomass in the apex: Atlantic rock crab (*Cancer irroratus*), horseshoe crab (*Limulus polyphemus*), American lobster (*Homarus americanus*), and longfin squid (*Loligo pealeii*). Five species comprised >75% of the finfish biomass: little skate (*Raja erinacea*), spring dogfish (*Squalus acanthias*), winter flounder (*Pseudopleuronectes americanus*), red hake

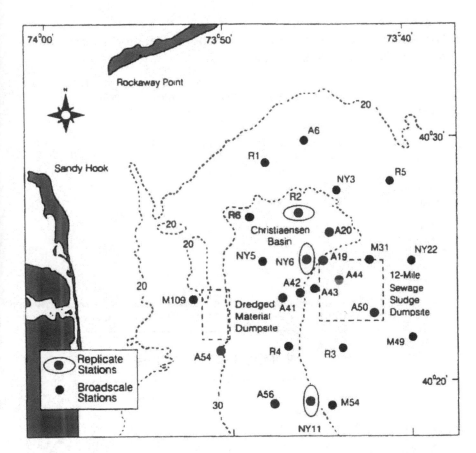

FIGURE 4   Locations of stations sampled on the replicate and broadscale surveys during the 12-mile dumpsite study. (From Pikanowski, R., in *Effects of the Cessation of Sewage Sludge Dumping at the 12-Mile Site*, Studholme, A. L., O'Reilly, J. E., and Ingham, M. C., Eds., NOAA tech. Rept. NMFS 124, NOAA Woods Hole, MA, 1995, 13.)

(*Urophycis chuss*), and ocean pout (*Macrozoarces americanus*). Demersal biomass was dominated seasonally by Atlantic rock crab (summer) and little skate and spring dogfish (winter).[51] According to Pikanowski,[12] three of the dominant species (i.e., Atlantic rock crab, little skate, and winter flounder) showed no statistically significant response to the cessation of sewage sludge dumping, whereas the American lobster increased in local abundance.

In summary, some benthic variables showed clear responses to the phaseout of sewage sludge dumping at the most sludge-altered site (NY6). The numbers of overall benthic species, as well as numbers of molluscan, crustacean, and amphipod species, increased significantly at station NY6 compared to stations R2 and NY11 subsequent to the termination of sludge dumping. Abundances of certain amphipod species (especially *Photis pollex* and *Uniciola irrorata*), indicators of improved environmental conditions, increased greatly at the heavily degraded site (NY6) after cessation of dumping.[52] With the decline in the abundance of *Capitella* spp., together

TABLE 1
Sediment and Bottom Water Characteristics at Stations R2, NY6, and NY11 in the New York Bight Apex

	R2 Before	R2 After	NY6 Before	NY6 After	NY11 Before	NY11 After
Depth[a]	29		31		29	
Sediment						
Grain size[b]	3.1 ± 0.1	3.2 ± 0.1	3.6 ± 0.1	3.5 ± 0.1	3.1 ± 0.1	3.1 ± 0.1
TOC[c]	0.9 ± 0.1	0.9 ± 0.1	4.6 ± 0.5	2.3 ± 0.3	0.3 ± 0.0	0.3 ± 0.0
Cr[d]	37.2	36.7	163.0	96.5	15.8	12.9
Bottom water						
minimum DO[e]	4.2	5.7	4.3	5.8	5.3	5.5

[a] meters.
[b] Phi units.
[c] % dry weight.
[d] ppm dry weight .
[e] mg/l.

From Steimle, F. W., Sewage sludge disposal and winter flounder, red hake, and American lobster feeding in the New York Bight, *Mar. Environ. Res.*, 37, 233, 1994. With permission.

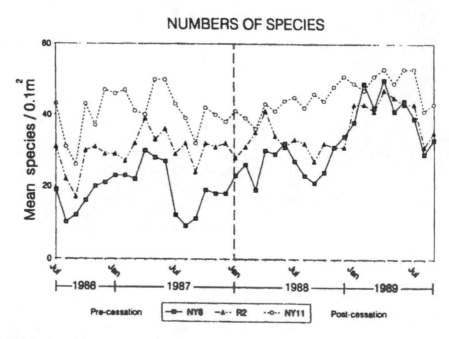

FIGURE 5   Mean numbers of species (all taxa combined) per 0.1 m² at sampling stations in the inner New York Bight. Station NY6 = square; station R2 = triangle; station NY11 = circle. (From Reid, R. N. et al., in *Effects of the Cessation of Sewage Sludge Dumping at the 12-Mile Site*, Studholme, A. L., O'Reilly, J. E., and Ingham, M. C., Eds., NOAA Tech. Rept. NMFS 124, Woods Hole, MA, 1995, 213.)

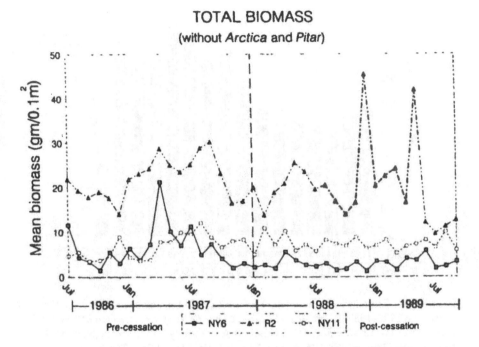

FIGURE 6   Mean biomasses of all taxa combined, excluding *Arctica islandica* and *Pitar morrhuanus*, per 0.1 m² at sampling stations in the inner New York Bight. Station NY6 = square; station R2 = triangle; station NY11 = circle. (From Fromm, S. A. et al., in *Effects of the Cessation of Sewage Sludge Dumping at the 12-Mile Site*, Studholme, A. L., O'Reilly, J. E., and Ingham, M. C., Eds., NOAA Tech. Rept. NMFS 124, Woods Hole, MA, 1995, 213.)

with the aforementioned changes, this area appears to have responded favorably to the removal of sewage sludge influences. Habitat quality appears to have improved strongly within two years of cessation of sludge dumping.

Studholme and O'Reilly[53] provide the following major conclusions of the 39-month study:

- Indicators of sewage pollution in sediments (e.g., rate of seabed oxygen consumption, redox potential, bottom dissolved oxygen gradients, total bacteria counts, coliform bacteria counts) responded rapidly to the cessation of sewage sludge dumping.
- Data collected on lead enrichment, *Clostridium perfringens* spores, and redox potential indicated a decrease in sewage contamination over a large area surrounding the dumpsite.
- At the most heavily polluted site (NY6), the abundance of pollution-tolerant benthic species decreased while the number of total species and number of crustaceans increased subsequent to the cessation of sludge dumping.
- The distribution of megafauna showed no appreciable changes attributable to the cessation of sludge dumping.

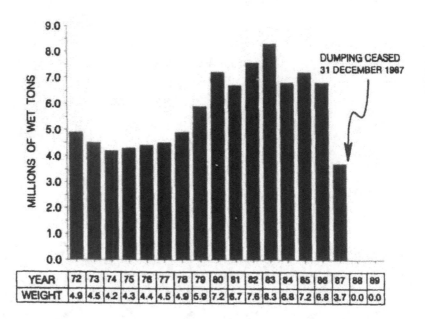

FIGURE 7   Estimated annual amounts of sewage sludge dumped in the New York Bight Apex from 1972–1989. (From Wilk, S. J. et al., in *Effects of the Cessation of Sewage Sludge Dumping at the 12-Mile Site*, Studholme, A. A., O'Reilly, J. E., and Ingham, M. C., Eds., NOAA Tech. Rept. NMFS 124, NOAA Woods Hole, MA, 1995, 173.)

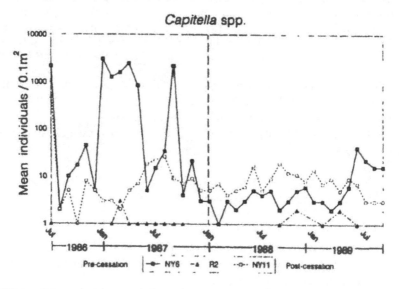

FIGURE 8   Mean abundances of *Capitella* spp. per 0.1 m² at sampling stations in the inner New York Bight. Station NY6 = square; station R2 = triangle; station NY11 = circle. (From Reid, R. N. et al., in *Effects of the Cessation of Sewage Sludge Dumping at the 12-Mile Site*, Studholme, A. L., O'Reilly, J. E., and Ingham, M. C., Eds., NOAA Tech. Rept. NMFS 124, Woods Hole, MA, 1995, 213.)

REFERENCES

1. O'Connor, J. S., Contaminant effects on biota of the New York Bight, in *Proceedings of the Gulf and Caribbean Fisheries Institute*, Higman, J. B., Ed., 28th Annual Session, University of Miami, Coral Gables, 1976, 50.
2. Swanson, R. L., Status of ocean dumping research in the New York Bight, *J. Waterway, Port, Coastal Ocean Div.*, 103 (WW1), 9, 1977.
3. Ketchum, B. H., Kester, D. R., and Park, P. K., Eds., *Ocean Dumping of Industrial Wastes*, Plenum Press, New York, 1981.
4. Mayer, G. F., Ed., *Ecological Stress and the New York Bight: Science and Management*, Estuarine Research Federation, Columbia, SC, 1982.
5. Swanson, R. L., Champ, M. A., O'Connor, T., Park, P. K., O'Connor, J., Mayer, G. F., Stanford, H. M., and Erdheim, E., Sewage-sludge dumping in the New York Bight Apex: a comparison with other proposed ocean dumpsites, in *Wastes in the Ocean*, Vol. 6, *Nearshore Waste Disposal*, Ketchum, B. H., Capuzzo, J. M., Burt, W. V., Duedall, I. W., Park, P. K., and Kester, D. R., Eds., John Wiley & Sons, New York, 1985, 461.
6. Santoro, E., Status report: phaseout of ocean dumping of sewage sludge in the New York Bight Apex, *Mar. Pollut. Bull.*, 18, 278, 1987.
7. NOAA, A plan for study: response of the habitat and biota of inner New York Bight to abatement of sewage sludge dumping, NOAA Tech. Memo. NMFS-F/NEC-55, Woods Hole, MA, 1988.
8. NOAA, Response of the habitat and biota of the inner New York Bight to abatement of sewage sludge dumping: second annual progress report — 1988. NOAA Tech. Memo. NMFS-F/NEC-67, Woods Hole, MA, 1989.
9. Reid, R., Responses of habitats and biota of the inner New York Bight to abatement of sewage sludge dumping — progress report, in *Conference Proceedings on Cleaning Up Our Coastal Waters: An Unfinished Agenda*, Southerland, M. T. and Swetlow, K., Eds., Manhattan College, Bronx, New York, March 12–14, 1990., U. S. Environmental Protection Agency Contract Rept. No. 68-C8-0052, 1990, 491.
10. Studholme, A. L., Ingham, M. C., and Pacheco, A., Eds., Response of the habitat and biota of the inner New York Bight to abatement of sewage sludge dumping: third annual progress report — 1989. NOAA Tech. Memo. NMFS-F/NEC-82, Woods Hole, MA, 1991.
11. Kennish, M. J., *Ecology of Estuaries: Anthropogenic Effects*, CRC Press, Boca Raton, FL, 1992.
12. Pikanowski, R. A., The effects of ocean disposal of sewage sludge on the relative abundance of benthic megafauna, *Chem. Ecol.*, 6, 199, 1992.
13. Steimle, F. W., Sewage sludge disposal and winter flounder, red hake, and American lobster feeding in the New York Bight, *Mar. Environ. Res.*, 37, 233, 1994.
14. Kennish, M. J., Ed., *Practical Handbook of Estuarine and Marine Pollution*, CRC Press, Boca Raton, FL, 1996.
15. Sindermann, C. J., *Ocean Pollution: Effects on Living Resources and Humans*, CRC Press, Boca Raton, FL, 1996.
16. Wolfe, D. A. and O'Connor, T. P., Eds., *Oceanic Processes in Marine Pollution*, Vol. 1, *Biological Processes and Wastes in the Ocean*, Robert E. Krieger Publishing, Malabar, FL, 1988.
17. Swanson, R. L. and Mayer, G. F., Ocean dumping of municipal and industrial wastes in the United States, in *Oceanic Processes in Marine Pollution*, Vol. 3, *Marine Waste Management: Science and Policy*, Camp, M. A. and Park, P. K., Eds., Robert E. Krieger Publishing Company, Malabar, FL, 1989, 35.

18. Mearns, A. J., Haines, E., Kleppel, G. S., McGrath, R. A., McLaughlin, J. J. A., Segar, D. A., Sharp, J. H., Walsh, J. J., Word, J. Q., Young, D. K., and Young, M. W., Effects of nutrients and carbon loadings on communities and ecosystems, in *Ecological Stress and the New York Bight: Science and Management*, Mayer, G. F., Ed., Estuarine Research Federation, Columbia, SC, 1982, 53.

19. Norton, M. G. and Champ, M. A., The influence of site-specific characteristics on the effects of sewage sludge dumping, in *Oceanic Processes in Marine Pollution*, Vol. 4, *Scientific Monitoring Strategies for Ocean Waste Disposal*, Hood, D. W. et al., Eds., Robert E. Krieger, Malabar, FL, 1989, 161.

20. Gross, M. G., Waste disposal, MESA New York Bight Atlas Monograph 26, New York Sea Grant Institute, Albany, New York, 1976.

21. Mueller, J. A., Jeris, J. S., Anderson, A. R., and Hughes, C. F., Contaminant inputs to the New York Bight, Tech. Mem. ERL MESA-6, U.S. National Oceanic and Atmospheric Administration, Boulder, CO, 1976.

22. Redfield, A. C. and Walford, L. A., A study of the disposal of chemical wastes at sea, Publication 201, National Academy of Science, Washington, D.C., 1951.

23. Steimle, F., Caracciola, J., and Pearce, J., Impacts of dumping on the New York Bight Apex benthos, in *Ecological Stress and the New York Bight: Science and Management*, Mayer, G., Ed., Estuarine Research Federation, Columbia, SC, 1982, 213.

24. Steimle, F. W., Biomass and estimated productivity of the benthic macrofauna in the New York Bight: a stressed coastal area, *Est. Coastal Shelf Sci.*, 21, 539, 1985.

25. Steimle, F. and Terranova, R., Trophodynamics of select demersal fishes in the New York Bight, U.S. Department of Commerce, NOAA Tech. Memo. NMFS-F/NEC 84, Woods Hole, MA, 1991.

26. Roberts, A. E., Hill, D. R., and Tifft, E. C., Jr., Evaluation of New York Bight lobsters for PCBs, DDT, petroleum hydrocarbons, mercury, and cadmium, *Bull. Environ. Contam. Toxicol.*, 29, 711, 1982.

27. Steimle, F. W., Boehm, P. D., Zdanowicz, V. S., and Bruno, R. A., Organic and trace metal levels in ocean quahog, *Arctica islandica Linné*, from the northwestern Atlantic, *Fish. Bull. (U.S.)*, 84, 133, 1986.

28. Young, J. S. and Pearce, J. B., Shell disease in crabs and lobsters from New York Bight, *Mar. Pollut. Bull.*, 6, 101, 1975.

29. Thomas, J. P., Phoel, W. C., Steimle, F. W., O'Reilly, J. E., and Evans, C. A., Seabed oxygen consumption — New York Bight Apex, *Am. Soc. Limnol. Oceanogr. Spec. Symp.* 2, 354, 1976.

30. O'Connor, J. S. and Segar, D. A., Pollution in the New York Bight: a case history, in *Impact of Man on the Coastal Environment*, Duke, T. W., Ed., Tech. Rept. EPA-600/8-82-021, U.S. Environmental Protection Agency, Washington, D.C., 1982, 47.

31. Tietjen, J., Population structure and species composition of the free-living nematodes inhabiting sand of the New York Bight Apex, *Est. Coastal Mar. Sci.*, 10, 61, 1980.

32. Gross, M. G., Ed., Middle Atlantic Continental Shelf and the New York Bight, *Am. Soc. Limnol. Oceanogr. Spec. Symp.* 2, Allen Press, Lawrence, KA, 1976.

33. Walsh, J. J., *On the Nature of Continental Shelves*, Academic Press, New York, 1989.

34. Malone, T. C., Phytoplankton productivity in the apex of the New York Bight: environmental regulation of phytoplankton/chlorophyll *a*, *Am. Soc. Limnol. Oceanogr. Spec. Symp.* 2, 260, 1976.

35. Malone, T. C., Esaias, W., and Falkowski, P., Plankton dynamics and nutrient cycling, in *Oxygen Depletion and Associated Benthic Mortalities in New York Bight, 1976*, Swanson, R. I. and Sindermann, C. J., Eds., NOAA Professional Paper 11, Rockville, MD, 1979, 193.

36. O'Connor, J. S., A perspective on natural and human factors, in *Oxygen Depletion and Associated Benthic Mortalities in New York Bight, 1976*, Swanson, R. I. and Sindermann, C. J., Eds., NOAA Professional Paper 11, Rockville, MD, 1979, 323.

37. Garside, C., Malone, T. C., Roels, O. A., and Sharfstein, B. A., An evaluation of sewage-derived nutrients and their influence on the Hudson River estuary and New York Bight, *Est. Coastal Mar. Sci.*, 4, 281, 1976.

38. Duedall, I. W., O'Connors, H. B., Parker, J. H., Wilson, R. W., and Robbins, A. S., The abundances, distribution, and flux of nutrients and chlorophyll *a* in the New York Bight Apex, *Est. Coastal Mar. Sci.*, 5, 81, 1977.

39. Duedall, I. W., O'Connors, H. B., Wilson, R. E., and Parker, J. H., The Lower Bay Complex. MESA New York Bight Atlas Monogr. No. 29, New York Sea Grant Institute, Albany, NY, 1979.

40. McLaughlin, J. J. A., Kleppel, G. S., Brown, M. P., Ingram, R. J., and Samuels, W. B., The importance of nutrients to phytoplankton production in New York Harbor, in *Ecological Stress and the New York Bight: Science and Management*, Mayer, G. F., Ed., Estuarine Research Federation, Columbia, SC, 1982, 469.

41. Malone, T. C., Factors influencing the fate of sewage-derived nutrients in the lower Hudson estuary and New York Bight, in *Ecological Stress and the New York Bight: Science and Management*, Mayer, G. F., Ed., Estuarine Research Federation, Columbia, SC, 1982, 389.

42. Capuzzo, J. M., Burt, W. V., Duedall, I. W., Park, P. K., and Kester, D. R., The impact of waste disposal in nearshore environments, in *Wastes in the Ocean*, Vol. 6, *Nearshore Waste Disposal*, Ketchum, B. H., Capuzzo, J. M., Burt, W. V., Duedall, I. W., Park, P. K., and Kester, D. R., Eds., John Wiley & Sons, New York, 1985, 3.

43. Reid, R. N., Radosh, D. J., Frame, A. B., and Fromm, S.A., Benthic macrofauna of New York Bight, 1979–1989, NOAA Tech. Rept. NMFS 103, Woods Hole, MA, 1991.

44. Pearce, J. B., The effects of solid waste disposal on benthic communities in the New York Bight, in *Marine Pollution and Sea Life*, Ruivo, M., Ed., Fishing News (Books) Ltd., Surrey, England, 1972, 401.

45. Pearce, J. B., Caracciolo, J. V., Halsey, M. B., and Rogers, L. H., Temporal and spatial distributions of benthic macroinvertebrates in the New York Bight, Am. Soc. Limnol. Oceanogr. Spec. Symp., 2, 394, 1976.

46. Carmody, D. J., Pearce, J. B., and Yasso, W. E., Trace metals in sediments of New York Bight, *Mar. Pollut. Bull.*, 4, 132, 1973.

47. Gross, M. G., Geologic aspects of waste solids and marine waste deposits, New York metropolitan region, *Geol. Soc. Am. Bull.*, 83, 3163, 1972.

48. Hatcher, P. G. and McGillivary, P. A., Sewage contamination in the New York Bight: coprostanol as an indicator, *Environ. Sci. Technol.*, 13, 1225, 1979.

49. Reid, R. N., O'Reilly, J. E., and Zdanowicz, V. S., Contaminants in New York Bight and Long Island Sound sediments and demersal species, and contaminant effects on benthos, summer 1980, NOAA Tech. Memo. NMFS-F NEC-16, 1982.

50. Reid, R. N., Ingham, M. C., and Pearce, J. B., Eds., NOAA's Northeast Monitoring Program (NEMP): a report on progress of the first five years (1979–1984) and a plan for the future, NOAA Tech. Memo. NMFS-F NEC-44, 1987.

51. Wilk, S. J., Pikanowski, R. A., Pacheco, A. L., McMillan, D. G., and Stehlik, L. L., Response of fish and megainvertebrates of the New York Bight Apex to the abatement of sewage sludge dumping — a overview, in *Effects of the Cessation of Sewage Sludge Dumping at the 12-Mile Site*, Studholme, A. L., O'Reilly, J. E., and Ingham, M. C., Eds., NOAA Tech. Rept. NMFS 124, Woods Hole, MA, 1995, 173.

52. Reid, R. N., Fromm, S. A., Frame, A. B., Jeffress, D., Vitaliano, J. J., Radosh, D. J., and Finn, J. R., Limited responses of benthic macrofauna and selected sewage sludge components to phaseout of sludge disposal in the inner New York Bight, in *Effects of the Cessation of Sewage Sludge Dumping at the 12-Mile Site*, Studholme, A. L., O'Reilly, J. E., and Ingham, M. C., Eds., NOAA Tech. Rept. NMFS 124, Woods Hole, MA, 1995, 213.
53. Studholme, A. L. and O'Reilly, J. E., Executive summary, in *Effects of the Cessation of Sewage Sludge Dumping at the 12-Mile Site*, Studholme, A. L., O'Reilly, J. E., and Ingham, M. C., Eds., NOAA Tech. Rept. NMFS 124, Woods Hole, MA, 1995, 251.

3 Case Study 2: Chesapeake Bay

I. INTRODUCTION

One of the most heavily utilized estuaries in the United States is Chesapeake Bay, a 300-km long shallow system comprised of a central mainstem and several major tributaries (e.g., Chester, Choptank, James, Nanticoke, Patapsco, Patuxent, Potomac, Rappahannock, Susquehanna, and York rivers). Primary urban centers located either along the mainstem of the estuary or its tributaries include Baltimore on the Patapsco River, Washington (D.C.) on the Potomac River, and Norfolk near the junction of the entrance of Hampton Roads and the Elizabeth River. Norfolk is a major naval operations center. The Susquehanna River at the far northern reach represents the largest influent system. It drains most of western Pennsylvania and adjacent areas of other states. Chesapeake Bay communicates freely with the Atlantic Ocean at its southern (Virginia) end.

During the past several decades, Chesapeake Bay has received increasing pollutant inputs from numerous municipal and industrial point sources, and more poorly defined nonpoint sources associated with a burgeoning population in nearby watersheds. Aside from being used for waste disposal, Chesapeake Bay and its tributaries are employed for electric power generation, commercial shipping, various recreational pursuits, and commercial harvesting of wildlife, finfish, and shellfish. In addition, two major industrial ports are located on the bay: Baltimore Harbor and Hampton Roads. Shipping activities at these ports contribute greatly to local anthropogenic impacts. As a consequence of continued human development in watershed areas and accelerated use of bay resources, environmental conditions in the estuary have deteriorated significantly, as manifested by repeated episodes of eutrophication, the reduction of sensitive habitat area (e.g., submerged aquatic vegetation), and the decline of finfish and shellfish resources. Decreases in abundance of oyster and striped bass populations, for example, may be attributable in part to rising levels of pollution. Some contaminants such as polychlorinated biphenyls (PCBs), phthalate esters, trace metals (e.g., copper, lead, and zinc), polycyclic aromatic hydrocarbons (PAHs), pesticides, and radionuclides are regionally dispersed in the bay. However, the cumulative impact of human activity and pollutants on baywide changes in biotic communities and habitats has not been clearly established.

With a volume of 5.2×10^{10} m^3 (excluding tributaries), Chesapeake Bay is the largest U.S. estuary (Figure 1). Historically, it ranks among the world's most productive estuarine systems in terms of recreational and commercial fisheries. During recent years, these fisheries have been threatened, primarily by eutrophication and toxic substances. Nutrient inputs from agricultural runoff and other nonpoint sources are largely responsible for eutrophication in the estuary and the development of

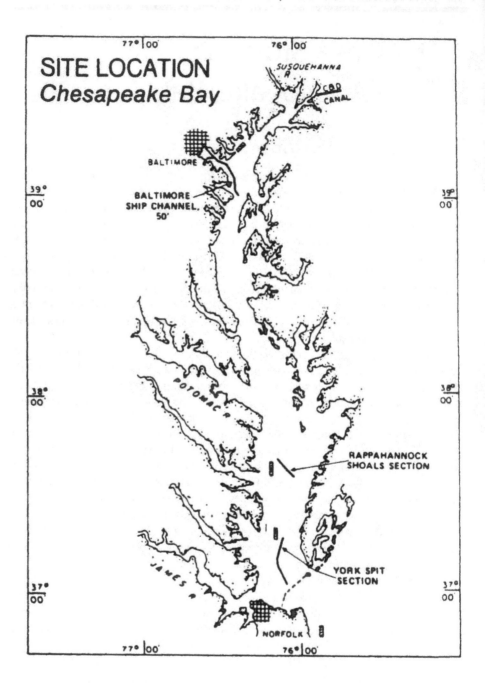

FIGURE 1   Map of Chesapeake Bay. Location of the Baltimore shipping channel and dredged-spoil disposal sites (hachured) are also shown. (From Nichols, M., Diaz, R. J., and Schaffner, L. C., *Environ. Geol. Water Sci.*, 15, 31, 1990. With permission.)

chronic algal blooms and low dissolved oxygen levels that can decimate commercially important, bottom-dwelling shellfish and finfish populations. Of mounting concern is the input and biotic effects of toxic organic substances from multiple sources, most notably the Susquehanna River. However, chemical contaminants also enter from runoff along the estuarine perimeter, from municipal and industrial wastewaters, and from atmospheric deposition. The Susquehanna River delivers about 3.46×10^{10} m^3 of water and 0.9×10^6 metric tons (mt) of sediment each year to the northern bay.[1] Toxic substances transported by the river tend to be particle-bound and accumulate in fine-grained sediments of the upper estuary. In the middle and lower reaches of the bay, agricultural runoff and shoreline erosion are major sources of contaminants and sediments. The bay receives an estimated annual input of 0.6×10^6 mt of clay and silt.[2] Chemical contaminants are commonly sorbed to these grains.

Because of marked decreases in the water quality and resources of Chesapeake Bay concomitant with escalating pollution problems, state and federal agencies formulated a cooperative agreement to assess environmental conditions in the bay. To this end, the U.S. Environmental Protection Agency (U.S. EPA), the State of Maryland, the Commonwealth of Pennsylvania and Virgina, and the District of Columbia implemented the Chesapeake Bay Agreements of 1983 and 1987.[3] The principal goal of these agreements is to establish a plan for a comprehensive, long-term program to revitalize the bay. Monitoring programs are underway, and environmental databases are being generated to discern long-term trends in estuarine conditions. A program of Best Management Practices applied throughout the estuary is designed to effectively manage the bay and its associated tributary systems by minimizing anthropogenic impacts.

II. PRINCIPAL POLLUTION PROBLEMS

A. EUTROPHICATION

One of the main factors decreasing the capacity of Chesapeake Bay to support living resources is the depletion of dissolved oxygen levels during the summer months.[4] Although oxygen-depleted bottom waters have been documented in Chesapeake Bay for at least the past 60 years, they have become more widespread and prolonged in recent years.[5] Eutrophication (excessive nutrient enrichment) appears to contribute greatly to the generation of hypoxia (<2 mg O_2/l) and anoxia (0 mg O_2/l) in the bay.[6-7] During the 1980s (1985–1989), substantially higher nitrogen to phosphorus ratios occurred in the upper mainstem of Chesapeake Bay, the Patuxent River estuary, and the Potomac River estuary than during earlier periods (e.g., the late 1960s to the late 1970s).[8] Such changes may have important implications for phytoplankton growth. Table 1 shows the concentrations of nitrogen and phosphorus in Chesapeake Bay compared to other estuarine systems. Pulses of primary production (e.g., phytoplankton blooms) stimulated by nutrient loading leads to high oxygen demand in the estuary.[9] Microbial decompositon of plankton and other organic matter accumulating on the estuarine bottom depletes dissolved oxygen in lower water layers.[10]

TABLE 1
Concentrations of Nitrogen and Phosphorus in Chesapeake Bay Compared to Those in Other Estuarine Systems[a]

| | Nutrient Concentration | |
Estuary	Nitrogen	Phosphorus
River dominated		
Pamlico River, North Carolina	1.5	8.0
Narrangansett Bay, Rhode Island	0.6	1.6
Western Wadden Sea, Netherlands	3.0	2.0
Eastern Wadden Sea, Netherlands	4.0	2.5
Mid-Patuxent River, Maryland	4.2	2.3
Upper Patuxent River, Maryland	10.0	2.0
Long Island Sound, New York	1.5	0.5
Lower San Francisco Bay, California	20.6	3.8
Upper San Francisco Bay, California	11.5	2.0
Barataria Bay, Louisiana	4.6	0.8
Victoria Harbor, British Columbia	11.5	2.0
Mid-Chesapeake Bay, Maryland	4.5	0.6
Upper Chesapeake Bay, Maryland	5.0	6.0
Duwamish River, Washington	60.0	3.0
Hudson River, New York	5.0	0.16
Apalachicola Bay, Florida	10.0	0.1
Embayments		
Roskeeda Bay, Ireland	0.4	2.2
Bedford Basin, Nova Scotia	0.6	0.5
Central Kaneohe Bay, Hawaii	0.8	0.3
S.E. Kaneohe Bay, Hawaii	1.0	0.5
St. Margarets Bay, Nova Scotia	1.1	0.5
Vostok Bay, Russia	1.0	0.05
Lagoons		
Beaufort Sound, North Carolina	0.5	0.5
Chincoteague Bay, Maryland	3.2	2.5
Peconic Bay, New York	1.9	1.3
High Venice Lagoon, Italy	2.4	0.05
Fjords		
Baltic Sea	1.3	0.1
Loch Etive, Scotland	1.1	0.06

[a] Nutrient concentrations in μg-atom/l.

From Boynton, W. R., Kemp, W. M., and Keefe, C. W., in *Estuarine Comparison*, Kennedy, V. S., Ed., Academic Press, New York, 1982, 69. With permission.

Water column respiration causes oxygen depletion that may rival or even exceed benthic respiration. Taft et al.[6] note that organic matter accumulating from the previous year (summer and fall) and settling through deep waters during winter provides most of the oxygen demand. However, new primary production in spring and summer, despite being a less significant driving force in oxygen dissipation, sinks and accelerates deep-water oxygen depletion in the bay.

Hypoxia and anoxia commonly develop in deeper channels at depths below the pycnocline during periods of increased stratification of the water column. For example, the increase in density stratification of the estuary between February and May, attributable to the spring freshet, strengthens water column stability, thereby minimizing advective transport of oxygen from the surface to the deep water layers. Hence, during summer, anoxia purportedly spreads along the bottom of the bay's channel in waters >10 m.[6] Wind-driven seiching can transport oxygen-depleted bottom waters into nearshore shallow areas where hypoxia and anoxia may also arise. Clearly, the combined effect of density stratification, plankton decomposition, and wind forcing can generate severe oxygen depletion during the summer months and can lead to the movement of hypoxic or anoxic waters onto the shallower flanks of Chesapeake Bay and its tributaries.[4]

Eutrophic conditions likewise occur in subestuaries.[11] Influent systems along the western shore of Chesapeake Bay characterized by deep basins at their mouth (e.g., Patuxent, Rappahannock, and York rivers) frequently exhibit hypoxic conditions. Other systems which receive even greater wastewater loadings but have greater gravitational circulation (e.g., James River) rarely display hypoxia or anoxia. The severity of dissolved oxygen depletion in these subestuaries is contingent upon a myriad of biological, chemical, and physical factors.

The most severe effects of low dissolved oxygen levels near the bottom are manifested in benthic communities.[12-14] Changes in species composition, abundance, diversity, and community biomass typify oxygen-deficient environments. Stressful conditions associated with extended periods of hypoxia and anoxia often adversely affect recruitment, growth, and survival of the benthos.

B. HYDROCARBONS

Chesapeake Bay is subject to chronic oil pollution from marine vessels, shoreline installations, nonpoint source runoff, as well as municipal and industrial waste discharges. Numerous chemical compounds found in oil can adversely affect organisms in the estuary. The susceptibility of estuarine and marine organisms to the damaging impacts of oil depends on the life-history stages exposed to the toxic compounds, the length of time of the exposure, and the concentration of the compounds. These organisms may take up petroleum hydrocarbons and associated chemical compounds by several routes, including active uptake of dissolved or dispersed substances, ingestion of petroleum-sorbed particles, and drinking or gulping (as in the case of fish) of water-soluble components.[1,15] When exposed to sufficiently high levels of toxic chemicals, sublethal or even lethal effects may be manifested. Sublethal effects, such as aberrant physiological and behavioral function, while not causing immediate mortality, usually predispose an organism to greater long-term risk of death. The growth, reproduction, and distribution of the affected organisms can be compromised, leading in some cases to gradual shifts in species composition, abundance, and diversity of communities. In extreme cases (e.g., local oil spills), the effects are often sudden and lethal.

In Chesapeake Bay, site-specific impacts of polluting oil are anticipated in areas most greatly influenced by anthropogenic activities. For example, Baltimore Harbor and the Hampton Roads area receive inputs of oil from various sources, most notably coastal installations and shipping operations. The northern reach of Chesapeake Bay, which is in close proximity to heavily populated watersheds, appears to be an important zone of entry of oil derived from various nonpoint sources.

Polycyclic aromatic hydrocarbons (PAHs), which tend to concentrate in bottom sediments of Chesapeake Bay, generally decrease in concentration downestuary along a north-to-south gradient. Highest concentrations, therefore, occur where human population densities peak. Unsubstituted PAH compounds (UPAH), in particular, attain highest levels in areas near heavy urban activity.[16] Figure 2 compares total PAH concentrations in bottom sediments of Chesapeake Bay to those of other estuarine and coastal marine systems.

At least 16 UPAH compounds have been recorded in mollusks and polychaete worms collected at sampling sites between Baltimore Harbor and the mouth of the Potomac River (Table 2). Total UPAH concentrations in biotic samples collected throughout the estuary range from 0.551 to 178 μg/g fraction organic carbon. Highest loadings are found at the entrance to Baltimore Harbor.

In bottom sediments, most PAH compounds originate from anthropogenic sources, principally the combustion or high temperature pyrolysis of carbonaceous fuels. The amount of PAHs from spills and land runoff is less important. Natural sources of PAHs account for less than 20% of the total PAH concentrations in bottom sediments.[17] The quantities of PAH compounds in bottom sediments generally ranges from about 1 to 400 μg/kg dry weight. The Elizabeth River, a subestuary of Chesapeake Bay, is heavily contaminated with PAHs; the concentrations in bottom sediments are as high as 390 μg/g dry weight.[18] A gradient in PAH levels is evident in bottom sediments, with concentrations gradually increasing upriver.[19]

C. ORGANOCHLORINE CONTAMINANTS

Like many other major estuaries worldwide, Chesapeake Bay receives an array of chlorinated hydrocarbon compounds that are potentially toxic to estuarine organisms and a threat to sensitive habitats. Because these compounds are extremely stable in aquatic environments, resist breakdown, and tend to bioaccumulate, they create acute as well as insidious pollution problems. The most notable contaminants in this group are the organochlorine biocides (insecticides, herbicides, fungicides, etc.) such as DDT, DDE, chlordane, dieldrin, lindane, mirex, and toxaphene, as well as a wide variety of hazardous industrial chemicals such as PCBs and numerous volatile organic compounds. Organochlorine contamination in Chesapeake Bay has existed for at least 50 years. Before 1980, most work on organochlorine contamination in the estuary involved studies of pesticides, particularly DDT and DDE. Since 1980, the focus of investigations has shifted to other contaminants as well, including industrial chemical compounds (e.g., PCBs).

Because of the biomagnification of organochlorines in the food chain of Chesapeake Bay and the potential harm to humans who consume contaminated finfish and shellfish from the estuary, great effort has been expended on monitoring

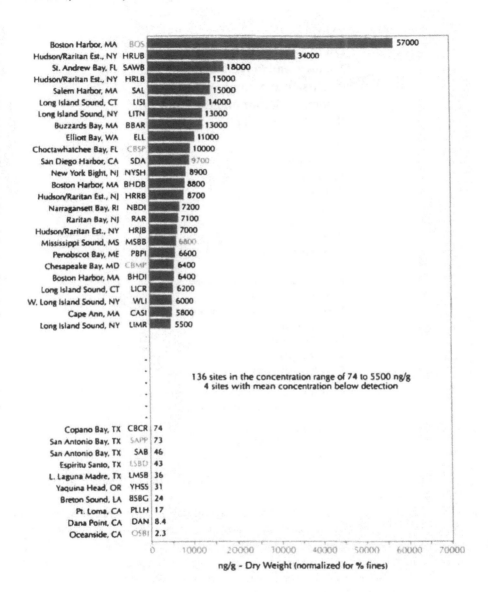

FIGURE 2   Total PAH concentrations in sediments of Chesapeake Bay compared to those of other estuarine and coastal marine systems of the United States. (From NOAA, *A Summary of Data on Individual Organic Contaminants in Sediments Collected During 1984, 1985, 1986, and 1987*, Tech. Mem. NOS OMA 47, National Oceanic and Atmospheric Administration, Rockville, MD, 1989.)

chlorinated hydrocarbon concentrations in biota of the bay. Data collected on the American oyster (*Crassostrea virginica*), blue crab (*Callinectes sapidus*), and spot (*Leiostomus xanthurus*) over a 20-year period indicate that at any point in time the concentrations of DDT and PCBs were higher in the blue crab and spot by a factor

---

TABLE 2
List of UPAH Compounds in
Mollusks and Polychaetes Sampled
in Chesapeake Bay between
Baltimore Harbor and the Mouth of
the Potomac River[a]

Naphthalene
Acenaphthylene
Acenaphthene
Fluorene
Phenanthrene/anthracene[b]
Fluoranthene
Pyrene
Benzo(a)anthracene
Chrysene
Benzo(b)- and benzo(a) anthracene
Chrysene/triphenylene
Benzo(b)- and benzo(k)fluoranthenes
Benzo(a)pyrene
Indo(123-cd)pyrene
Benzo(ghi)perylene
Dibenz(ah)anthracene

[a] Listed in order of increasing molecular weight.
[b] Unresolved chromatographic retention.

Data from Foster, G. D. and Wright, D. A., *Mar. Pollut. Bull.*, 19, 459, 1988.

---

of 4–5 than in the American oyster (Figure 3). This difference is largely attributable to biomagnification.

Much research on organochlorine contamination in Chesapeake Bay biota has targeted upper-trophic-level organisms, especially bird and finfish populations which may have elevated contaminant levels due to biomagnification. For example, reductions in the thickness of bald eagle and osprey eggs observed along the Chesapeake Bay shoreline during the 1960s and 1970s were coupled to high levels of DDT and DDE concentrations. Osprey eggs measured in 1977 and 1978 were 10–24% thinner than those measured during the pre-DDT era. DDE levels in osprey eggs collected along the eastern and western shores of Chesapeake Bay in 1977 and 1978 ranged from 1.9–2.8 μg/g wet weight and 1.4–3.5 μg/g wet weight, respectively.[20] During 1978 and 1979, the mean DDE concentration of bald eagle eggs in the Chesapeake Bay region amounted to about 10 μg/g wet weight compared to a mean DDE concentration of about 0.1 μg/g wet weight in black duck eggs.[21,22]

Studies have also been conducted on DDT contamination in benthic invertebrates of the estuary. For example, surveys of DDT concentrations in mussels (*Mytilus edulis*) and oysters (*Crassostrea virginica*) have revealed declining levels of

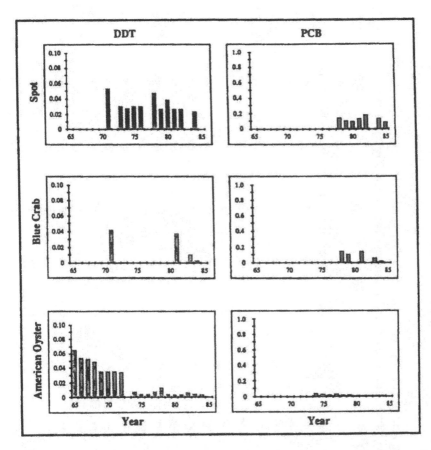

FIGURE 3   Concentrations (μg/g dry weight) of DDT and PCBs in all samples of the American oyster (*Crassostrea virginica*), blue crab (*Callinectes sapidus*), and spot (*Leisotomus xanthurus*) collected from Chesapeake Bay from 1965–1985. (From Mearns, A. J., Matta, M. B., Simecek-Beatty, D., Buchman, M. F., Shigenaka, G., and Wert, W. A., PCB and Chlorinated Pesticide Contamination in U.S. Fish and Shellfish: A Historical Assessment Report, NOAA Tech. Mem. NOS OMA 39, National Oceanic and Atmospheric Administration, Seattle, WA, 1988.)

DDT. Table 3 ranks the trends of DDT contamination in shellfish of Chesapeake Bay relative to those of other estuarine and coastal marine systems.

Concern also exists with regard to finfish populations. During the 1970s, the total organochlorine residues in striped bass from the Potomac River ranged from 0.21–0.40 μg/g wet weight. The levels of DDT, DDE, and chlordane were less than 0.06 μg/g wet weight.[23] PCBs have been the most persistent and prevalent organic contaminant residue in striped bass from Chesapeake Bay.[24]

Surveys of PCB contamination in commercially important fish and shellfish populations inhabiting Maryland and Virginia waters reveal values approximately one order of magnitude less than the U.S. Food and Drug Administration action level of 2 μg/g wet weight.[25] Sediments appear to be an important source of PCB

TABLE 3

National Status and Trends Sites Ranking Among the Highest 20 in 1986, 1987, and 1988 and Overall for Total DDT in Mussels (*Mytilus edulis, Mytilus californianus*) and Oysters (*Crassostrea virginica*) from Various Estuarine and Coastal Marine Systems

Code	Location	State	Species	Ranking 1986	1987	1988	Overall	Significant Difference?	Trend
HRLB	Hudson/Raritan Estuary	NY	me	1t	7	10	5	yes	d
SPFP	San Pedro Harbor	CA	me	1t	4	1	2	no	no
PVRP	Palos Verdes	CA	mc	1t	1	2	1	no	no
ABWJ	Anaheim Bay	CA	mc	4	3	4	4	yes	no
SFDB	San Francisco Bay	CA	me	5	11t	v	16	no	no
CBSP	Choctawhatchee Bay	FL	cv	6	2	3	3	no	no
BBAR	Buzzards Bay	MA	me	7t	v	v	v	yes	d
NYSH	New York Bight	NJ	me	7t	19t	16t	14t	no	no
IBNJ	Imperial Beach	CA	mc	9	10	7	10	no	no
NYSR	New York Bight	NJ	me	10	v	v	v	no	no
LITN	Long Island Sound	NY	me	11	11t	v	17	yes	no
HRUB	Hudson/Raritan Estuary	NY	me	12t	14t	8	11	no	no
CBCC	Chesapeake Bay	VA	cv	12t	v	v	v	yes	d
SAWB	St. Andrew Bay	FL	cv	12t	v	v	v	no	no
SLSL	San Luis Ob. Bay	CA	mc	12t	v	v	v	no	no
HRJB	Hudson/Raritan Estuary	NY	me	16t	v	13	20	yes	no
DBBD	Delaware Bay	DE	cv	16t	8	v	12	no	no
PDPD	Pt. Dume	CA	mc	18	6	12	9	no	no
LIMR	Long Island Sound	NY	me	19t	v	v	v	no	no
MBCP	Mobile Bay	AL	cv	19t	v	18t	v	no	no

Code	Location	State	Species						
GBYC	Galveston Bay	TX	cv	19t	v	v	v	no	no
MBSC	Monterey Bay	CA	mc	19t	v	v	v	no	no
SFEM	San Francisco Bay	CA	me	—	5	9	7	—	—
MDSJ	Marina Del Rey	CA	me	v	9	5	8	yes	i
DBAP	Delaware Bay	DE	cv	v	13	v	v	no	no
BHDB	Boston Harbor	MA	me	v	14t	v	v	no	no
NBWJ	Newport Beach	CA	mc	v	14t	v	v	no	no
SFSM	San Francisco Bay	CA	me	v	14t	v	v	no	no
SDHI	San Diego Bay	CA	me	v	15t	v	v	no	no
PGLP	Pacific Grove	CA	mc	v	15t	v	v	no	no
DBKI	Delaware Bay	DE	cv	v	19t	v	v	no	no
MBHI	Mobile Bay	AL	cv	—	—	6	6	—	—
OSBJ	Oceanside	CA	me	v	v	11	v	yes	i
MRPL	Mississippi River	LA	cv	—	—	14	13	—	—
GBSC	Galveston Bay	TX	cv	—	—	15	14t	—	—
APDB	Apalachicola Bay	FL	cv	v	v	16t	v	no	no
MSPB	Mississippi Sound	MS	cv	v	v	18t	v	yes	i
BRFS	Brazos River	TX	cv	—	—	18t	18t	—	—
SANM	San Miguel Island	CA	mc	—	—	18t	18t	—	—

Note: DDT concentrations in Chesapeake Bay samples exhibit a decreasing trend for the survey period. me = *Mytilus edulis*; mc = *Mytilus californianus*; cv = *Crassostrea virginica*; t = two or more concentrations were equal; d = decreasing trends; i = increasing trends.

From NOAA, *A Summary of Data on Tissue Contamination from the First Three Years (1986–1988) of the Mussel Watch Project*, NOAA Tech. Mem. NOS OMA 49, National Oceanic and Atmospheric Administration, Rockville, MD, 1989.

contamination in estuarine biota. Localized "hot spots" of sediment contamination have been delineated in several areas of the Chesapeake Bay system. For instance, in Baltimore Harbor, Munson[26] and Tsai et al.[27] documented PCB concentrations >2 mg/kg in surface sediments. Other "hot spot" locations have been discerned in the Patapsco and Elizabeth Rivers.[28] Some organisms sampled near highly populated areas around Hampton Roads have exhibited a general increase in PCB levels.[29] Table 4 lists the ranges of PCBs in sediments of Chesapeake Bay compared to those of other estuarine and coastal marine systems.

Aside from the principal organochlorine contaminants found in Chesapeake Bay biota — DDT, DDE, and PCBs — chlordecone (Kepone), hexachlorobenzene, and a number of other compounds have also been detected.[1,24,30] Among the most notable chemical contamination in the system was the release of Kepone (an organochlorine pesticide) from a manufacturing plant on the James River estuary. Entering the estuary mainly via wastewater from a municipal sewage system at Hopewell, Virginia, 120 km upriver of Chesapeake Bay, and secondarily via leaching and erosion of contaminated soils and discharges of solid waste, Kepone began to accumulate in bottom sediments of the middle estuary in 1966 and reached peak concentrations in 1974. It contaminated an estimated 37×10^6 mt of sediment over a distance of 120 km to the mouth of the James River and spread through the food chain.[31] Highest concentrations (>200 ng/g) occurred in near-source sediments and rapidly increased in biota. During the mid-1970s, Kepone concentrations exceeded 7.0 μg/g wet weight in some fish populations. As a result, all forms of fishing in the James River were banned in December 1975. Contaminant levels gradually declined, and by the mid-1980s, concentrations of Kepone in fish and crabs ranged from 0.2–0.8 μg/g wet weight, and they were less than 0.1 μg/g wet weight in oysters. Nonetheless, Kepone continued to be monitored in bottom sediments and biota, particularly in the lower James River, where it remained an important contaminant. Natural sedimentation has provided the most effective means of Kepone decontamination in the estuary.[31]

Chesapeake Bay has been the target of numerous monitoring and survey programs for chlorinated hydrocarbon pesticides, being the most heavily sampled U.S. estuary for chemical contamination of fish and shellfish. Between 1965 and 1986 alone, at least 38 surveys measured or monitored chlorinated hydrocarbon pesticides and PCBs in biota of Chesapeake Bay.[32] Chief among the biota investigated were the American oyster Crassostrea virginica and the blue mussel Mytilus edulis, both extremely important biomonitors of chemical contamination in U.S. waters. However, fish tissue has also proven to be valuable in monitoring chlorinated hydrocarbon contamination in the estuary and elsewhere (Figure 4). Such monitoring and survey programs have continued to this day.[15]

Based on surveys of chlorinated hydrocarbons in Chesapeake Bay biota conducted during the past three decades, some general conclusions can be drawn.[32] First, the concentrations of some chlorinated hydrocarbon compounds (e.g., DDT) have appreciably decreased in biota of the estuary during this period. Second, while biota of the eastern shore are now relatively free of DDT and PCBs, this is not the case for other organochlorine contaminants in these organisms, such as chlordane,

---

TABLE 4
PCB Concentrations (ng/g dry weight)
in Sediments of Chesapeake Bay
Compared to Those of Other Major
Estuarine and Coastal Marine Systems

Area	PCB Conc.
Mediterranean Sea	0.8–9.0
Gulf of Mexico	0.2–35.0
Chesapeake Bay	4–400
Tiber estuary	28–770
Rhine-Meuse estuary	50–1000
New York Bight	0.5–2200.0
Palos Verdes Peninsula	30–7900
Hudson River	tr–6700
Escambia Bay	190–61,000

Note: tr - trace. See original source for particular studies.

From Phillips, D. J. H., in *PCBs and the Environment*, Vol. 2, Waid, J. S., Ed., CRC Press, Boca Raton, FL, 1986, 132. With permission.

---

heptaclor, and toxaphene. Third, chlorinated hydrocarbon pesticide and PCB contamination in the American oyster is mainly confined to the far northern reaches (Susquehanna River and the Chesapeake-Delaware Canal), the northwest region (Patapsco River at Baltimore), and the southern shore (Hampton Roads-James River region). However, because of the great persistence of these compounds in sediments and biota of the estuary, it is necessary to continue to monitor and assess their impacts on the entire system.

D. HEAVY METALS

Chesapeake Bay receives heavy metal inputs primarily from the Susquehanna River, and secondarily from shore erosion, atmospheric deposition (especially lead and zinc), as well as municipal and industrial wastewaters. Bottom sediments represent a major repository for heavy metals in the estuary (Table 5). Major industrial ports (Baltimore Harbor and Hampton Roads) are "hot spots" of heavy metal accumulation, but they contribute little contamination to the main bay, as does the transport of metals out of the northern bay. Baltimore Harbor acts as a sediment trap and thereby retains much of the heavy metal load. Lead and zinc have the highest enrichment values in all areas, which infers an atmospheric source.[33] Helz et al.[34] asserted that zinc enrichment in the middle and upper bay has occurred for more than a century. Cadmium and lead enrichment likewise takes place in the upper bay. In general, trace metal concentrations decrease downestuary, as the influence of the

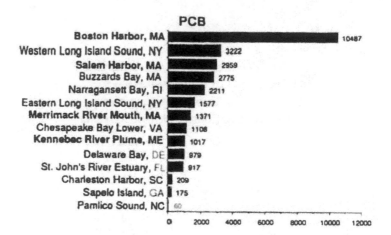

FIGURE 4 Concentrations (ng/g wet weight) of PCBs in Atlantic Coast fish liver tissue sampled in Chesapeake Bay and other estuarine systems by the NOAA Status and Trends Program. (From Larsen, P. F., *Rev. Aquat. Sci.*, 6, 67, 1992.)

Susquehanna River declines seaward. Trace metals in the lower bay are largely derived from offshore, which is consistent with the continental-shelf source of sediment in this area of the estuary.[32,35]

Table 6 compares enrichment factors of heavy metals for Chesapeake Bay to those of Delaware Bay, San Francisco Bay, and elsewhere. The enrichment factors are determined by the following equation:

$$EF = \frac{(X/Fe) \text{ sediment}}{(X/Fe) \text{ Earth's crust}}$$

where X/Fe is the ratio of the concentration of element X to iron, the element of normalization. An EF value of unity signifies neither enrichment nor depletion in relationship to the average Earth's crust. As is evident in Table 6, the highest enrichment factors in the Chesapeake Bay ecosystem occur in the Susquehanna River, upper bay, Baltimore Harbor, and Hampton Roads. By comparison, enrichment factors are higher for Delaware Bay, a heavily industrialized estuary, and generally lower (1 or 2) for San Francisco Bay. In contrast to Delaware Bay which is flanked by numerous industrial sources of heavy metals along its shoreline, Chesapeake Bay has industrial activity concentrated in Baltimore Harbor and Hampton Roads, as noted previously.

The effects of heavy metals on biota of Chesapeake Bay have been the focus of a number of investigations during the past 10–15 years. For example, Mehrle et al.[36] examined the toxicity of several heavy metals (arsenic, cadmium, copper, lead, and selenium) to striped bass larvae. Hall et al.[37] detailed the chemical toxicity of striped bass larvae in the Potomac River estuary. Sinex and Wright[33] described the use of the benthic clam *Macoma balthica* as a biomonitor of trace metals in Chesapeake Bay. Phelps et al.[38] chronicled trace metal accumulation in the oyster *Crassostrea virginica*. Bender and Huggett[39] discussed the general impact of heavy metals, as

TABLE 5
Heavy Metal Composition in Sediments of Chesapeake Bay and Tributary Systems

Area	Elements	Author(s)/Date of Publication[a]
Main stem	Al, C, Ca, Co, Cr, Cu, Fe, H, K, Mg, Mn, Na, Ni, Pb, S, Si, Ti, V	Sommer and Pyzik, 1974
	Cr, Cu, Ni, Pb	Schubel and Hirschberg, 1977
	Cd, Cu, Fe, Mn, Ni, Zn	Cronin et al., 1974
	Ag, Al, Cd, Co, Cr, Cu, Fe, Mn, Ni, Pb, V, Zn	Goldberg et al., 1978
	Al, Co, Cr, Cu, Fe, Ga, Mn, Ni, Org C, Org N, Pb, Si, Ti, V, Zn	Sinex, 1981
	As, Cd, Cu, Fe, Hg, Mn, Ni, Pb, Sn, Zn	Harris et al., 1980
Back River	Cd, Cu, Pb, Zn	Helz et al., 1975
Patapsco	Cd, Cr, Cu, Hg, Mn, Ni, Pb, Zn	Villa and Johnson, 1974
	As, Cd, Cr, Cu, Hg, Mn, Ni, Pb, Zn	Tsai et al., 1979
	Al, Co, Cr, Fe, Mn, Ni, Si, Ti, V, Zn	Sinex and Helz, 1982
Rhode River	Cd, Cr, Fe, Mn, Zn	Frazier, 1976
Patuxent	Cd, Co, Cr, Cu, Fe, Mn, Ni, Pb, Zn	Ferri, 1977
Potomac	Ag, Ba, Cd, Co, Cr, Cu, Fe, Li, Mn, Ni, Pb, Sr, V, Zn	Pheiffer, 1972
	Ba, Co, Cr, Cu, Fe, Mn, Pb, Sr, Ti, V, Zi, Zn	Mielke, 1974
Rappahannock, York	Cu, Zn	Huggett et al., 1975
Elizabeth	Al, Cd, Cr, Cu, Fe, Hg, Pb, Zn	Johnson and Villa, 1976
	Al, Co, Cr, Fe, Mn, Ni, Si, Ti, V, Zn	Helz et al., 1983
	Cd, Co, Cr, Cu, Fe, Mn, Ni, Pb, Zn	Rule, 1986

[a] See original source for particular studies.

From Helz, G. R. and Huggett, R. J., in *Contaminant Problems and Management of Living Chesapeake Bay Resources*, Majumdar, S. K., Hall, L. W., Jr., and Austin, H. M., Eds., Pennsylvania Academy of Science, Easton, 1987, 170. With permission.

well as other contaminants, on shellfish of the bay. Despite these studies, many questions remain unanswered with regard to the effects of heavy metals on organisms in the estuary and its tributaries.

III. OTHER ANTHROPOGENIC IMPACTS

A. DREDGING AND DREDGED-SPOIL DISPOSAL

Chesapeake Bay is not only North America's largest estuary, it is also one of the most valuable in terms of marine resources and shipping. The estuary produces more than $80 million in commercial fisheries annually, and the worth of cargo on ships docking at Norfolk and Baltimore alone exceeds $2.7 billion.[40] Deep draft ships, such as bulk cargo carriers, require deeper channels for routine shipping. Hence,

TABLE 6
Enrichment Factors (average or typical range) for Heavy Metals in Chesapeake Bay Compared to Those for Other Major Estuarine and Coastal Marine Systems

Area	Cr	Mn	Ni	Cu	Zn	Pb	Ref.[a]
Chesapeake Bay (whole)	1.0	2.0	1.0	1.0	5	5	
Upper (0–75 km)	1.0	3.0–6.0	—	2.0	6–8	4–7	Sinex, 1981
Middle (75–200 km)	1.0	1.0	—	1.0	4–6	3–4	Sinex and Helz, 1981
Lower (200–300 km)	1.0	<1.0	—	1.0	2–4	2–4	
Baltimore Harbor	4.0	1.0	1.0	—	8	—	Sinex and Helz, 1982
Hampton Roads	1.0	1.0	1.0	2.0–4.0	5–10	7–21	Rule, 1986
Susquehanna River	0.6	6.0	3.0	2.0	8	7*	Sinex and Helz, 1981
Offshore[b]	0.7	0.8	0.6	0.2	2	3	Rule, 1986
Delaware Bay	3.0	—	12.0	2.0	10	16	Bopp and Biggs, 1973
San Francisco Bay	—	<1.0	2.0	2.0	2	—	Eaton, 1979

[a] See original source for particular studies.
[b] Proposed disposal site on the inner Virginian continental shelf approximately 27 km off the Chesapeake Bay mouth.

From Sinex, S. A. and Wright, D. A., *Mar. Pollut. Bull.*, 19, 425, 1988. With permission.

approach channels to Hampton Roads and Norfolk, as well as the main shipping channel between Baltimore and the sea, must be periodically dredged. With the trend of larger ships and anticipated port expansion, dredging is expected to remove more than 2.0×10^8 m^3 of sediment from Chesapeake Bay during the next two decades.[40]

An array of biotic impacts is coupled to dredging and dredged-spoil disposal, the most acute being those on benthic communities. Operation of the dredge increases mortality of the benthos by mechanical injury, smothering of sediment, and destruction of the bottom habitat. Resuspension of bottom sediment during dredging and high turbidity levels during dredged-spoil disposal often reduce primary production and may alter water quality by the release of organic matter, nutrients, and chemical contaminants from the bottom. The disposal of high sediment loads usually blankets extensive areas and buries the benthos below a thick layer of material often derived from remote environments and, hence, usually with different physical-chemical properties.

More studies have been conducted in Chesapeake Bay on the effects of dredged-spoil disposal than on dredging impacts. For example, the consequences of turbidity and sedimentation have been investigated primarily at disposal sites, with much less attention given to resuspension and accumulation of sediments at dredging sites. Short-term turbidity effects and biotic impacts of these operations have been examined in the upper,[41,42] middle,[40] and lower bay.[43] Two major conclusions can be drawn from these studies. First, turbidity is confined to a 4–5 km^2 area at sediment disposal sites. Second, abundance of benthic populations decreases markedly at these sites, but the populations typically recolonize and reestablish within less than 18 months.[40]

Nichols et al.[40] assessed the effects of redeposition of dredged-induced suspended sediment on benthic populations at Rappahannock Shoals in central Chesapeake Bay (Figure 1). While dredging and dredged-spoil disposal clearly cause environmental perturbations in the area, effects on the benthic community are minimal and transient. Infaunal assemblages are typically resilient and the life history strategies and population dynamics of the benthos enable them to successfully cope with these instabilities.

Evaluation of the benthic community at Rappahannock Shoals Channel revealed similar macrofaunal species composition before and after dredging. Four polychaete species (i.e., *Paraprionospio pinnata*, *Pseudeurythoe paucibranchiata*, *Streblospio benedicti*, *Mediomastus ambiseta*) and one gastropod species (i.e., *Cyclostremiscus pentagona*) dominated the benthic community after dredging. Abundance of taxonomic groups were as follows: polychaetes (67% of all individuals), gastropods (17%), nemerteans (5%), bivalves (4%), amphipods (2%), and oligochaetes (1%). In addition, no consistent pattern of increasing or decreasing abundance and biomass could be discerned with increasing distance from the channel. Community structure was similar prior to dredging, with the three burrowing polychaetes *Nephytes* cf. *cryptomma*, *P. paucibranchiata*, and *Sigambra tentaculata* and the tubicolous spionid polychaete *P. pinnata* dominating the community.

Nichols et al.[40] proposed several reasons why macrobenthic assemblages in the vicinity of the Rappahannock Channel were not significantly affected by deposition of dredged sediments: (1) the species characterizing this area of the bay were generally short-lived forms that have flexible life history strategies and high motilities which enable them to cope with the rate of new sediment supply; (2) the dredged sediments were not contaminated; (3) the properties of the dredged material were the same as those of natural background sediments; and (4) the rate of accumulation of dredged sediments on channel flanks was low and occurred over a period of several months. In conclusion, despite elevated suspended sediment concentrations in the water column during dredging and dredged-spoil disposal, which can exceed some water quality standards, benthic communities are highly resilient to these impacts.

B. ELECTRIC GENERATING STATIONS

A number of large electric generating stations sited along shorelines of the Chesapeake Bay ecosystem adversely affect estuarine organisms by thermal loading or calefaction of receiving waters, the release of biocides and other chemical substances, the impingement of individuals on intake screens, and entrainment of various life forms in cooling water systems.[44] Thermal discharges from these power plants commonly decrease primary productivity but accelerate the metabolic rate of faunal inhabitants. The heated effluent elicits behavioral and physiological responses in animals, such as avoidance or attraction reactions as well as altered enzyme activity and respiration, that can lead to reduced growth, increased mortality, and ultimately changes in community structure. Sublethal impacts including decreased reproduction, diminished hatching success of eggs, and inhibition of larval development, exacerbate these effects. Attraction of finfishes to thermal plumes in near-field regions of the power plants periodically results in heat- and cold-shock mortality.

Biocides used to control biofouling of heat exchanger surfaces in the power plants are toxic to many estuarine and marine organisms. Phytoplankton and zooplankton are particularly susceptible to the toxic effects of chlorine, as manifested by acute decreases in abundance of these diminutive forms in outfall waters. As a consequence of power plant chlorination, primary and secondary production is often much less in near-field than far-field regions.[45]

The corrosion of copper-nickel alloy condensers in power plants has been implicated in the release of heavy metals that may impact estuarine organisms. For example, condenser corrosion at the Chalk Point power station and the subsequent discharge of copper (~600 kg/yr) into the Patuxent River caused greening and copper accumulation in oysters (*Crassostrea virginica*). Condenser replacement in 1982 with copper-nickel alloy (70/30) condensers did not improve conditions. Hence, copper discharges from the plant after condenser replacement were approximately twice those estimated before condenser replacement.[46] However, the biotic impacts of heavy metal releases from the plant to receiving waters were not determined.

The most significant impacts of the power plants on estuarine populations are those associated with impingement and entrainment. Thousands of fish and macroinvertebrates impinged on intake screens are killed each year, as are millions of eggs, larvae, and juveniles of some species passively drawn into the plants and entrained in the cooling water systems. Mortality results from mechanical damage, thermal stresses, chemical toxicity, and pressure changes incurred during in-plant passage.

Examples of thermal and cooling water effects are those observed at the Chalk Point, Calvert Cliffs, and Surrey power plants. Morgan and Stross[45] reported a depression in rates of phytoplankton photosynthesis when thermal discharges from the Chalk Point power plant raised ambient water temperatures in the Patuxent River estuary above 20°C. Abbe[47] reported more rapid growth of oysters (*Crassostrea virginica*) exposed to thermal discharges from the Calvert Cliffs power plant. At the Surry power plant on the James River estuary, White and Brehmer[48] examined impingement mortality and conveyed how the retrofitting of Ristroph vertical traveling screens onto the intake structure reduced impingement mortality of organisms an average of 93.3% over an 18-month study period. The application of revised operating strategies and new engineering features has mitigated environmental impacts of power plants in the Chesapeake Bay ecosystem.

IV. BIOTIC COMMUNITIES

A. TROPHIC STRUCTURE

Baird and Ulanowicz[49] have described the trophic relationships of organisms in the mesohaline region of Chesapeake Bay, and much of the following discussion derives from their work. The biomass of many taxa peak during the summer months. Two exceptions are phytoplankton and mesozooplankton biomasses which attain highest levels in the spring. Phytoplankton productivity, which is highest during the summer when zooplankton biomass is low, ranges from ~75–850 g C/m²/yr. Production rates in the mid-bay region (~165–850 g C/m²/yr) far exceed those in the upper bay

($\sim$75–85 g C/m^2/yr).[50] Several species seasonally dominate the phytoplankton community along the length of the estuary. *Skeletonema costatum* and *Asterionella glacialis* are winter dominants; *S. costatum, A. glacialis, Cerataulina pelagica,* and *Rhizosolenia fragilissima,* spring dominants; *Coscinodiscus marginatus, R. calcar avis,* and *Ceratium furca,* summer dominants; and *S. costatum* and *Chaetoceros sociale,* fall dominants. More than 200 phytoplankton species have been identified in lower Chesapeake Bay, with more than half of them being diatoms.[51] An annual pattern of phytoplankton abundance and species composition is discernible in the bay. A successional series of phytoplankton takes place throughout the year, with populations derived from coastal oceanic waters or present as indigenous forms in the estuary and its tributary systems.

Zooplankton graze heavily on phytoplankton in spring, with $\sim$35% of the phytoplankton net production being consumed at this time. However, the cropping of zooplankton by ctenophores, sea nettles, and fish may exceed zooplankton production in summer, thereby causing a depression of zooplankton standing stock during the warmest months of the year. The calanoid copepods *Acartia tonsa, A. hudsonica,* and *Eurytemora affinis* tend to dominate the zooplankton community. *Acartia tonsa* dominates the community in summer and fall, while *Eurytemora affinis* dominates the community in fall and late winter. Cladocerans, barnacle nauplii, polychaete larvae, and other meroplankton become quantitatively important at specific times of the year and at certain salinity regimes. ·

The benthic invertebrate community consists of three main components: suspension feeders, deposit feeders, and predator/scavengers. Important suspension feeders include *Crassostrea virginica, Mya arenaria, Mulinia lateralis,* and *Rangia cuneata.* Polychaetes (e.g., *Heteromastus filiformes, Nereis succinea, Scololepides viridis*), tellinid bivalves (*Macoma* spp.), amphipod crustaceans (*Corophium lacustrae, Leptocheirus plumulosus*), and meiofauna are the major deposit feeders. The foremost benthic scavenger/predator in the bay is the blue crab *Callinectes sapidus.* Several species of diving ducks (i.e., canvasback, scaup, goldeneye, bufflehead, and scoters) feed heavily on some of these invertebrates.

Similarly, the finfish community may be subdivided into three broad trophic categories: suspension feeders, benthic feeders, and piscivorous fish. The first group, suspension feeders, is principally comprised of fish larvae, alewives and herring, bay anchovy, menhaden, and shad. Benthic macroinvertebrate feeders of significance are the Atlantic croaker, hogchoker, spot, and white perch. Piscivorous fish (bluefish, weakfish, summer flounder, striped bass), in turn, feed primarily on bay anchovy and menhaden. During a 14-month sampling program, Orth and Heck[52] found that spot, northern pipefish, bay anchovy, and silver perch composed 63%, 14%, 9%, and 5% of all fish collected, respectively. According to Horwitz,[53] bay anchovy, spot, and Atlantic croaker either consistently or frequently dominate fish communities in the estuary.

B. BENTHIC COMMUNITIES

Investigations of pollution impacts on the biota of Chesapeake Bay have focused on benthic communities whose populations are particularly vulnerable to environmental disturbances because they tend to be either immobile or have very limited mobility. While anthropogenic effects on benthic communities of the estuary can be profound, it is often difficult to differentiate natural from man-induced changes in the communities. Natural seasonal patterns are a major source of variation for the Chesapeake Bay macrobenthos.[54-56] In addition to large year-to-year fluctuations in abundance, spatial distributions of the benthos vary at both regional and local scales. Quantification of macrobenthic responses to the myriad of pollutants and anthropogenic activities occurring in the estuary using empirical data requires an extended collection phase that is generally labor intensive.

Salinity and sediment type are major controlling factors on the benthic communities of Chesapeake Bay.[57] Over a 14-year period, for example, macrobenthic sampling in the mesohaline region of Chesapeake Bay (1971–1984) and the estuarine portion of the Potomac River (1980–1984) revealed five distinct macrobenthic assemblages corresponding to major sediment types and salinity zones: (1) nearshore sand; (2) transitional muddy sand; and (3) deep-water mud assemblage in the mid-bay and lower Potomac River; (4) nearshore sand; and (5) a deep-water mud assemblage in the mid-Potomac River.[58] The following euryhaline marine species were most abundant in these areas: *Acteocina canaliculata, Gemma gemma, Glycinde solitaria, Eteone heteropoda, Haminoe solitaria, Heteromastus filiformis, Macoma mitchelli, Micrura leidyi, Monoculodes edwardsi, Mulinia lateralis, Mya arenaria, Nereis succinea, Paraprionospio pinnata, Streblospio benedicti*, and *Tagelus plebeius*. Several estuarine endemic species (e.g., *Cyathura polita, Leptocheirus plumulosus, Macoma balthica, Rangia cuneata, Scolecolepides viridis*, and *Tubificoides* sp.) were commonly found in oligohaline habitats, and they attained peak abundance in the middle segment of the Potomac estuary.

In the Calvert Cliffs area, polychaetes and mollusks dominated the benthic community in terms of numerical abundance and biomass, respectively. While resident species (e.g., *Glycinde solitaria, Heteromastus filiformis, Macoma balthica, M. mitchelli, Micrura leidyi, Mya arenaria, Paraprionospio pinnata, Retusa canaliculata*, and *Scolecolepides viridis*) did not exhibit long-term increases or decreases in abundance during the 14-year period, opportunistic species (e.g., *Mulinia lateralis, Nereis succinea*, and *Streblospio benedicti*) fluctuated markedly in abundance from year to year due to high fecundity and large recruitment pulses. Eurytolerant marine or polyhaline species (e.g., *Glycera dibranchiata* and *Tagelus plebeius*) largely comprised the nonresident benthos in the area.

C. BENTHIC COMMUNITY IMPACTS

From 1985–1991, Dauer and Alden[3] conducted long-term monitoring of macrobenthic communities in the lower Chesapeake Bay. The purpose of this benthic monitoring program was twofold: (1) to characterize the health of regional areas of the lower bay as indicated by the structure of the benthic communities; and (2) to relate spatial and temporal trends of the benthic communities to changes in water and sediment quality in the lower bay. Five parameters characterizing macrobenthic community structure were tested for trends using a nonparametric trend analysis procedure including community biomass, species richness, abundance of individuals, proportion of biomass composed of opportunistic species (opportunistic biomass composition), and proportion of biomass composed of equilibrium species (equilibrium biomass composition).

The relationship between benthic community structure and different levels of pollution (e.g., eutrophication and/or sediment contamination) has been a focal point of many basic research and monitoring programs. Highly stressed estuarine and marine macrobenthic communities are typically characterized by: (1) low levels of species diversity (or species richness), abundance (number of individuals), and biomass; (2) dominance by species that are short-lived, (opportunistic, pioneering, r-selected, stress tolerant), shallow-dwelling, and primarily annelids; and (3) the absence or rarity of species that are long-lived (equilibrium, k-selected) and often deep-dwelling within the sediment.[3,14,59-61] Low and intermediate levels of stress on benthic communities associated with specific pollutants or human activities are typically more difficult to interpret and characterize.[62,63]

Dauer and Alden[3] summarized the distribution of 36 significant trends in macrobenthic community parameters detected in the Chesapeake Bay estuarine ecosystem. Based on trend analysis of the data, improving conditions were noted for the benthos of the James and York Rivers and deteriorating conditions for the benthos of the Rappahannock River and the mainstem of Chesapeake Bay. Low dissolved oxygen events in the mainstem of Chesapeake Bay appear to have increased the opportunistic species composition at three stations and decreased the equilibrium species composition at one station, resulting in the deteriorating classification for this area of the bay. A mixed pattern of data was evident for the Rappahannock River, with all trends in abundance and equilibrium biomass composition declining at one station and opportunistic biomass composition increasing at one station. Strong positive trends were apparent for the James and York Rivers, which both showed five trends of increasing community biomass, five of six trends of increasing species richness, and two trends of increasing equilibrium biomass composition. No trends of data calculated for these two influent systems were indicative of deteriorating conditions.

Benthic community trends in the James, York, and Rappahannock subestuaries appear to reflect the patterns of water quality. The trends of the benthos in the James and York Rivers reflect increasing community health, whereas those of the Rappahannock River are more indicative of stress. Eutrophication and reduced bottom-dissolved oxygen concentrations have become more marked in recent years in the Rappahannock River. Similar to the influent systems, there appears to be inferential agreement between the patterns of water quality conditions and the long-term trends in the benthic communities of the mainstem of Chesapeake Bay.

Orth and Heck[52] and Orth and Moore[64] observed a significant baywide decline of eelgrass (*Zostera marina*) beds, as well as other submerged aquatic plants in Chesapeake Bay. The decline of eelgrass was noted as early as 1973. Orth and Heck[52] also discerned decreasing fish abundances over an annual period of sampling, attributing it in part to a reduction in submerged aquatic plant habitat. Some investigators have ascribed the loss of eelgrass habitat in the estuary to anthropogenic nutrient enrichment, although other factors have not been discounted. The same principal factor has been implicated in seagrass losses in other systems.[65] The long-term health and viability of the Chesapeake Bay is closely coupled to that of critical habitats such as seagrasses.

Eelgrass beds have multiple biotic functions in Chesapeake Bay. They are important nursery areas for a large number of invertebrates and fish, and offer many estuarine organisms a refuge from predators.[66] Some commercially and recreationally important finfish and shellfish rely heavily on them. In addition, eelgrass plants provide attachment sites for epiphytes that support food webs in the bay, thereby playing a vital role in the energy flow of this coastal system. The diet of numerous invertebrates includes epiflora attached to eelgrass. Furthermore, eelgrass blades appear to promote habitat complexity which increases the density and/or diversity of motile seagrass epifauna.

Eelgrass also affects chemical and physical processes in the estuary. For example, it removes nutrients from sediments and overlying waters, and due to this activity, influences nutrient cycling processes. Acting as nutrient pumps, eelgrass may assist in the regulation of water quality in shallow areas. The removal of nutrients from bottom sediments by eelgrass and their subsequent release to surrounding waters enhance the productivity of the system.

As protective cover in shallow areas, eelgrass protects the estuarine bottom and shoreline from the devastating impact of major storms. It buffers the estuary from serious wind and wave erosion. By stabilizing bottom sediments and mitigating erosion, eelgrass is both ecologically and economically important in the bay.

In conclusion, complex riverine-estuarine systems such as exist along the Chesapeake Bay are generally difficult to assess in regard to pollution effects. The interaction of a rigorous physical environment with benthic communities of the mainstem bay and its subestuaries imparts a high degree of biological variability. Large natural variation in the Chesapeake Bay ecosystem tends to mask pollution impacts and confounds assessment programs.

REFERENCES

1. Kennish, M. J., *Ecology of Estuaries: Anthropogenic Effects*, CRC Press, Boca Raton, FL, 1992.
2. Schubel, J. R. and Carter, H. H., Suspended sediment budget for Chesapeake Bay, in *Estuarine Processes*, Vol. 2, *Circulation, Sediments, and Transfer of Material in the Estuary*, Wiley, M., Ed., Academic Press, New York, 1976, 48.
3. Dauer, D. M. and Alden, R. W., III, Long-term trends in the macrobenthos and water quality of the lower Chesapeake Bay (1985–1991), *Mar. Pollut. Bull.*, 30, 840, 1995.
4. Breitburg, D. L., Nearshore hypoxia in the Chesapeake Bay: patterns and relationships among physical factors, *Est. Coastal Shelf Sci.*, 30, 593, 1990.
5. Kuo, A. Y. and Neilson, B. J., Hypoxia and salinity in Virginia estuaries, *Estuaries*, 10, 277, 1987.
6. Taft, J. L., Taylor, W. R., Hartwig, E. O., and Loftus, R., Seasonal oxygen depletion in Chesapeake Bay, *Estuaries*, 4, 242, 1980.
7. Officer, C. B., Biggs, R. B., Taft, J. L., Cronin, L. E., Tyler, M. A., and Boynton, W. R., Chesapeake Bay anoxia: origin, development, and significance, *Science*, 223, 22, 1984.

8. Magnien, R. E., Summers, R. M., and Sellner, K. G., External nutrient sources, internal nutrient pools, and phytoplankton production in Chesapeake Bay, *Estuaries*, 15, 497, 1992.

9. Cooper, S. R. and Brush, G. S., A 2500-year history of anoxia and eutrophication in Chesapeake Bay, *Estuary*, 16, 617, 1993.

10. Seliger, H. H., Boggs, J. A., and Biggley, S. H., Catastrophic anoxia in Chesapeake Bay in 1984, *Science*, 228, 70, 1985.

11. Day, J. W., Jr., Hall, C. A. S., Kemp, W. M., and Yanez Arancibia, A., *Estuarine Ecology*, John Wiley & Sons, New York, 1989.

12. Holland, A. F., Shaughnessy, A. T., and Hiegel, H., Long-term variation in mesohaline Chesapeake Bay macrobenthos: spatial and temporal patterns, *Estuaries*, 10, 227, 1987.

13. Dauer, D. M., Rodi, A. J., Jr., and Ranasinghe, J. A., Effects of low dissolved oxygen events on the macrobenthos of the lower Chesapeake Bay, *Estuaries*, 15, 384, 1992.

14. Dauer, D. M., Biological criteria, environmental health and estuarine macrobenthic community structure, *Mar. Pollut. Bull.*, 26, 249, 1993.

15. Kennish, M. J., Ed., *Practical Handbook of Estuarine and Marine Pollution*, CRC Press, Boca Raton, FL, 1997.

16. Foster, G. D. and Wright, D. A., Unsubstituted polynuclear aromatic hydrocarbons in sediments, clams, and clam worms from Chesapeake Bay, *Mar. Pollut. Bull.*, 19, 459, 1988.

17. Huggett, R. J., de Fur, P. O., and Bieri, R. H., Organic compounds in Chesapeake Bay sediments, *Mar. Pollut. Bull.*, 19, 454, 1988.

18. Roberts, M. H., Jr., Hargis, W. J., Jr., Strobel, C. J., and De Lisle, P. F., Acute toxicity of PAH contaminated sediments to the estuarine fish, *Leiostomus xanthurus*, *Bull. Environ. Contam. Toxicol.*, 42, 142, 1989.

19. Weeks, B. A. and Warinner, E., Effects of toxic chemical on macrophage phagocytosis in two estuarine fishes, *Mar. Environ. Res.*, 14, 327, 1984.

20. Ohlendorf, H. M. and Fleming, W. J., Birds and environmental contaminants in San Francisco and Chesapeake Bays, *Mar. Pollut. Bull.*, 19, 487, 1988.

21. Ohlendorf, H. M., The Chesapeake Bay's birds and organochlorine pollutants, *Trans. North Am. Widl. Nat. Res. Conf.*, 46, 259, 1981.

22. Halestine, S. D., Mulhern, B. M., and Stafford, C., Organochlorine and heavy metal residues in black duck eggs from the Atlantic Flyway, 1978, *Pestic. Monit. J.*, 14, 53, 1980.

23. Mehrle, P. M., Haines, T. A., Hamilton, S., Ludke, J. L., Mayer, F. L., and Ribick, M. A., Relationship between body contaminants and bone development of East Coast striped bass, *Trans. Am. Fish. Soc.*, 111, 231, 1982.

24. Klauda, R. J. and Bender, M. E., Contaminant effects on Chesapeake Bay finfishes, in *Contaminant Problems and Management of Living Chesapeake Bay Resources*, Majumdar, S. K., Hall, L. W., Jr., and Austin, H. M., Eds., Pennsylvania Academy of Science, Easton, PA, 1987, 321.

25. Eisenberg, M., Mallman, R., and Tubiash, H. S., Polychlorinated biphenyls in fish and shellfish in the Chesapeake Bay, *Mar. Fish. Rev.*, February 21–25, 1980.

26. Munson, T. O., in Upper Bay Survey, Vol. 2, Final Report to the Maryland Department of Natural Resources, Westinghouse Electric Corporation, Ocean Sciences Division, Annapolis, MD, 1976.

27. Tsai, C. F., Welch, J., Kewi-yang, C., Schaeffer, J., and Cronin, L. E., Bioassay of Baltimore Harbor sediments, *Estuaries*, 2, 141, 1979.

28. Bieri, R., Hein, C., Huggett, R., Shou, P., Slone, H., Smith, C., and Su, C., Toxic Organic Compounds in Surface Sediments from the Elizabeth and Patapsco Rivers and Estuaries, Final Report to U.S. Environmental Protection Agency, EPA-600/53-83-012, EPA, Washington, D.C., 1983.

29. Helz, G. R. and Huggett, R. J., Contaminants in Chesapeake Bay: the regional perspective, in *Contaminant Problems and Management of Living Chesapeake Bay Resources*, Majumdar, S. K., Hall, L. W., Jr., and Austin, H. M., Eds., Pennsylvania Academy of Science, Easton, PA, 1987, 270.

30. Gossett, R. W., Brown, D. A., and Young, D. R., Predicting the bioaccumulation of organic compounds in marine organisms using octanol/water partition coefficients, *Mar. Pollut. Bull.*, 14, 387, 1983.

31. Nichols, M. M., Sedimentologic fate and cycling of Kepone in an estuarine system: example from the James River estuary, *Sci. Total Environ.*, 97/98, 407, 1990.

32. Mearns, A. J., Matta, M. B., Simecek-Beatty, D., Buchman, M. F., Shigenaka, G., and Wert, W. A., PCB and Chlorinated Pesticide Contamination in U.S. Fish and Shellfish: A Historical Assessment Report, NOAA Tech. Mem. NOS OMA 39, National Oceanic and Atmospheric Administration, Seattle, WA, 1988.

33. Sinex, S. A. and Wright, D. A., Distribution of trace metals in the sediments and biota of Chesapeake Bay, *Mar. Pollut. Bull.*, 19, 425, 1988.

34. Helz, G. R., Sinex, S. A., Setlock, G. H., and Cantillo, A. Y., Chesapeake Bay Trace Elements, U.S. Environmental Protection Agency, EPA-600/53-83-012, EPA, Washington, D.C., 1983.

35. Officer, C. B., Lynch, D. R., Setlock, G. H., and Helz, G. R., Recent sedimentation rates in Chesapeake Bay, in *The Estuary as a Filter*, Kennedy, V., Ed., Academic Press, New York, 1984, 131.

36. Mehrle, P. M., Buckler, D., Finger, S., and Ludke, L., Impact of contaminants on striped bass, Interim Tech. Rept., U.S. Fish and Wildlife Service, Columbia National Fisheries Research Laboratory, Columbia, MO, 1984.

37. Hall, L. W., Jr., Hall, W. S., Bushong, S. J., and Herman, R. L., *In situ* striped bass (*Morone saxatilis*) contaminant and water quality studies in the Potomac River, *Aquat. Toxicol.*, 10, 73, 1987.

38. Phelps, H. L., Wright, D. A., and Mihursky, J. A., Factors affecting trace metal accumulation by estuarine oysters *Crassostrea virginica*, *Mar. Ecol. Prog. Ser.*, 22, 187, 1985.

39. Bender, M. E. and Huggett, R. J., Contaminant effects on Chesapeake Bay shellfish, in *Contaminant Problems and Management of Living Chesapeake Bay Resources*, Majumdar, S. K., Hall, L. W., Jr., and Austin, H. M., Eds., Pennsylvania Academy of Science, Easton, PA, 1987, 373.

40. Nichols, M., Diaz, R. J., and Schaffner, L. C., Effects of hopper dredging and sediment dispersion, Chesapeake Bay, *Environ. Geol. Water Sci.*, 15, 31, 1990.

41. Biggs, R. B., Environmental effects of overboard spoil disposal, *J. Sanit. Eng. Div.*, 94, 477, 1968.

42. Cronin, L. E., Biggs, R. B., Flemer, D. A., Pfitzmeyer, G. T., Goodwin, F., Jr., Dovel, W. L., and Richie, D. E., Jr., Gross Physical and Biological Effects of Overboard Spoil Disposal in Upper Chesapeake Bay, Natural Resources Institute Special Report No. 3, Chesapeake Biological Laboratory, University of Maryland, Solomons, MD, 1970.

43. Nichols, M. and Diaz, R., Plume Monitoring of Rappahannock and York Spit Channels, Baltimore Harbor and Channels (Phase 1), Contribution Report to the U.S. Army Corps of Engineers, Baltimore District, Baltimore, MD, 1989.

44. Hocutt, C. H., Stauffer, J. R., Jr., Edinger, J. E., Hall, L. W., Jr., and Morgan, R. P., II, Eds., *Power Plants: Effects on Fish and Shellfish Behavior*, Academic Press, New York, 1980.

45. Morgan, R. P., II and Stross, R. G., Destruction of phytoplankton in the cooling water supply of a steam electric station, *Chesapeake Sci.*, 10, 165, 1969.

46. Abbe, G. R. and Sanders, J. G., Condenser replacement in a coastal power plant: copper uptake and incorporation in the American oyster, *Crassostrea virginica*, *Mar. Environ. Res.*, 19, 93, 1986.

47. Abbe, G. R., Thermal and other discharge-related effects on the bay ecosystem, in *Ecological Studies in the Middle Reach of Chesapeake Bay: Calvert Cliffs*, Heck, K. L., Jr., Ed., Springer-Verlag, Berlin, 1987, 270.

48. White, J. C. and Brehmer, M. L., Eighteen-month-evaluation of the Ristroph traveling fish screens, in *Proceedings of the 3rd National Workshop on Entrainment and Impingement: 316(b) Research and Compliance*, Jensen, L. D., Ed., Ecological Analysts, Melville, NY, 1976, 367.

49. Baird, D. and Ulanowicz, R. E., The seasonal dynamics of the Chesapeake Bay ecosystem, *Ecol. Monogr.*, 59, 329, 1989.

50. Sellner, K. G., Phytoplankton in Chesapeake Bay: role in carbon, oxygen and nutrient dynamics, in *Contaminant Problems and Management of Living Chesapeake Bay Resources*, Majumdar, S. K., Hall, L. W., Jr., and Austin, H. M., Eds., Pennsylvania Academy of Science, Easton, 1987, 134.

51. Marshall, H. G., The composition of phytoplankton within the Chesapeake Bay plume and adjacent waters off the Virginia coast, U.S.A., *Est. Coastal Shelf Sci.*, 15, 29, 1982.

52. Orth, R. J. and Heck, K. L., Jr., Structural components of eelgrass (*Zostera marina*) meadows in the lower Chesapeake Bay — fishes, *Estuaries*, 3, 278, 1980.

53. Horwitz, R. J., Fish, in *Ecological Studies in the Middle Reach of Chesapeake Bay Calvert Cliffs*, Heck, K. L., Jr., Ed., Springer-Verlag, Berlin, 1987, 167.

54. Boesch, D. F., Classification and community structure of macrobenthos in the Hampton Roads area, Virginia, *Mar. Biol.*, 21, 226, 1973.

55. Holland, A. F., Mountford, N. K., and Mihursky, J. A., Temporal variation in upper bay mesohaline benthic communities. I. The 9-m mud habitat, *Chesapeake Sci.*, 18, 370, 1977.

56. Holland, A. F., Long-term variation of macrobenthos in a mesohaline region of Chesapeake Bay, *Estuaries*, 8, 93, 1985.

57. Diaz, R. J., Benthic resources of the Chesapeake Bay estuarine system, in *Contaminant Problems and Management of Living Chesapeake Bay Resources*, Majumdar, S. K., Hall, L. W., Jr., and Austin, H. M., Eds., Pennsylvania Academy of Science, Easton, 1987, 97.

58. Holland, A. F., Shaughnessy, A. T., and Hiegel, M. H., Long-term variation in mesohaline Chesapeake Bay macrobenthos: spatial and temporal patterns, *Estuaries*, 10, 227, 1987.

59. Pearson, T. H. and Rosenberg, R., Macrobenthic succession in relation to organic enrichment and pollution of the marine environment, *Ann. Rev. Oceanogr. Mar. Biol.*, 16, 229, 1978.

60. Rhoads, D. C. and Boyer, L. F., The effects of marine benthos on physical properties of sediments: a successional perspective, in *Animal-Sediment Relations*, McCall, P. L. and Tevesz, M. J. S., Eds., Plenum Press, NY, 1982, 3.

61. Warwick, R. M., A new method for detecting pollution effects on marine macrobenthic communities, *Mar. Biol.*, 92, 557, 1986.

62. Scott, K. J., Effects of contaminated sediments on marine benthic biota and communities, in *Contaminated Marine Sediments — Assessment and Remediation*, National Research Council, National Academy Press, Washington, D.C., 1989, 132.

63. Gray, J. S., Eutrophication in the sea, in *Marine Eutrophication and Population Dynamics*, Giuseppe, C., Ferrari, I., Ceccherelli, V. U., and Rossi, R., Eds., 25th European Marine Biology Symposium, Olsen and Olsen, Fredensgorg, 1992, 3.

64. Orth, R. J. and Moore, K. A., Chesapeake Bay: an unprecedented decline in submerged aquatic vegetation, *Science*, 222, 51, 1983.

65. Tomasko, D. A., Dawes, C. J., and Hall, M. O., The effects of anthropogenic nutrient enrichment on turtle grass (*Thalassia testudinum*) in Sarasota Bay, Florida, *Estuaries*, 19, 448, 1996.

66. Heck, K. L., Jr. and Thoman, T. A., The nursery role of seagrass meadows in the upper and lower reaches of the Chesapeake Bay, *Estuaries*, 7, 70, 1984.

4 Case Study 3: Southern California Bight

I. INTRODUCTION

The coastal waters of the Southern California Bight are heavily utilized by industry and a large residential population. Sources of contamination in these waters in approximate order of importance include municipal wastes, river runoff, atmospheric fallout, harbor discharges, thermal discharges, and marine transportation. The effects of municipal waste outfalls far outweigh all other sources of pollution in the Southern California Bight. Nevertheless, urban, industrial, and agricultural runoff deliver various loads of the contaminants to the bight that can impact coastal marine habitats and communities.

A series of submarine outfalls discharges billions of liters of municipal waste-waters each day into Southern California shelf waters.[1] Effluent treatment ranges from advanced primary and partial secondary (50:50) to full secondary treatment. One small municipal wastewater facility (Summerland) provides tertiary treatment. Four major sanitation districts (i.e., City of Los Angeles, Los Angeles County, Orange County, and City of San Diego) operate the largest outfalls. About 90% of municipal wastewaters released directly to the Southern California Bight derives from effluents of the Hyperion Treatment Plant (HTP) of the City of Los Angeles, the Joint Water Pollution Control Plant (JWPCP) of the County Sanitation Districts of Los Angeles County, Wastewater Treatment Plants 1 and 2 of the County Sanitation Districts of Orange County (CSDOC), and the Point Loma Wastewater Treatment Plant (PLWTP) of the City of San Diego (Figure 1). Wastewaters of the HTP, JWPCP, and CSDOC enter coastal waters via outfalls at a depth of ~60 m. The PLWTP disposes of effluent from an extended outfall at a depth of 93 m, although prior to 1993 the plant discharged wastewaters at a depth of ~60 m. The collective effluent flow rate of these large treatment plants amounted to $1278-1656 \times 10^9$ l/yr between 1971 and 1994. Aside from these major treatment plants, there are 15 smaller municipal wastewater facilities that also release effluent to the Southern California Bight. From 1971 to 1994, their cumulative rate of effluent flow ranged from approximately $95-189 \times 10^9$ l/yr.

Despite continued high flow rates at the treatment plants, the water quality of wastewaters entering the Southern California Bight has improved steadily over the past 25 years due to increased source control and land disposal of biosolids, upgraded sludge and primary treatment, and increased secondary treatment. The decrease of mass emissions from the major wastewater treatment plants has been significant. These changes have led to a gradual upgrade of environmental conditions in the bight and positive responses of marine communities.

149

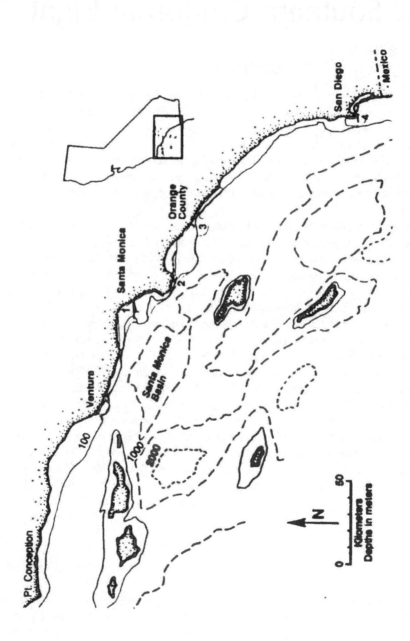

FIGURE 1   Map of the Southern California coast showing the areas affected by the four largest municipal wastewater treatment plants: (1) Hyperion Treatment Plant; (2) Joint Water Pollution Control Plant; (3) County Sanitations Districts of Orange County; and (4) Point Loma Wastewater Treatment Plant. (Modified from Bascom, W., *Environ. Sci. Technol.*, 16, 226A, 1982.)

Investigations of the ecological effects of the municipal wastewater discharges on the coastal waters of Southern California have centered around efforts of the Southern California Coastal Water Research Project (SCCWRP). Founded in 1969 when five local government agencies (i.e., the county sanitation districts of Los Angeles, Orange, and Ventura counties and the cities of Los Angeles and San Diego) signed a joint agreement to sponsor environmental studies, the SCCWRP also has been dedicated to understanding the changing ecology of the coastal waters of the bight due to other anthropogenic activities. For example, multiple nonpoint sources of pollution and extensive water diversion projects likewise stress the coastal zone. The shelf waters of the bight have been affected for years by an array of anthropogenic factors, such as discharges from power plants, pollutants from oil refineries, recreational and commercial fisheries, marine transportation, numerous industrial installations, and various tourist activities in the coastal zone. However, the principal focus of the SCCWRP is to delineate the effects of municipal waste discharges on marine life in the bight.

This chapter examines the environmental impacts of wastewater discharges in the Southern California Bight. It focuses on the effects of the major municipal wastewater treatment plants. It also deals with long-term studies of marine communities affected by wastewater discharges from these facilities.

II. PHYSICAL DESCRIPTION

The Southern California Bight is a large open embayment covering an area of approximately 100 km^2. It extends seaward from the Southern California coastline between Point Conception and Cabo Colnet to the abrupt continental slope about 200-250 km offshore. A basin-and-range submarine topography characterizes the bight, and rocky terrain occasionally rises above the sea surface to form islands. The system of basin and ranges separating the coastal shelf from the deep ocean greatly influences water circulation throughout the region. The network of shelf basins serves as an efficient sediment trap. In addition, a series of submarine canyons (e.g., Santa Monica Canyon, San Pedro Canyon, Redondo Canyon, etc.) channels the seaward movement of particulates. Bottom sediments in the bight primarily consist of richly organic, fine-grained silts.[2]

The southward-flowing California Current passes over the western part of the bight. A portion of this current turns eastward along the Mexican border and becomes part of a counterclockwise eddy that mixes with northward-flowing warm water and returns through the bight. During most of the year (except late winter), waters are stratified, with a shallow (10–30 m) surface layer being isolated by the thermocline-pycnocline. Sub-thermocline currents transport seawater and entrained wastewaters offshore in a northwest direction, often oriented along depth contours. Current velocities over the shelf range from 7–10 cm/s; the net transport is only about 3 cm/s.[2] Oscillating tidal currents are superimposed on these sub-thermocline currents. The upwelling of cooler, nutrient-rich deep water onto the shelf develops during periods of strong and persistent easterly winds (Santa Ana Winds).

III. WASTE DISPOSAL

The Southern California Bight has a long history of sewage sludge and municipal wastewater disposal. For example, the HTP has discharged sewage effluent into Santa Monica Bay since 1894. From September 1957 to November 1987, the HTP also discharged sewage sludge through a 11.2-km pipeline into the bay.[3] The Los Angeles County Sanitation Districts has operated the JWPCP since 1937. The submarine outfall system of this facility currently extends about 3 km offshore from Whites Point and releases effluent onto the Palos Verdes Shelf.[4] The CSDOC and PLWTP have likewise discharged wastewaters onto the shelf for decades.

The SCCWRP publishes annual reports of effluent flow and mass emission estimates of constituents for the four aforementioned large treatment plants.[5] The SCCWRP also compiles effluent characteristics of smaller treatment plants. Estimates of mass emissions of constituents from these smaller municipal wastewater facilities were summarized by the SCCWRP in 1971, 1987, 1989, 1993, and 1994.[6] The following discussion reviews the effluent characteristics of both large and small wastewater treatment plants that discharge to the Southern California Bight.

A. LARGE WASTEWATER TREATMENT PLANTS

Effluent quality of the largest municipal treatment plants (i.e., HTP, JWPCP, CSDOC, and PLWTP) changed markedly over the period of 1971 to 1994. Although the cumulative flow from these plants increased 15% during this 24-year period, mass emissions declined substantially. For example, the combined mass emissions of chlorinated hydrocarbons and trace metals decreased 99% and 95%, respectively, within this time interval. In addition, the combined mass emissions of suspended solids, oil and grease, and biochemical oxygen demand (BOD) dropped 77%, 69%, and 53%, respectively. Figure 2 shows the combined mass emission values of DDT and PCBs from 1971 to 1994, and Figure 3 illustrates the combined mass emission values of BOD during the same period. As is evident, most of the decline in concentrations occurred between 1985 and 1990. Table 1 compares constituent mass emissions of these wastewater treatment plants during 1994. Reductions in constituent mass emissions in recent years are ascribed mainly to improved primary treatment, increased secondary treatment, and greater source control.[5]

Of particular note was the precipitous drop in chlorinated hydrocarbon concentrations from 1971 to 1994. Much of this decrease was due to federal bans on the use of DDT and PCBs in the 1970s. Hence, in 1971 the combined mass emissions of DDT and PCBs for the four major treatment plants amounted to 21,527 kg and 8730 kg, respectively. In 1994, the emissions of DDT had dropped to 7.9 kg, and they were below detection limits for PCBs. Between 1947 and 1971, the JWPCP sewer system processed DDT wastes from the Montrose Chemical Corporation, the largest DDT manufacturer in the world. Small amounts of DDT are still detected in plant effluents in some months. Despite the diminution of DDT and PCB concentrations in plant effluents, the levels of these contaminants in bottom sediments and biota of the Southern California Bight remain a concern (Janet Stull, County Sanitations Districts of Los Angeles County, personal communication, 1996).

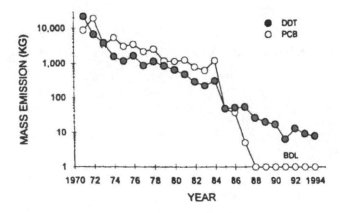

FIGURE 2   Combined mass emissions of DDT and PCB from the four largest wastewater treatment plants in Southern California (BLD = below detection limits). (From Raco-Rands, V. E., in *Southern California Coastal Water Research Project, Annual Report 1994–95,* Southern California Coastal Water Research Project, Long Beach, CA, 1995, 10.)

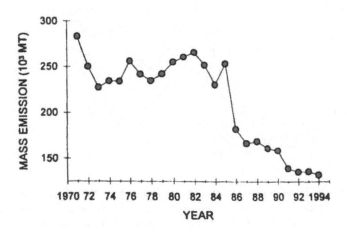

FIGURE 3   Combined mass emissions of biochemical oxygen demand from the four largest wastewater treatment plants in Southern California (MT = metric tons). (From Raco-Rands, V. E., in *Southern California Coastal Water Research Project, Annual Report 1994–95,* Southern California Coastal Water Research Project, Long Beach, CA, 1995, 10.)

B. SMALL WASTEWATER TREATMENT PLANTS

The SCCWRP summarized constituent mass emissions from 15 small (<9.5 × 10⁷ l/d) wastewater treatment plants in 1971, 1987, 1989, 1993, and 1994 (Table 2). Most of these plants provide secondary treatment of wastes (Table 3). From 1971 to 1994, the annual combined volume of effluent discharged by these smaller facilities into the Southern California Bight increased by 90%. However, during this period the mass emissions of oil and grease, BOD, suspended solids and cyanide decreased 91%, 80%, 79%, and 73%, respectively. Because of the higher rate of decrease in

TABLE 1
Estimated Constituent Mass Emissions from the Largest Municipal Wastewater Treatment Plants in Southern California in 1994

Constituent	HTP	JWPCP	CSDOC	PLWTP	Total
Flow[a] (L x 10⁹)	454	452	331	237	1474
Suspended solids (mt)	13,471	29,068	14,712	10,875	68,126
BOD (mt)	37,009	44,776	23,396	27,076	132,257
Oil and grease (mt)	5371	5254	4418	3491	18,534
Nitrate-N (mt)	78	118	—	15	211
Nitrite-N (mt)	—	20	—	—	20
Ammonia-N (mt)	11,334	15,925	7754	6093	41,106
Organic N (mt)	2526	2963	—	—	5489
Phosphate (mt)	—	—	—	202	202
Total phosphorus (mt)	1821	1627	—	—	3448
Cyanide (mt)	6.7	3.8	0.2	0.9	12
Silver (mt)	2.4	2.6	0.7	—	5.7
Arsenic (mt)	1.8	1.4	0.4	0.4	4.0
Cadmium (mt)	0.2	0.2	0.2	0.07	0.7
Chromium (mt)	0.6	4.1	1.6	0.4	6.7
Copper (mt)	16	10	9.2	14	49
Mercury (mt)	0.008	—	—	0.02	0.03
Nickel (mt)	5.1	16	5.8	0.6	28
Lead (mt)	—	0.6	0.7	—	1.3
Selenium (mt)	0.04	6.7	0.4	0.3	7.4
Zinc (mt)	21	33	12	6.2	72
Phenols[b] (mt)	—	230	9.6	—	240
Chlorinated[c]	—	2.1	—	—	2.1
Nonchlorinated[c]	1.4	66	1.2	3.1	72
Total DDT (kg)	4.8	2.9	0.2	—	7.9
Total PCB[d] (kg)	—	—	—	—	—

Note: HTP = Hyperion Treatment Plant, City of Los Angeles. JWPCP = Joint Water Pollution Control Plant, County Sanitation Districts of Los Angeles County. CSDOC = County Sanitation Districts of Orange County. PLWTP = Point Loma Wastewater Treatment Plant, City of San Diego. mt = metric tons.

[a] Annual flow volumes were the sum of mean daily flow per month times the number of days in each month.
[b] EPA method 420.2 (Colorimetric method).
[c] EPA method 604 or 625 (GCMS method).
[d] Total PCB = PCB 1016, 1221, 1232, 1242, 1248, 1254, and 1260.

From Raco-Rands, V. E., in *Southern California Coastal Water Research Project, Annual Report 1994–95*, Southern California Coastal Research Project, Long Beach, CA, 1995, 10.

mass emissions at large treatment plants relative to small treatment plants in recent years, the percent contribution of certain constituents (e.g., chromium and lead) discharged from small plants to the Southern California Bight has exceeded 50% (Table 4). As a result, the concentrations of some effluent constituents released to

TABLE 2
Combined Mass Emissions of Constituents from the 15 Small (<9.5 × 10⁷ l/d) Wastewater Treatment Plants that Discharged to the Southern California Bight from 1971–1994

Constituent	Mass Emissions						Percent Change		
	1971	1987	1989	1993	1994	1971–1994	1987–1994	1993–1994
Flow	0.26	0.50	0.52	0.51	0.50	90	−1	−3
Suspended solids (mt)	8200	4193	2984	2297	1737	−79	−59	−24
BOD (mt)	11,000	5178	4751	2285	2207	−80	−57	−3
Oil and Grease (mt)	4200	708	460	425	377	−91	−47	−11
Ammonia-N (mt)	1600	1757	2716	3668	3118	95	77	−15
Cyanide (mt)	8	1.7	0.67	3.6	2.2	−73	29	−39
Arsenic (mt)	—	0.43	0.84	0.32	0.44	—	2	38
Cadmium (mt)	—	1.7	0.53	1.2	0.87	—	−49	−28
Chromium (mt)	—	2.3	0.84	1.5	1.6	—	−30	7
Copper (mt)	—	6.9	3.4	4.5	3.2	—	−54	−29
Lead (mt)	—	6.5	2.9	4.6	2.8	—	−57	−39
Mercury (mt)	—	0.18	0.23	0.01	0.008	—	−96	−20
Nickel (mt)	—	5.5	2.8	4.2	4.9	—	−11	17
Silver (mt)	—	0.87	0.58	0.71	1.4	—	61	97
Zinc (mt)	—	16	12	11	11	—	−31	0
DDT (kg)	—	nd[a]	nd	0.91	0.07	—	—	−92
PCB (kg)	—	nd	nd	0.09	0.09	—	—	0

Note: Flow = l/d × 10⁹; mt = metric tons.

[a] nd = no data.

From Raco-Rands, V. E. in *Southern California Coastal Water Research Project, Annual Report 1994–95*, Southern California Coastal Research Project, Long Beach, CA, 1995, 21.

TABLE 3
Volume and Level of Effluent Treatment for the Small Municipal
Wastewater Treatment Facilities that Discharge to the Southern
California Bight

Facility	Flow[a]	Treatment Level
Goleta	1.74	Primary/secondary
Santa Barbara	2.65	Secondary
Montecito	0.33	Secondary
Summerland	0.05	Tertiary
Carpinteria	0.49	Secondary
Oxnard	6.59	Secondary
Terminal Island	6.09	Secondary
Avalon	0.20	Secondary
San Clemente Island	0.007	Secondary
Aliso Water Management Agency	6.70	Secondary
South East Regional Reclamation Authority	6.13	Secondary
Oceanside	4.58	Secondary
Encina	7.76	Secondary
San Elijo and Escondido	6.51	Secondary

[a] l/day × 10^7.

From Raco-Rands, V. E., in *Southern California Coastal Water Research Project, Annual Report
1994–95*, Southern California Coastal Research Project, Long Beach, CA, 1995, 21.

the the Southern California Bight from small treatment plants have become more
significant during the 1990s.[6]

C. AFFECTED WATERS

1. Santa Monica Bay

As mentioned previously, Santa Monica Bay has been affected by both sewage sludge
and wastewater disposal from the HTP in El Segundo. It is also influenced by
wastewater discharges from the JWPCP located about 12.5 km southeast of Point
Palos Verdes. Some wastewaters from JWPCP enter the southern end of Santa
Monica Bay, where they affect water quality. However, the greatest impacts on water
quality in the bay are due to discharges from the HTP.

Santa Monica Bay, located west of Los Angeles, extends for ~40 km between
headlands, and it covers an area of ~500 km². Although the submarine topography
is basically flat, several low-relief rocky outcrops rise above the seafloor. The bay
opens into a series of alternating basins with depths to 2 km, and the Santa Monica
submarine canyon forms a steep-sided trench in the shelf seaward of the outfalls.[7]
Muddy sediments predominate in the basins.

The velocity and direction of currents vary considerably, and one or two gyres
occasionally develop in Santa Monica Bay. While the residence time of water for
mixed layers of the Southern California Bight is about 3 months, it is much shorter

TABLE 4
**Constituent Mass Emissions from Large (>56.8 × 10⁹ l/d) and Small
(<9.5 × 10⁷ l/d) Municipal Wastewater Treatment Plants that
Discharged to the Southern California Bight in 1994**

	Mass Emissions			Percent of Total	
Constituent	Large[a]	Small	Total	Large	Small
Flow	4.05	0.50	4.55	89	11
Suspended solids (mt)	68,126	1737	69,863	98	2
BOD (mt)	132,257	2207	134,464	98	2
Oil and Grease (mt)	18,534	377	18,911	98	2
Ammonia-N (mt)	41,106	3118	44,224	93	7
Cyanide (mt)	12	2.2	14	85	15
Arsenic (mt)	4.0	0.4	4.4	91	9
Cadmium (mt)	0.7	0.9	1.6	44	56
Chromium (mt)	6.7	1.6	8.3	81	19
Copper (mt)	49	3.2	52	94	6
Lead (mt)	1.3	2.8	4.1	32	68
Mercury (mt)	0.03	0.008	0.04	79	21
Nickel (mt)	28	4.9	33	85	15
Silver (mt)	5.7	1.4	7.1	80	20
Zinc (mt)	72	11	83	87	13
DDT (kg)	7.9	0.07	8	99	1
PCB (kg)	nd[b]	0.09	0.09	—	—

Note: Flow = l/d × 10⁹; mt = metric tons.

[a] Hyperion Wastewater Treatment Plant (City of Los Angeles), Joint Water Pollution Control Plant of the County Sanitation Districts of Los Angeles County, Wastewater Treatment Plants 1 and 2 of County Sanitation Districts of Orange County, and Point Loma Wastewater Treatment Plant (City of San Diego).
[b] nd = no data.

From Raco-Rands, V. E., in *Southern California Coastal Water Research Project, Annual Report 1994–95*, Southern California Coastal Research Project, Long Beach, CA, 1995, 21.

(~1 month) in the waste field of the bay. The density of seawater along the shelf controls the height to which waste plumes rise in outfall areas. It varies seasonally. During most of the year, coastal waters in Santa Monica Bay are stratified. As a consequence, wastes discharged through the 8.3-km pipeline rarely reach the sea surface, but tend to concentrate at a depth of ~20–40 m. During several weeks in winter (albeit not every year), the thermocline/pycnocline stratification breaks down and the water column becomes well mixed. This condition enables some of the waste to reach the sea surface. When wastewaters are discharged during most of the year, a dilution of 100–200 is rapidly achieved, and the resulting plume rises to mid-depth.

The sludge outfall of the City of Los Angeles, when active, discharged nearly 20 × 10⁶ l/d of digested sludge at the head of Santa Monica Canyon for 30 years. This effluent consisted of three principal components: digested primary sludge, secondary sludge, and secondary effluent. Sewage solids (96% pass through a

0.25 mm screen) comprised less than 1% of the discharge. Over an annual cycle, 5.7×10^4 mt (dry weight) of solids with an approximate volume (wet) of 2.85×10^5 cu m were discharged at the terminus of the 11.2-km sludge pipeline.[7] Long-term sludge disposal caused a significant increase in the concentrations of organic matter and chemical contaminants over an area ~3 km² at the bottom of the Santa Monica Canyon, where changes in biological communities were also noted (see below).

Sludge material that accumulated directly in front of the discharge contained plant fibers, bits of leaves, cigarette filters, watermelon and tomato seeds, plastics, and other items. Particles larger than 1 mm only comprised about 1% of the solids discharged and those between 0.25 and 1 mm in size, about 3%. These larger particles settled soon after release from the pipeline and accounted for much of the sludge remaining in the canyon.[8] An estimated 10% of the discharged solids fell within an oval area ~2 km around the outfall.[7] The finer solid particles remained suspended for days or weeks, being dispersed by currents far offshore generally northwest of the discharge point.

Between 1971 and 1979, total suspended solids discharged through the 11.2-km outfall ranged from 4400–10,300 mg/l and averaged 8700 mg/l dry weight. Among the heavy metals analyzed, copper (13.2 mg/l) and zinc (21.0 mg/l) had the highest mean concentrations. Mean values of oil and grease amounted to 710 mg/l. DDT and PCBs had mean concentrations of 3.7 and 20.4 µg/l, respectively.

Although the 11.2-km sewage sludge outfall was closed in 1987 when sludge combustion and handling facilities became operational, the 8.3-km wastewater outfall has remained open. A 1.6-km outfall is also used during certain emergency conditions. Unchlorinated liquid effluent is discharged from the 3.7-m diameter, 8.3-km long pipeline through a "Y"-shaped diffuser at a depth of 57 m. Wastewater flow from the HTP increased from 988×10^6 l/d in 1960 to 1571×10^6 l/d in 1985. Effluent solids reached a peak level of 155 mg/l in 1985, and have fallen to approximately 30 mg/l at present.

The quality of wastewaters discharged through the 8.3-km pipeline was enhanced in 1986 with initiation of the Hyperion Interim Improvements Projects and construction of full secondary treatment facilities. Due to the termination of sewage sludge discharge from the 11.2-km pipeline, combined sewage emissions from the four largest municipal wastewater treatment plants decreased 40% from 1987 to 1988.[5] In addition, reduced flows to the HTP which occurred during the early 1990s due to water conservation and drought conditions (e.g., 1188×10^6 l/d during 1992–1993) contributed to improved effluent quality. For example, suspended solids in wastewaters during 1992–1993 reached low levels, averaging only 33 mg/l.[3]

In 1994, the 8.3-km outfall discharged 454×10^9 l of liquid effluent. Suspended solids emissions amounted to 13,471 mt. It currently discharges 1.21×10^9 l/d of mixed primary and secondary effluent.[3] Much lower concentrations of heavy metals have been released from the 8.3-km wastewater outfall than were released from the 11.2-km sewage sludge outfall. With the exception of copper and zinc, heavy metal mass emissions were less than 10 mt in 1994. Oil and grease values totaled 5371 mt. Chlorinated hydrocarbons were low, with total DDT being 4.8 kg and total PCBs undetectable.[5] At certain times in the past when floods or equipment failures have occurred, some solids released from the 8.3-km outfall accumulated on the shelf and in the canyon.

2. Palos Verdes Shelf

Located seaward of an isolated coastal promontory, the Palos Verdes Shelf forms a narrow terrace. It is bounded on the north by the Redondo Canyon and on the south by the San Pedro Sea Valley. During most of the year, subsurface currents generally flow toward the northwest. As a result, contaminants released in wastewater discharges (e.g., heavy metals and organochlorine compounds), as well as organic matter, have accumulated in highest concentrations in an area within 2 km northwest of the outfalls.[9]

The JWPCP of the Los Angeles County Sanitation Districts, located 10 km inland, discharges more than 1.0×10^9 l/d of partial secondary treated wastewaters through two outfall pipes which extend 2.8 km and 3.6 km, respectively, offshore of Whites Point to a water depth of ~60 m (Figure 4). The wastewaters were initially discharged in 1937 through smaller pipes that were replaced in 1956 and 1966 by "Y"-shaped (2.3 m in diameter) and "L"-shaped (3.05 m in diameter) outfalls, respectively. Effluent quality has improved significantly over the past two decades. The implementation of greater source control of trace constituents among industries tributary to the plant, advances in treatment and operational strategies such as the addition of polymer to sedimentation tanks, upgrades of facilities for dewatering solids, the application of primary effluent screening, and increases in land use and volatization of solids have greatly reduced mass emissions to the sea.[2,4,10] Hence, between 1971 and 1981, the emissions of suspended solids decreased by 50%, DDT and PCBs declined by >95%, and chromium dropped by 77%.[2] In 1971, the solids mass emission rate to the ocean was five times greater than that in 1992. By 1994, the emissions of cadmium, copper, chromium, lead, and zinc were only a small fraction of 1971 values.[4] Even greater reductions were evident for chlorinated hydrocarbons. The decrease in JWPCP solids emissions from 1971 and 1994 was responsible for 69% of the reduction in the combined mass emissions of suspended solids (77% total reduction), oil and grease (69%), and biochemical oxygen demand (53%) from the four largest municipal wastewater treatment plants in Southern California (i.e., HTP, JWPCP, CSDOC, and PLWTP).[5] Figure 5 illustrates the flow of treated wastewaters and mass emissions of suspended solids discharged to the Palos Verdes Shelf from 1937 to 1992.

Contaminant burdens of surface sediments along the Palos Verdes Shelf have also declined over time owing to progressive improvements in effluent quality. For years, the flux of effluent particulates to the seafloor altered bottom sediments and benthic communities of the Palos Verdes Shelf.[2] However, from 1972 to 1992 considerable environmental and biological recovery was observed along the Palos Verdes Shelf concomitant with reduced discharges of solids and contaminants from the outfalls (see below).[2,4]

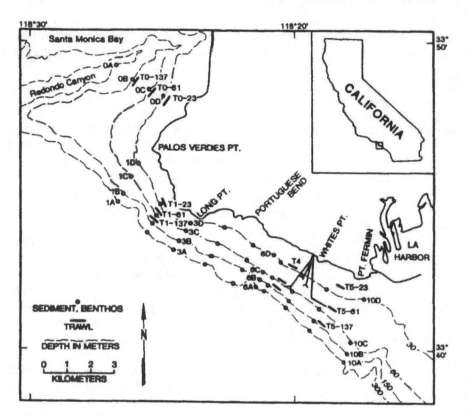

FIGURE 4   Biotic sampling sites in the Palos Verdes shelf and slope during the 1972–1992 period. (From Stull, J. K., *Bull. Southern California Acad. Sci.*, 94, 21, 1995.)

3. San Pedro Shelf

Wastewater Treatment Plants 1 and 2 of the CSDOC discharge treated domestic and industrial effluent (50% advanced primary and 50% secondary) from a 3-m diameter, 7.25-km pipe at a depth of 60 m on the shelf. A multiport diffuser along the final 1.6-km segment of the pipeline results in a 175:1 dilution of wastewaters at the diffuser ports. The total flow from the CSDOC in 1994 amounted to 4.52×10^{11} l. Long-term flow from the Orange County outfall has altered a 10 km² area of the seafloor. However, coefficient of pollution values calculated by Maurer et al.[11] for various outfalls in the Southern California Bight revealed generally lower values for the Orange County outfall than elsewhere.

Current flow near the outfall is principally aligned along bottom contours, with the Southern California Eddy exerting strong influence on current variations in the area. At the outfall, currents primarily flow northward at velocities up to 35 cm/s and significantly affect the movement of wastewater plumes and sediments. The San Gabriel Canyon and Newport Canyon, located on the western and eastern flanks of the shelf, respectively, serve as major repositories for fine grained (silt-clay) sediments.

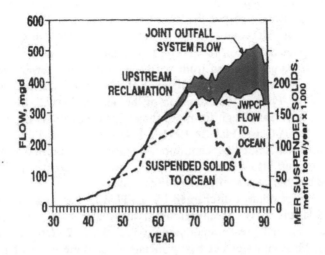

FIGURE 5 Flow of treated wastewaters and mass emissions of suspended solids discharged to the ocean off Palos Verdes by the Los Angeles County Sanitation Districts, 1937–1992. mgd = million gallons per day; MER = mass emission rate. Upper line is the Districts' entire Joint Outfall System Flow. (From Stull, J. K., *Bull. Southern California Acad. Sci.*, 94, 21, 1995.)

The outfall of the CSDOC discharged a total of 14,712 mt of suspended solids in 1994. Mass emissions of biochemical oxygen demand and oil and grease at this time amounted to 23,396 mt and 4418 mt, respectively. Mass emissions of heavy metals were <1 mt for arsenic, cadmium, lead, selenium, and silver. Somewhat higher values were recorded for chromium (1.6 mt), nickel (5.8 mt), copper (9.2 mt), and zinc (12 mt). Total DDT emissions in 1994 equalled 0.2 mt (Table 1).

From 1971 to 1994, the volume of wastewater discharged by the CSDOC nearly doubled. In addition, effluent acute toxicity (to fathead minnows *Pimephales promelas*) significantly increased at the CSDOC from 1990 to 1994. However, major decreases in constituent mass emissions at the CSDOC during the past two decades have been noted. These are ascribed to improved effluent treatment and greater source control.[5]

4. Point Loma

The PLWTP at San Diego discharges wastewaters on the shelf at a depth of 93 m, although it discharged wastewaters at a depth of 60 m prior to November 24, 1993. The length of the outfall from shore was 3.6 km prior to November 24, 1993 and 7.285 km after this date. The plant currently processes about 7.0×10^8 l/d of sewage generated by more than 1.7×10^6 people in San Diego and 15 other nearby cities and sewer districts. In 1994, the total wastewater flow amounted to 2.37×10^{11} l. The San Diego Metropolitan Sewerage System made several improvements in sewage treatment operations between 1979 and 1992. In 1988, an upgrade of the system from primary to chemically enhanced primary treatment increased the efficiency of solids removal by more than 20%. The total wastewater flow also increased from about 4.5×10^8 to 7.0×10^8 l/d. The total suspended solids that were discharged did not change significantly despite the increased flow.

Because of the exceedence of bacteriological standards in the Point Loma kelp beds and the potential exposure of divers to pathogenic organisms associated with wastewater discharges released from the 60-m outfall, the sewage plant extended the pipeline in November 1993 to a distance of more than 7 km offshore and a depth of nearly 100 m. Prior to the extension of the outfall, the U.S. EPA had filed suit against the city of San Diego for more than 20,000 violations of the Clean Water Act and California Ocean Plan. In 1991, a federal court found the city in violation of the Clean Water Act almost continuously since the statute was enacted in 1972. However, the court also declared that the wastewater discharges had not adversely affected the benthic community around the outfall.[12]

The quality of effluent discharged by the PLWTP has steadily improved since 1980. In 1994, the mass concentrations of suspended solids, biochemical oxygen demand, and oil and grease amounted to 10,875 mt, 27,076 mt, and 3491 mt, respectively. Mass emissions of heavy metals at this time were <1 mt for arsenic, cadmium, chromium, lead, mercury, nickel, and selenium. They were somewhat higher for zinc (6.2 mt) and copper (14 mt). DDT and PCBs were not detected in the effluent (Table 1).

IV. BIOTIC EFFECTS

A. SANTA MONICA BAY

Bascom,[1,7,8] Dorsey et al.,[3] Thompson,[13] and Anderson et al.[14] have examined in detail the effects of sewage sludge disposal and sewage wastewater discharges on biological communities in Santa Monica Bay. Daily et al.[15] also reported on sewage impacts on biological communities elsewhere in the Southern California Bight. Long-term studies have been conducted on the composition of the sewage sludge discharged, where it accumulates on the seafloor, the extent of its effects on benthic communities and fish assemblages, and the influences of associated contaminants on the food web and human health. The accumulation of particulates and associated contaminants on the bottom near the 11.2-km sewage sludge discharge point was largely responsible for the biological impacts observed during the 1970s. During the period of active sewage sludge dumping, an area ~3 km² around the outfall, representing about 0.4% of the bay, was described as "degraded."[7]

The area most seriously impacted by sewage waste disposal occurs entirely in the Santa Monica Basin at depths greater than 100 m. While benthic community structure in this degraded area was clearly altered by the sludge disposal, some members of the community proliferated as evidenced by substantial increases in their biomass. The additional organic carbon represented by this biomass provided a new food supply that supported a greater number of organisms. The larger than average size of many individuals in the area also contributed significantly to the increased biomass values.[1] Some populations, however, disappeared from the sewage sludge disposal grounds. Hence, the sludge-lined canyon habitats appear to have had both beneficial and detrimental effects on the benthos.

As effluent quality improved at the 8.3-km wastewater outfall during the past two decades, the benthic community shifted in composition from pollution-tolerant assemblages to assemblages more typical of nonpolluted areas of the bight. Within the old sludge field of the terminated 11.2-km outfall, the benthic infaunal community has changed gradually from highly degraded to an early transitional state in terms of species compostion and abundance. The benthic community at the 8.3-km outfall is expected to recover further as the HTP moves toward full secondary treatment in 1998.[3] It is anticipated that the benthic community surrounding the 11.2-km outfall will likewise continue to be upgraded as clean shelf sediments settle to the seafloor and bury the sludge deposits to greater depths.

1. Benthic Communities

Hartman[16] surveyed the benthic infauna of Santa Monica Bay during the 1952–1956 period. He conducted these surveys to characterize infaunal assemblages around the 1.6-km outfall and (at that time) the proposed 8.3-km and 11.2-km outfalls. This work focused on descriptive associations of the benthos.

Since 1972, benthic communities have been monitored consistently to assess spatial and temporal changes in infaunal assemblages due to waste discharges from the HTP outfalls. Species composition and abundance patterns at the outfall sites were compared to the predictive model of Pearson and Rosenberg[17] which defines four zones of infaunal communities affected by organic enrichment:

1. A degraded zone nearly devoid of infauna;
2. A polluted zone of low diversity, but high biomass owing to proliferation of opportunistic species;
3. A transition zone typified by high abundance and diversity of species due to a mix of opportunistic and natural occurring populations;
4. An unaffected zone characterized by a natural assemblage of organisms.

Results of the benthic monitoring programs in Santa Monica Bay are summarized below.

The benthic communities of Santa Monica Bay, consisting of relatively sedentary hard-substrate and soft-sediment dwellers, tend to integrate the effects of long-term exposure to low-level contaminants and therefore have been used to assess the magnitude and character of effluent-related environmental stresses. At sewage outfalls, biotic gradients develop from polluted to unpolluted areas. More pollution-tolerant opportunistic species tend to fluorish in stressed habitats near the outfalls and are displaced by a natural assemblage of organisms, including many highly pollution-sensitive forms, away from the stressed areas.

Observations by Bascom[7,8] of the terminal structure and last few hundred meters of the 11.2-km effluent pipeline revealed a variety of marine animals typically encountered on rocky bottoms. Of greatest abundance was the sea anemone *Metridium senile* which nearly covered the terminal structure. On the soft-sediment bottom nearby, gastropods (*Kelletia kellite*, *Megasercula carpenteriana*, and *Terebra* sp.) and starfishes (*Astropectin verrilli* and *Luidia foliolata*) were common. An Infaunal

Trophic Index (ITI) used as a pollution index and formulated for the general outfall area indicated that the greatest changes in benthic faunal composition occurred along the upper part of the Santa Monica Canyon. Here, ITI values were often <30, inferring significant degradation of the benthic communities. Averaged among all sampling stations, the ITI index was 76.9, demonstrating normal conditions throughout most of the bay.[18] In heavily impacted upper canyon areas, the number of benthic species was greatly reduced, perhaps to as few as 10, during active sludge disposal. Included in this group were three species of mollusks (*Axinopsida serricata, Macoma carlotensis*, and *Parvalucina tenuisculpta*), as well as dorvellid and polychaete worms. Although the number of infaunal species was low in the upper canyon, the total abundance of individuals and the biomass at most canyon stations were about double those at noncanyon reference sites, reflecting the overwhelming importance of the high food supply derived from sludge disposal and the few species that could utilize it. Highly unstable, stressful environments, such as the upper Santa Monica Basin, favor the successful colonization by relatively few benthic infaunal species.

Trawl samples collected in the Santa Monica Basin directly beyond the sludge discharge point revealed high abundances of invertebrates on and immediately above the sludge-covered seafloor. Not only were more epibenthic invertebrate species (mean number = 15) taken in the Santa Monica trawls compared to control stations (mean number = 11), but also more individual animals (mean number = 734) were found at sludge-affected areas than at controls (mean number = 455). During the period of 1976–1979, 40 species of macroinvertebrates were recovered by trawl sampling at the outfall.[7,8]

Ongoing studies have been conducted on the benthic infaunal community in the area of the 11.2-km outfall to document the recovery rate subsequent to the termination of sewage sludge disposal in November 1987. Thompson,[13] applying ordination analyses to infaunal data, disclosed that the highly degraded assemblage near the outfall initially exhibited rapid recovery to more natural conditions soon after termination of sludge disposal. However, by 1990 the rate of recovery had slowed substantially (Figure 6). During this recovery period, key indicator species of degraded, polluted, and transitional assemblages attained peak abundances and then decreased. For example, the polychaete *Capitella capitata*, a pollution indicator species, became the most abundant species within a year after sludge termination, but was nearly absent in the old sludge field by 1990. *Pectinaria californiensis* replaced *C. capitata* as a dominant form. Another polychaete, *Ophryotrocha* sp. C, was the most common species at the time of sludge disposal termination in 1987, but it had nearly disappeared in the old sludge field by the summer of 1990.

Dorsey et al.[3] asserted that a mix of opportunistic and normal benthic populations (an average of about 50 species) now inhabit the old sludge field. As cleaner shelf sediments continue to accumulate in the area, more pollution-sensitive species are likely to settle along the upper Santa Monica Canyon. Thompson[13] predicts that assemblages in the sludge field will reach control site conditions by the year 2002.

As in the case of the benthic infaunal community at the 11.2-km outfall, that at the 8.3-km outfall has shifted in species composition. However, some important differences between the two communities still remain. For example, Hyperion benthic surveys conducted during the summer of 1992 revealed two different outfall

INFAUNAL ORDINATION

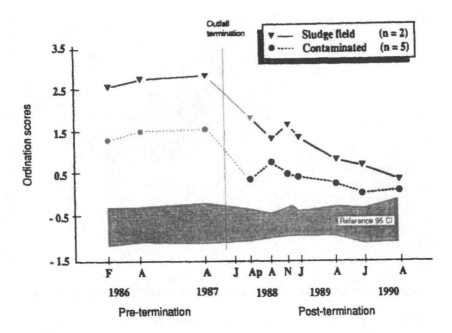

FIGURE 6   Ordination values of sites sampled from 1986 through 1990 during the sludge termination study by Southern California Wastewater Research Project. n = number of stations in strata denoted by symbols. (From Thompson, B. E., Recovery of Santa Monica Bay from Sludge Discharge, Southern California Coastal Water Research Project, Tech. Rept. No. 349, Long Beach, CA.)

groups. The 11.2-km outfall group was characterized by pollution-tolerant species, such as the polychaete *Pectinaria californiensis* and bivalves *Axinopsida serricata*, *Parvilucina tenuisculpta*, and *Tellina carpenteri*, along with a mix of normal populations. The 8.3-km outfall group was typified by the polychaetes *Acmira catherinae*, *Aphelochaeta* sp., *Capitella capitata*, and *Mediomastus* spp., and the ostracod *Euphilomedes carcharodonta*, together with a variety of other species. *Capitella capitata* and *E. carcharodonta* are pollution-tolerant forms. The aforementioned pollution-tolerant, opportunistic species are capable of attaining extremely high population abundances in organically enriched sediments, but they are very unstable and subject to rapid shifts in population sizes.

Based on benthic sampling during the 1990s,[3,19] the infaunal community within the old sludge field at the 11.2-km outfall has changed along a pollution gradient from a degraded to early transitional assemblage, while that at the 8.3-km outfall has shifted from a polluted to a late transitional, nearly normal assemblage. The area of the seafloor affected by the combined sewage outfall system has declined from about 51 km² in 1987 to 16 km² in 1992 (Figure 7). If the trend continues, both

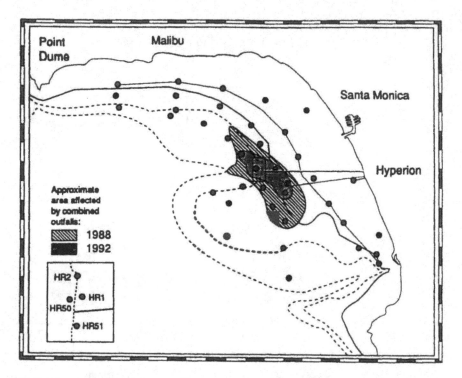

FIGURE 7   Approximate extent of infaunal assemblages affected by discharges from the Hyperion outfall system in 1988 and 1992. (From Dorsey, J. H., Phillips, C. A., Dalkey, A., Roney, J. D., and Deets, G. B., *Bull. Southern California Acad. Sci.*, 94, 46, 1995.)

outfall communities should recover further from the cumulative impacts of the pollution stresses.

2. Finfish Communities

Schools of fish commonly occur along the effluent pipelines. Bascom[7,8] reported high abundances of bocaccio *Sebastes paucispinis* and vermilion rockfish *Sebastes miniatus* near the terminus of the 11.2-km pipeline. The shortbelly rockfish *Sebastes jordani* frequently swam above the pipe, and the Dover sole *Microstomus pacificus* was observed near the pipe and along the soft bottom.

The density of fishes inshore of the Santa Monica Canyon ($0.05/m^2$) was significantly less than that of fishes in the sewage sludge field ($0.45/m^2$). The dominant species outside the sludge field included longspine combfish *Anoiolepis latinpinnis* ($0.02/m^2$), pink sea perch *Zalembius rosaceus* ($0.01/m^2$), and speckled sanddab *Citarichthys stigmaeus* ($0.01/m^2$). In the sludge field, the dominant species were the white croaker *Genyonemus lineatus*, shiner perch *Cymatogaster aggregata*, and Pacific electric ray *Torpedo californica*, with densities of $0.18/m^2$, $0.06/m^2$, and $0.04/m^2$, respectively.

Trawl samples collected between 1976 and 1979 in the canyon sludge area at depths ranging from 124–183 m revealed more species of fish (mean = 17.4), greater biomass (mean = 40.7 kg), and larger individuals (mean = 58.9 g) than at control stations in San Pedro Canyon (8.3, 6.3 kg, 31.0 g) and Redondo Canyon (17.0, 28.8 kg, 38.4 g). Catches by trawl at the sludge outfall during this period were dominated by the following species of fish: Dover sole *Microstomus pacificus*, slender sole *Lyopsetta exilis*, Pacific sanddab *Citharichthys sordidus*, and white croaker *Genyonemus lineatus*. Overall, more than 40 species of fish were collected by trawl sampling at the sewage sludge outfall.

Diseased fish were frequently found in Santa Monica Bay and other areas of the Southern California Bight during the early 1970s. For example, external tumors commonly appeared in white croakers and flatfishes, and fin erosion in numerous species. Along the Palos Verdes Shelf alone, fin erosion was recorded in more than 30 fish species, but most often in Dover sole *Microstomus pacificus*, rex sole *Errex zachirus*, calico rockfish *Sebastes dalli*, and greenstriped rockfish *Sebastes elongatus*. Dover sole was the species most frequently affected by fin erosion disease along the shelf and in vicinity of the 11.2-km outfall some 20 km to the northwest. Bascom[1,7] attributed fin erosion in Dover sole and several other species with similar disease distribution patterns to sediment contaminant exposure in the bight. More specifically, he correlated fin erosion disease in Dover sole in the region of Santa Monica Bay and San Pedro Bay to the JWPCP outfall. However, the extent to which migration could be responsible for the presence of diseased Dover sole in sewage sludge and wastewater outfalls was not determined.

B. PALOS VERDES SHELF

Extensive ecological surveys of the benthic macroinvertebrate community on the Palos Verdes Shelf have been conducted by the Los Angeles County Sanitation Districts for more than 25 years. Stull et. al.[2,9] investigated benthic community structure and successional changes in proximity to submarine outfalls on the shelf. They focused on benthic successional communities and shelf sediments along a unidepth pollution gradient. Results of their work over a 10-year period (1972–1982) clearly showed declining wastewater-related impacts of the JWPCP on the benthos of the 60-m isobath. Stull[4] provided a review of two decades (1972–1992) of Palos Verdes surveys on benthic infauna, epibenthic megainvertebrates, and demersal fish.

1. Benthic Communities

Long-term wastewater discharges from the Whites Point outfalls at a depth of 60 m have historically affected extensive areas of the Palos Verdes Shelf. In the early 1970s, prior to improved source control and waste treatment practices, the environment of the Palos Verdes Shelf was significantly impacted by effluents from the JWPCP. Bottom sediments near the outfalls were degraded, being characterized by organic and sulfide enrichment, low redox potential, and high concentrations of toxic chemical contaminants, such as organochlorine compounds (e.g., DDT and PCBs) and heavy metals. Concentrations of these contaminants peaked in surface sediments (down to 20-cm depth) near the diffusers and decreased with increasing distance away from the point of discharge. Benthic communities in the impacted area had reduced species diversity, high abundances of pollution-tolerant species, missing groups of pollution-sensitive forms, and large changes in species composition over time. While organic loading tended to eliminate pollution-sensitive species, it provided an abundant food source for opportunists which commonly flourished in the organically rich sediments. Both sediment and biotic impacts were most severe at, and immediately northwest of, the outfalls.[2,9]

During the early period, polychaetes dominated the shelf benthic infaunal assemblages. Near the outfalls, species such as *Capitella* spp. and *Schistomeringos longicornis* were abundant. Many infaunal species at unimpacted sites of the shelf were conspicuously absent near the outfalls. Crustaceans and echinoderms were rare.[4]

Stull et al.[2,9] investigated benthic community structure and successional changes in proximity to the outfalls between 1972 and 1982. When outfall influences predominated (i.e., 1972–1973, 1978–1980) and when sediment conditions as measured by pH, Eh, and H_2S were most severe, pollution-tolerant polychaete worms (*Capitella* spp. and *Schistomeringos longicornis*) were most abundant. Due to relatively poor effluent quality, benthic species diversity decreased substantially in 1972–1974 and 1978–1980, although the area of reduced diversity was less extensive in 1972–1974 than in 1978–1980. Short-term settlement of large numbers of the echiuran *Listriolobus pelodes* in shelf sediments between 1974 and 1977 helped to mitigate some adverse outfall effects by aerating the sediments and reducing porewater hydrogen sulfide via its burrowing, respiratory, and feeding activities. These biological alterations of the habitat enabled colonization by a more diverse fauna, particularly in outfall-impacted regions. Hence, during the 3-year period of echiuran dominance of the community, natural events rather than the wastewater discharges directed structural changes in the benthos. The largest population of *L. pelodes* (~1500/m² with biomass of 2 kg/m²) was recorded in 1975 near the points of discharge.[9] The echiuran began a steady decline in 1976 and was rare in the area by 1978.

Discharge effects once again became the controlling influence on faunal succession after the collapse of the echiuran population in the late 1970s. Opportunistic species recolonized the newly available habitat. The polychaete *Capitella capitata* was the most numerous early invader following the decline of *L. pelodes*. The benthic community soon reverted toward former conditions after the disappearance of the echiuran, with outfall stress-gradient succession being principally dependent on the magnitude of the wastewater discharge impact. Therefore, over the 10-year study period (1972–1982) when benthic community succession was investigated at 11 sites along a unidepth pollution gradient (i.e., 60 m) extending from the outfalls, both wastewater discharges and natural biological perturbations (i.e., short-term abundance of echiuran worms) played significant roles.[2,9]

As water quality of wastewaters improved during the 1980s, benthic assemblages near the outfalls became more balanced. The numerical dominance by only a few species, evident during the early 1970s, began to change. The number of species increased, as did the dominance index. Enhanced biomass diminished. More taxa also were documented in nearshore than offshore areas of the outfall.

Stull[4] recorded a distinct pattern of species replacement between 1972 and 1992. Small deposit-feeding annelids (e.g., *Schistomeringos longicornis*), which actively rework subsurface sediments, were abundant in 1972. During the 1973–1977 period, the echiuran *Listriolobus pelodes*, the clam *Mysella tumida*, and the polychaete genera *Mediomastus*, *Prionospio*, and *Tharyx* reached high numbers. As *L. pelodes* declined in the late 1970s, the polychaetes *Capitella capitata* and *Nereis procera* and the clams *Parvilucina tenuisculpta* and *Tellina carpenteri* began to thrive. Strong El Niño conditions and storms in the early 1980s were accompanied by pulses of *Euphilomedes carcharodonta* and *Spiophanes missionensis*. High abundances of *E. carcharodonta* occurred again in the 1990s, together with locally elevated abundances of *Pectinaria californiensis*.

2. Epibenthic Macroinvertebrates

Stull[4] examined Palos Verdes epibenthic macroinvertebrate samples collected by otter trawl sampling from the inner shelf (23 m), outer shelf (61 m), and upper slope (137 m) between 1978 and 1992. Results of 10-minute otter trawls at 12 sites during this 15-year period revealed the dominance of arthropods and echinoderms (Figure 8). The dominant species included *Sicyonia ingentis* (ridgeback rock shrimp, 40% of the 15-year invertebrate catch), *Lytechinus pictus* (white sea urchin, 30%), and *Pleuroncodes planipes* (pelagic red crab, 25%). Arthropods were most abundant in the 1980s, and echinoderms, most numerous in the 1990s. Among arthropod taxa, the northern ocean shrimp *Pandalus jordani* was common in the area of Redondo Canyon from 1978–1982. The Xantus swimming crab reached high numbers near the JWPCP outfalls on the inner shelf in 1983. Along the upper slope outfall area, *P. planipes* was most abundant in 1984 and 1985 and *S. ingentis* from 1983–1986. Other abundant forms in the 1980s were *Pyromaia tuberculata* (inner shelf, 23 m) and *Mursia gaudichaudii* and *Pleurobranchaea californica* (outer shelf, 61 m). Peak arthropod abundances occurred in 1984 and 1985.

Population abundances of echinoderms were lowest in 1984 and 1985 and highest during the 1990s. Abundant forms were the California sand star *Astropecten verrilli* and the fragile sea urchin *Allocentrotus fragilis*, commonly found on the shelf and upper slope (137 m), respectively. The white sea urchin *Lytechinus pictus* dominated trawl catches on both the inner shelf (23 m) and outer shelf (61 m) and was common inshore of the outfalls.

3. Finfish Communities

a. Composition

Stull and Tang[20] analyzed 21 years (1973–1993) of demersal fish trawl data from the Palos Verdes Shelf and upper slope. Flatfish (Pleuronectiformes) and rockfish (Scorpaenidae) dominated otter trawl catches during this period, with the most abundant species being Dover sole *Microstomus pacificus*, stripetail rockfish *Sebastes saxicola*, slender sole *Eopsetta exilis*, Pacific sanddab *Citharichthys sordidus*, plainfin midshipman *Porichthys notatus*, yellowchin sculpin *Icelinus quadriseriatus*, and speckled sanddab *Citharichthys stigmaeus* (Figure 9). As in the case of benthic and epibenthic invertebrate taxa on the shelf, populations of soft-bottom demersal fish fauna exhibited considerable interannual variation in abundance, usually

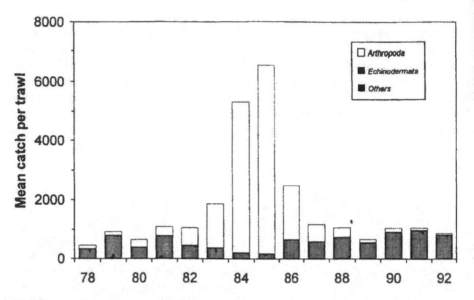

FIGURE 8   Palos Verdes epibenthic invertebrate catches, 1978–1992. Catch is the mean of quarterly 10-minute otter trawls at 12 sampling sites. (From Stull, J. K., *Bull. Southern California Acad. Sci.*, 94, 21, 1995.)

over the entire Palos Verdes study area. While temperature shifts exerted the greatest influence on fish populations through time, water depth was the primary factor controlling spatial distribution patterns. Other factors affecting fish distributions included wastewater discharges from the JWPCP outfall, sediment type and quality (grain size, concentrations of contaminants, and organic matter), and topography (shelf vs. canyon areas). Fish abundances were also habitat dependent.

Impacts of wastewater discharges on the demersal fish fauna were greatest during the early 1970s.[21] In close proximity to the outfalls, low species diversity, abundance, and biomass were recorded.[22,23] Some members of the fish community were missing from this area. In addition, high incidences of fin erosion were observed in certain species (e.g., Dover sole *Microstomus pacificus*) at sampling sites near the outfall.[20]

Several temperate and cold-temperate species were characteristic of the Palos Verdes outfall areas in the early 1970s. Most notable were the shiner perch *Cymatogaster aggregata*, Dover sole *Microstomus pacificus*, English sole *Pleuronectes vetulus*, curlfin sole *Pleuronichthys decurrens*, and white seaperch *Phanerodon furatus*. As these species decreased in the mid-1970s, others (e.g., blackbelly eelpout *Lycodopsis pacifica*, California halibut *Paralichthys californicus*, hornyhead turbot *Pleuronichthys verticalis*, and California tonguefish *Symphurus atricauda*) became more abundant near the outfalls.

A major change in the composition of demersal fish assemblages occurred across the Palos Verdes Shelf in the 1980s and 1990s, owing to a warming trend. Water temperatures in the Southern California Bight during the 1970s were several degrees cooler than during the 1980s and 1990s. Significant warming of coastal waters caused by the 1982–1983 El Niño had a dramatic effect on demersal fish populations. The

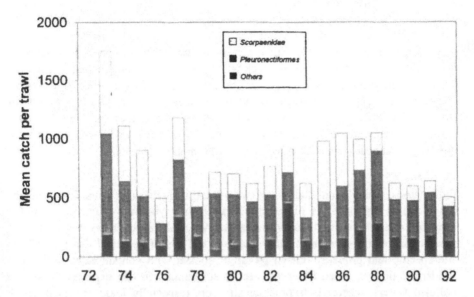

FIGURE 9   Palos Verdes demersal fish catches, 1973–1992. Catch is the mean of semiannual or quarterly 10-minute otter trawls at 12 sampling sites. (From Stull, J. K., *Bull. Southern California Acad. Sci.*, 94, 21, 1995.)

El Niño of 1983–1986 also influenced fish assemblages on the shelf. The warming trend of the 1980s and 1990s resulted in the northward movement of warm-temperate species inhabiting southerly waters. It also caused the displacement of some species farther offshore or to the north and altered recruitment patterns. Trawl catches of the following species increased during or after the 1982–1983 El Niño: Pacific hake *Merluccius productus*, fantail sole *Xystreurys liolepis*, queenfish *Seriphus politus*, roughback sculpin *Chitonotus pugettensis*, greenblotched rockfish *Sebastes rosenblatti*, bigmouth sole *Hippoglossina stomata*, hornyhead turbot *Pleuronichthys verticalis*, gulf sanddab *Citharichthys fragilis*, pink seaperch *Zalembius rosaceus*, longspine combfish *Zaniolepis latispinnis*, California tonguefish *Symphurus atricauda*, California lizardfish *Synodus lucioceps*, splitnose rockfish *Sebastes diploproa*, white croaker *Genyonemus lineatus*, plainfin midshipman *Porichthys notatus*, and yellowchin sculpin *Icelinus quadriseriatus*.

Species showed particular preferences for inner-shelf (23 m), outer-shelf (61 m), and upper-slope (137 m) areas at certain times. At the 23-m isobath, California halibut *Paralichthys californicus*, fantail sole *Xystreurys liolepis*, longspine combfish *Zaniolepis latipinnis*, and hornyhead turbot *Pleuronichthys verticalis* increased in abundance in the 1980s and 1990s, while speckled sanddab *Citharichthys stigmaeus*, English sole *Pleuronectes vetulus*, curfin sole *Pleuronichthys decurrens*, and white seapearch *Phanerodon furcatus* decreased. At 61 m, California tonguefish *Symphurus atricauda*, longspine combfish *Zaniolepis latipinnis*, hornyhead turbot *Pleuronichthys verticalis*, bigmouth sole *Hippoglossina stomata*, and longfin sanddab *Citharichthys xanthostigma* increased. In contrast, stripetail rockfish *Sebastes saxicola*, white croaker *Genyonemus lineatus*, calico rockfish *Sebastes dalli*, and shiner perch *Cymatogaster aggregata* decreased. At 137 m, increases in abundance of slender sole *Eopsetta exilis*, plainfin midshipman *Porichthys notatus*, blackbelly eelpout *Lycodopsis pacifica*, greenblotched rockfish *Sebastes rosenblatti*, and shortspine combfish *Zaniolepis frenata* occurred in the 1980s and 1990s; Dover sole *Microstomus pacificus*, stripetail rockfish *Sebastes saxicola*, splitnose rockfish *Sebastes diploproa*, rex sole *Errex zachirus*, and sablefish *Anoplopoma fimbria* decreased at this time. Although not all of the factors causing the fluctuations in catches of the aforementioned species have been determined, water temperature, water depth, and reduced mass emissions of suspended solids and contaminants from the JWPCP appear to have been of paramount importance.

b. Diseases

Fin erosion and epidermal tumors are two predominant diseases found in fish inhabiting polluted environments. Fin erosion or fin rot is a syndrome coupled to degraded estuarine and coastal marine environments. When observed, it is often most pronounced in bottom-dwelling fish. Damage to the fins of demersal fish typically is site specific and probably related to direct contact with contaminated sediments. Epidermal tumors, when observed, are most common in fish exposed to heavily polluted waters, where environmental stressors (especially toxic chemicals, high microbial populations, and low dissolved oxygen) are present.[23] Despite the strong association between environmental contaminants and the occurrence of fin erosion and epidermal tumors in susceptible fish, the specific causal factors have largely been elusive.

A high incidence of fin erosion was evident in fish exposed to wastewater discharges in the Southern California Bight during the early 1970s. Nearly half of the 72 species of fish collected off Palos Verdes Peninsula at this time exhibited fin erosion.[24] Between 1971 and 1983, more than 15,000 demersal fish (29 species) sampled along the Palos Verdes Shelf — representing 8.5% of all individuals collected — had fin erosion. The incidence of epidermal lesions in these fish was much lower, occurring in 0.3% of all individuals and in 12% of all species collected. These two external diseases were most prevalent in Dover sole *Microstomus pacificus*. Dorsal and anal fins in frequent contact with sediment displayed most fin erosion effects.[20] This species, which comprised 23% of the fish collected, accounted for 93% of all the individuals with epidermal tumors and 89% of the individuals with fin erosion.[25] Both diseases appeared to be greater in fish near the JWPCP wastewater outfalls and were uncommon at more distant sites.[4] Cross[25] correlated the occurrence of these diseases to sediment contamination by the effluent flows.

With improved water quality and the upgrade of general environmental conditions in the Southern California Bight in the 1980s and 1990s, fin erosion and epidermal lesions in demersal fish declined markedly.[4,20] Both diseases disappeared first at more distant sites from the outfalls, and subsequently nearer the outfalls. In addition, the severity of fin rot syndrome diminished steadily into the mid-1980s. During the early 1980s, minor fin anomalies were observed; typically, only a small section of the mid-dorsal fin of Dover sole was afflicted at this time.[4] Fin erosion in small Dover sole (<9 cm) has not been documented since 1980, and in larger Dover sole (>14 cm), not since the mid-1980s. Epidermal pseudotumors, however, are more frequent (<5%) near wastewater outfalls than elsewhere on the Palos Verdes Shelf, but have also been reported in flatfish far removed from the discharges.[20]

C. SAN PEDRO SHELF

Maurer et al.[26] provided an overview of benthic infaunal sampling conducted on the San Pedro Shelf between 1985 and 1989. During this sampling period, the number of species, abundance, and biomass of infauna, as well as the total organic carbon were determined at each sampling site. Species, abundance, and biomass curves (SAB) for the CSDOC ocean outfall were then compared to those of the Pearson-Rosenberg model (PRM) which focuses on the response of soft-bottom benthic infauna along a gradient of organic enrichment.[17]

The San Pedro Shelf is characterized by rich benthic infaunal assemblages. Approximately 1200 species of soft-bottom infauna have been chronicled in the San Pedro Shelf compared to about 4500 species across the entire California shelf. The mean number of species found at sampling sites in the San Pedro Shelf from 1985–1989 ranged from 46–106. Mean abundance ranged from $3060/m^2$ to $12,700/m^2$, and mean biomass from 43 g/m^2 to 89 g/m^2. The polychaete *Capitella capitata*, the ostracod *Euphilomedes carcharodonta*, and the bivalve *Parvilucina tenuisculpta* were the dominant (abundance and biomass) forms. These three species can greatly influence SAB curves.

The PRM represents a major paradigm for marine benthic pollution ecology.[27] It predicts a slight increase in the number of species as an organic pollutant source is approached and a rapid increase in biomass. Closer to the pollutant source, the number of species and biomass decrease. Abundance increases as the number of species begins to decline. Similarities between the PRM and the test SAB curves for the CSDOC ocean outfall involve enhancement in benthic infaunal abundance and biomass approaching the pollution source. Departure of the test SAB curves from the PRM include: (1) no sharp decline in SAB curves to azoic conditions; (2) displacement of SAB curves away from the source of organic enrichment rather than occurrence in the immediate vicinity; and (3) dominant (abundant) species did not exclude or eliminate rare species. The PRM may more accurately reflect organic pollutant conditions in semi-enclosed estuarine systems characterized by reduced flushing rates and higher sedimentation than open shelf environments. The San Pedro Shelf experiences a wide variety of physical phenomena that can significantly affect organic enrichment in outfall areas (e.g., El Niños, storms, upwellings, internal waves, and frontal systems).[17] These phenomena may account, in part, for the departure of test SAB curves for the San Pedro Shelf from those of the PRM.

V. CONCLUSIONS

Areas of the Southern California Bight which have been affected by municipal wastewater discharges for many years continue to show strong environmental and biological recovery. As the quality of wastewater discharges from municipal treatment plants improved between 1972 and today, the areal extent of organic enrichment and sediment contamination in coastal habitats decreased. During the early 1970s, the temporal patterns of disease prevalence in fish, body burdens of contaminants in various biota, high mass emissions of contaminants from municipal treatment plants, and contamination of extensive areas of shelf sediments in the Southern California Bight suggested a direct coupling to wastewater and sewage sludge disposal. Most of these problems have been ameliorated through time due to stringent source control and land disposal of biosolids, advanced primary treatment, partial secondary treatment, and improved solids handling (e.g., centrifuges), all of which have decreased the solid mass emission rates.

Water quality in shore areas in the vicinity of treatment plant outfalls has improved concomitant with the upgrade in effluent quality discharged by the municipal facilities. For example, long-term monitoring of water quality near outfall areas shows that coliform bacterial counts have dropped despite increased flows and waste loads. Daily monitoring of the shoreline and weekly testing of nearshore water for total coliform bacteria in Santa Monica Bay have yielded a large bacteriological database useful for gauging water quality and identifying onshore activities that can adversely affect coastal waters.[28]

Although there have been substantial reductions in mass emissions and chemical contaminants from municipal wastewater treatment plants discharging to the Southern California Bight over the past 25 years, some problems persist in outfall areas. Venkatesan and Kaplan[29] estimated that the wastewater treatment plants release into the Southern California Bight 260 mt/yr of fecal sterols and 5×10^4 mt/yr of sewage carbon. It is necessary to continue monitoring the distribution and fate of this waste and any associated contaminants, as well as their potential impact on the local environment.

Some California shelf biota still carry body burdens of DDT and PCBs, and the concentrations of these contaminants tend to be higher in biota nearer the wastewater outfalls. The source of these contaminants is bottom sediment on the shelf which serves as a reservoir for the compounds.[30] PCBs have not been detected in plant effluents for many years, and DDT concentrations are very near detection limits.[4] Nevertheless, the great persistence of residues of these contaminants in tissues of shelf biota and their potential effect on the food web requires continued monitoring of shelf habitats (Janet Stull, County Sanitation Districts of Los Angeles County, personal communication, 1996). Over the years, discharges from the JWPCP outfall system have been the principal source of DDT wastes to the Southern California Bight, whereas PCB inputs are more generally distributed throughout the region from multiple sources.[30]

Stull et al.[31] reported that sediment cores from the Palos Verdes Shelf exhibit temporal and spatial patterns in organic matter, chlorinated hydrocarbon, and heavy metal distributions that are clearly linked to wastewater discharges from municipal treatment plants. These cores provide a chronological record of waste flow. They reflect the rising urbanization and industrialization of Los Angeles County from 1937 to the 1960s, followed by decreased emissions of solids and trace contaminants from the 1970s to the 1990s. During the past several decades, the Palos Verdes Shelf has been an area characterized by elevated levels of certain contaminants — such as DDT and PCBs — compared to those of other systems (Figures 10 and 11). However, as less contaminated wastewater particulate matter and industrial wastes are deposited and compacted on the shelf, the historical sediment reservoir with high chemical contaminant concentrations should be even farther removed from the surface of the seafloor and the biotic communities inhabiting the area. Hence, contaminant impacts of these persistent chemicals are expected to decline into the next century.

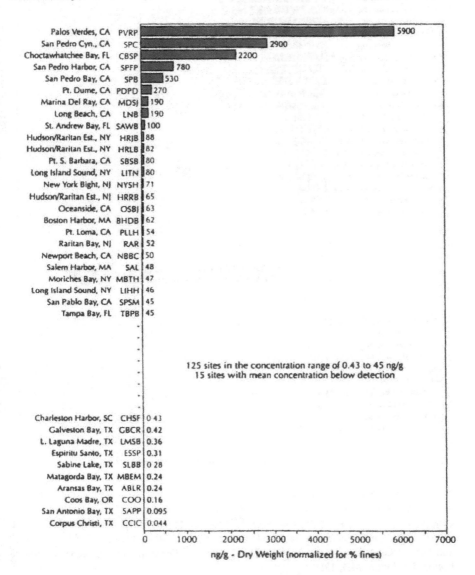

FIGURE 10 Total DDT in bottom sediments of various estuarine and coastal marine systems of the United States during the mid1980s. Note elevated concentrations at Palos Verdes relative to the other systems. (From NOAA, A Summary of Data on Individual Organic Contaminants in Sediments Collected During 1984, 1985, 1986, and 1987, NOAA Tech. Mem. NOS OMA 47, National Oceanic and Atmospheric Administration, Rockville, MD, 1989.)

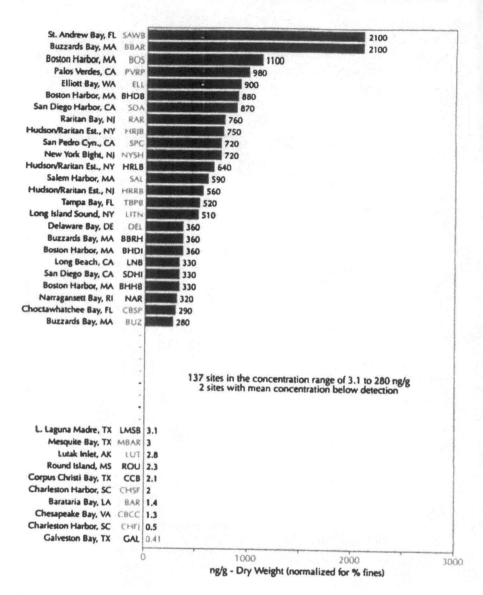

FIGURE 11 Total PCB concentrations in bottom sediments of various estuarine and coastal marine systems of the United States during the 1984–1987 survey period. Note elevated concentrations at Palos Verdes relative to the other systems. (From NOAA, A Summary of Data on Individual Organic Contaminants in Sediments Collected During 1984, 1985, 1986, and 1987, NOAA Tech. Mem. NOS OMA 47, National Oceanic and Atmospheric Administration, Rockville, MD, 1989.)

REFERENCES

1. Bascom, W., The effects of waste disposal on the coastal waters of Southern California, *Environ. Sci. Technol.*, 16, 226A, 1982.
2. Stull, J. K., Haydock, C. I., and Montagne, D. E., Effects of *Listriolobus pelodes* (Echiura) on coastal shelf benthic communities and sediments modified by a major California wastewater discharge, *Est. Coastal Shelf Sci.*, 22, 1, 1986.
3. Dorsey, J. H., Phillips, C. A., Dalkey, A., Roney, J. D., and Deets, G. B., Changes in assemblages of infaunal organisms around wastewater outfalls in Santa Monica Bay, California, *Bull. Southern California Acad. Sci.*, 94, 46, 1995.
4. Stull, J. K., Two decades of marine biological monitoring, Palos Verdes, California, 1972 to 1992, *Bull. Southern California Acad. Sci.*, 94, 21, 1995.
5. Raco-Rands, V. E., Characteristics of effluents from large municipal wastewater treatment facilities in 1994, in *Southern California Coastal Water Research Project*, Annual Report 1994–95, Southern California Coastal Research Project, Long Beach, CA, 1995, 10.
6. Raco-Rands, V. E., Characteristics of effluents from small municipal wastewater treatment facilities in 1994, in *Southern California Coastal Water Research Project*, Annual Report 1994–95, Southern California Coastal Research Project, Long Beach, CA, 1995, 21.
7. Bascom, W., The effects of sludge disposal in Santa Monica Bay, in *Impact of Marine Pollution on Society*, Tippie, V. K. and Kester, D. R., Eds., Praeger, New York, 1982, 217.
8. Bascom, W. N., Disposal of sewage sludge via ocean outfalls, in *Oceanic Processes in Marine Pollution*, Vol. 3, *Marine Waste Management: Science and Policy*, Camp, M. A. and Park, P. K., Eds., Robert E. Krieger Publishing, Malabar, FL, 1989, 25.
9. Stull, J. K., Haydock, C. I., Smith, R. W., and Montague, D. E., Long-term changes in the benthic community on the coastal shelf of Palos Verdes, Southern California, *Mar. Biol.*, 91, 539, 1986.
10. Moshiri, R., Luthy, R. F., Jr., and Hansen, B. E., Upgrading a large treatment plant — problems and solutions, *J. Wat. Pollut. Control Fed.*, 54, 1270, 1982.
11. Maurer, D., Robertson, G., and Haydock, I., Coefficient of pollution (p): the Southern California Shelf and some ocean outfalls, *Mar. Pollut. Bull.*, 22, 141, 1991.
12. National Research Council, Managing Wastewater in Coastal Urban Areas, National Academy Press, Washington, D.C., 1993.
13. Thompson, B. E., Recovery of Santa Monica Bay from Sludge Discharge, Southern California Coastal Water Research Project, Tech. Rept. No. 349, Long Beach, CA, 1991.
14. Anderson, J. W., Reish, D. J., Spies, R. B., Brady, M. E., and Segelhorst, E. W., Human impacts, in *Ecology of the Southern California Bight*, Daily, M. E., Reish, D. J., and Anderson, J. W., Eds., University of California Press, Berkeley, CA, 1993, 682.
15. Daily, M. D., Reish, D. J., and Anderson, J. W., Ed., *Ecology of the Southern California Bight*, University of California Press, Berkeley, CA, 1993.
16. Hartman, O., Contributions to a Biological Survey of Santa Monica Bay, California, Tech. Rept., Geology Department, University of Southern California, Los Angeles, CA, 1956.
17. Pearson, T. H. and Rosenberg, R., Macrobenthic succession in relation to organic enrichment and pollution of the marine environment, *Ann. Rev. Oceanogr. Mar. Biol.*, 16, 229, 1978.

18. Bascom, W., Mearns, A. J., and Word, J. Q., Establishing boundaries between normal, changed, and degraded areas, in *Coastal Water Research Project, Annual Report for the Year 1978*, Bascom, W., Ed., Southern California Coastal Water Research Project, Long Beach, CA, 1978, 81.

19. City of Los Angeles — Environmental Monitoring Division, Marine Monitoring in Santa Monica Bay: Annual Assessment Report for the Period July 1992 through June 1993, Department of Public Works, Bureau of Sanitation, Los Angeles, CA, 1994.

20. Stull, J. K. and Tang, C-L., Demersal fish trawls off Palos Verdes, Southern California, 1973–1993, *Calif. Coop. Oceanic Fish. Invest. Rept.*, 37, 211, 1996.

21. Mearns, A. J., Allen, M. J., Word, L. S., Word, J. Q., Greene, C. S., Sherwood, M. J., and Myers, B., Quantitative Responses of Demersal Fish and Benthic Invertebrate Communities to Coastal Municipal Wastewater Discharges, Final Report to U.S. Environmental Protection Agency, National Marine Water Quality Laboratory, Southern California Coastal Water Research Project, El Segundo, CA, 1976.

22. Allen, M. J., Pollution-related alterations of Southern California demersal fish communities, *Am. Fish. Soc., Cal-Neva Wildl. Trans.*, 1977, 103, 1977.

23. Sindermann, C. J., *Ocean Pollution: Effects on Living Resources and Humans*, CRC Press, Boca Raton, FL, 1996.

24. Southern California Coastal Water Resource Project, The Ecology of the Southern California Bight: Implications for Water Quality Management, Ref. No. SCCWRP TR 104, Southern California Coastal Water Research Project, El Segundo, CA, 1973.

25. Cross, J. N., Fin erosion and epidermal tumors in demersal fish from Southern California, in *Oceanic Processes in Marine Pollution*, Vol. 5, *Urban Wastes in Coastal Marine Environments*, Wolfe, D. A. and O'Connor, T. P., Eds., Robert E. Krieger, Malabar, FL, 1988, 57.

26. Maurer, D., Robertson, G., and Gerlinger, T., San Pedro Shelf California: testing the Pearson-Rosenberg Model (PRM), *Mar. Environ. Res.*, 35, 303, 1993.

27. Simboura, N., Zenetos, A., Panayotidis, P., and Makra, A., Changes in benthic community structure along an environmental pollution gradient, *Mar. Pollut. Bull.*, 30, 470, 1995.

28. Garber, W. F. and Wada, F. F., Water quality in Santa Monica, as indicated by measurements, in *Oceanic Processes in Marine Pollution*, Vol. 5, *Urban Wastes in Coastal Marine Environments*, Wolfe, D. A. and O'Connor, T. P., Eds., Robert E. Krieger, Malabar, FL, 1988, 49.

29. Venkatesan, M. I. and Kaplan, I. R., Sedimentary coprostanol as an index of sewage addition in Santa Monica Basin, Southern California, *Environ. Sci. Technol.*, 24, 208, 1990.

30. Young, D. R., Gossett, R. W., and Heesen, T. C., Persistence of chlorinated hydrocarbon contamination in a California marine ecosystem, in *Oceanic Processes in Marine Pollution*, Vol. 5, *Urban Wastes in Coastal Marine Environments*, Wolfe, D. A. and O'Connor, T. P., Eds., Robert E. Krieger, Malabar, FL, 1988, 49.

31. Stull, J., Baird, R., and Heezen, T., Relationship between declining discharges of municipal wastewater contaminants and marine sediment core profiles, in *Oceanic Processes in Marine Pollution*, Vol. 5, *Urban Wastes in Coastal Marine Environments*, Wolfe, D. A. and O'Connor, T. P., Eds., Robert E. Krieger, Malabar, FL, 1988, 23.

5 Case Study 4: San Francisco Bay

I. INTRODUCTION

Covering an area of approximately 1125 km² and extending for more than 100 km along the north-central coastline of California, San Francisco Bay is the largest estuary in the western United States. It is a highly urbanized system bordered by several major cities (e.g., Oakland, Palo Alto, San Jose, and San Francisco). A population of approximately six million people surrounds the bay and greatly influences its environment.

San Francisco Bay is located at the mouth of the Sacramento-San Joaquin river system, and it receives runoff from an extensive drainage basin (~153,000 km²) representing 40% of California's surface area (Figure 1).[1] Secondary flows enter via the Petaluma and Napa Rivers which drain major wine growing areas. A complex delta characterized by interconnected embayments, sloughs, marshes, channels, and rivers forms the easternmost boundary of the estuary. It originates from the confluence of the Sacramento River on the north and the San Joaquin River on the south. Suisun and San Pablo Bays lie directly west of the delta, while Central and South San Francisco Bays trend northwest-southeast.

Based on hydrographic and geographic characteristics, the bay can be subdivided into three distinct areas: (1) the northern reach consisting of Suisun Bay and San Pablo Bay; (2) Central San Francisco Bay; and (3) South San Francisco Bay. From the confluence of the Sacramento and San Joaquin Rivers to the Golden Gate Bridge, salinity ranges from approximately 0-32 practical salinity units (PSU). In the northern bays, the Sacramento and San Joaquin Rivers are the major sources of nutrients and freshwater inflow. Central San Francisco Bay connects the Pacific Ocean with South San Francisco Bay, a lagoonal system with a salinity range of about 26–30 PSU.[2] South San Francisco Bay receives negligable natural freshwater inputs (1×10^{11} m³/yr) relative to that of the entire system (2.1×10^{16} m³/yr). Wastewater discharges in the southern reach (5×10^{11} m³/yr) exceed natural freshwater inflow and, therefore, have greater impact on this area of the estuary than elsewhere.[3]

Suisun, San Pablo, Central, and South San Francisco Bays are all shallow, with a mean depth of 6 m at mean low water. Deepest areas consist of narrow (natural and dredged) channels that typically incise the bays to depths of 10–20 m, but reach 27 m and 110 m at Carquinez Strait and the Golden Gate, respectively.[4] The northern reach of the estuary — Suisun, San Pablo, and Central San Francisco Bays — is partially mixed. In contrast, South San Francisco Bay is a tidally oscillating lagoon with low freshwater inflow and a long residence time.[5] Strong bottom water inflow occurs at Central San Francisco Bay, and waters south of San Francisco are usually well mixed.[6]

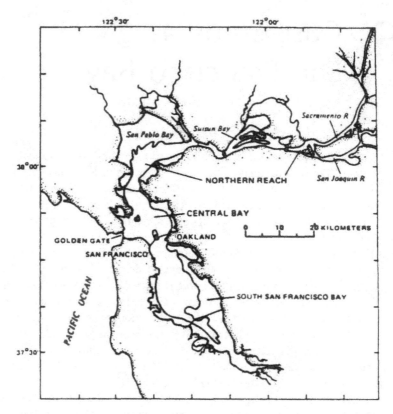

FIGURE 1   Map of San Francisco Bay showing the south, central, and northern segments as well as the delta region. (Modified from Walters, R. A. and Gartner, J. W., *Est. Coastal Shelf Sci.*, 21, 17, 1985.)

Anthropogenic activities have played a major role in altering the ecology of San Francisco Bay during the 20th Century. The bay is more heavily impacted by anthropogenic activity than most estuaries because of the long history of urban and industrial development and mining along its shores and a myriad of human-induced stresses in watershed areas upstream of the mouth of influent systems. The Suisun Bay/delta has been identified as one of the most susceptible U.S. estuarine areas to pollution effects based on its physical characteristics and pollution inputs from the general population, industry, and agriculture (Table 1).

The introduction of exotic plants and animals, some of which are pest species, has dramatically altered the composition of estuarine communities. The diversion of water for agricultural, municipal, and industrial uses also has had significant ecological consequences. Land reclamation has substantially reduced the area of freshwater and saltwater marshland. Specifically, diking and filling of wetlands bordering the bay have destroyed habitat for fish and waterfowl, thereby exacerbating the impact of low freshwater inflow. Only about 125 km^2 of undiked marsh remains from an original tidal marsh area of 2200 km^2.[7]

Runoff from urban and agricultural lands and wastewaters from municipal and industrial outfalls have transported an array of chemical contaminants for years to the estuarine system. The Sacramento and San Joaquin Rivers receive substantial amounts of agricultural chemicals (herbicides, pesticides, etc.) in runoff from the Central Valley of California, and they discharge them into San Francisco Bay. A wide range of halogenated hydrocarbon compounds has been found in water, sediment, and biotic samples from the estuary.[8] In addition, polycyclic aromatic hydrocarbons originating from petrogenic, pyrogenic, and urban sources, occur throughout the system.

Heavy metals represent a potentially significant source of biological stress in the bay, being derived from municipal and industrial point discharges as well as nonpoint source runoff. However, most modern anthropogenic inputs of heavy metals originate from sources on the estuary, rather than from riverine influx.[9] Luoma and Phillips[9] and Luoma et al.[10] report very high concentrations of heavy metals in benthic biota in several reaches of the bay. Nutrient concentrations are periodically high, with waste-derived nutrients more evident in South San Francisco Bay, where storm drains and waste treatment plants collect and release large quantities of nutrients.[1] To set the problem in perspective, 50 municipal waste treatment plants and 18 industrial dischargers released approximately 2.9×10^9 and 4×10^5 l/d of effluent, respectively, into the bay during the 1980s.[9] This waste input delivered high concentrations of nutrients, heavy metals, and other contaminants into the estuary. As a result of municipal and industrial wastewater discharges as well as other intense and widely diverse human activities, San Francisco Bay is now one of the most heavily modified major estuarine systems in the United States.[11]

II. PRINCIPAL ANTHROPOGENIC IMPACTS

A. Freshwater Diversion

There is considerable seasonal variation in freshwater inflow into San Francisco Bay; maximum levels occur in winter ($\sim$12,000 m³/s) when precipitation and runoff peak, and minimum levels take place in the summer and fall (100–300 m³/s) when dry conditions predominate. As a result, the residence time of water in northern San Francisco Bay ranges from about 1 d during the period of peak winter freshwater inflow to approximately 2 months during the time of restricted summer inflow.[12] Protracted drought conditions, such as during the 1986–1992 period, greatly reduce riverine inputs.[2] Because of the pressing need for reliable water supplies, state and federal agencies have constructed canals, dams, and reservoirs to increase storage capacity. Reservoir releases provide the principal source of freshwater inflow into the bay during the summer months. Hence, water management in the central and northern regions of California greatly affects biological communities in the estuary.

The interaction of freshwater inflow and tidal exchange modulates the salinity regime of the bay. As noted previously, salinity ranges from near 0 PSU at the delta to about 32 PSU at the Golden Gate Bridge. A strong gradient from salt to freshwater is evident when proceeding from San Pablo Bay to Suisun Bay. Periods of high river inflow, most conspicuous during the winter months, cause an increase in nontidal

TABLE 1
Summary of Pollutant Concentration Susceptibility in Estuaries

Most Susceptible Systems	Least Susceptible Systems
General population	General population
Brazos River**	Willapa Bay**
Ten Thousand Islands	St. Catherines/Sapelo Sound**
San Pedro Bay**	Penobscot Bay**
North-South Santee Rivers**	Humboldt Bay**
Galveston Bay**	Broad River**
Suisun Bay**	Hood Canal**
Sabine Lake**	Coos Bay**
St. Johns River*	Casco Bay**
Apalachicola Bay	Grays Harbor
San Antonio Bay**	Chincoteague Bay*
Connecticut River*	Bogue Sound**
Great South Bay*	St. Andrew/St. Simons Sound*
Merrimack River	Sheepscot Bay*
Atchafalaya/Vermillion Bays*	Apalachee Bay
Matagorda Bay*	Rappahannock River
Heavy industry	St. Helena Sound
Brazos River**	Puget Sound*
North/South Santee Rivers**	Heavy industry
Galveston Bay**	St. Catherines/Sapelo Sound**
Sabine Lake**	Hood Canal**
San Pedro Bay**	Penobscot Bay**
Connecticut River*	Casco Bay**
Calcasieu Lake	Humboldt Bay**
Hudson River/Raritan Bay	Buzzards Bay
Charleston Harbor	Boston Bay
Perdido Bay	Coos Bay**
Potomac River	Broad River**
San Antonio Bay**	Willapa Bay**
Mobile Bay	Bogue Sound**
Suisun Bay**	Puget Sound*
Great South Bay*	Narragansett Bay
Baffin Bay	Santa Monica Bay
Chesapeake Bay	Saco Bay
Agricultural activities	St. Andrew/St. Simons Sound*
Brazos River**	Agricultural activities
Suisun Bay**	Humboldt Bay**
North/South Santee Rivers**	Hood Canal**
St. Johns River*	Penobscot Bay**
Matagorda Bay*	Coos Bay**
Atchafalaya/Vermillion Bays*	St. Catherines/Sapelo Sound**
San Pedro Bay**	Chincoteague Bay*
Sabine Lake**	Bogue Sound**

TABLE 1 (CONTINUED)
Summary of Pollutant Concentration Susceptibility in Estuaries

Most Susceptible Systems	Least Susceptible Systems
Corpus Christi Bay	Long Island Sound
Galveston Bay**	Casco Bay**
San Antonio Bay**	Willapa Bay**
Winyah Bay	Broad River**
Albemarle Sound	Sheepscot Bay*
Neuse River	Klamath River
Laguna Madre	

Note: Suisun Bay among the most susceptible systems.

*,** Systems that are present in all three categories are marked with two asterisks; systems present in two categories are marked by one asterisk.

From Biggs, R. B., DeMoss, T. B., Carter, M. M., and Beasley, E. L., *Rev. Aquat. Sci.*, 1, 203, 1989.

---

currents that promote greater exchange of water between the semienclosed embayment of South San Francisco Bay and Central San Francisco Bay. Concomitantly, the water residence time in South San Francisco Bay decreases substantially following episodes of high river discharges.

Changes in river-driven circulation appear to affect the concentrations of some contaminants in South San Francisco Bay. For example, the levels of heavy metals (e.g., copper and silver) in benthic organisms near a waste outfall have been shown to decrease subsequent to winter floods.[12] Changes in circulation and salinity accompanying pulses of freshwater inflow may also affect the biological availability of other contaminants.

Freshwater inflow to San Francisco Bay is a key factor influencing the abundance and distribution of estuarine organisms. Consumptive uses and diversions of freshwater in and upstream from the delta for agricultural irrigation have gradually reduced the amount of freshwater reaching the bay. While natural flow through the estuary should average about 34 km^3/yr, human uses have decreased this amount by more than 60%. Even more freshwater is expected to be diverted by the year 2000, with estimates of decreased natural flow being as high as 70%.[1]

Much of the diversion of freshwater from the Sacramento and San Joaquin Rivers and their tributaries is coupled to irrigation needs of Central Valley farmlands during the dry season (May through October). Some of the water is exported in aqueducts for agricultural and municipal consumption in southern California. Because 90% of the freshwater entering the bay passes through the delta and Carquinez Strait and only 10% originates from the local bay watershed, the large volumes of water diverted for human needs have great potential impact on biological communities in the estuary.[6]

As recounted by Nichols et al.,[1] biological communities of the estuary are affected by both the physical process of diverting water as well as the changes in the natural flow patterns. The upstream flow of water caused by pumping operations in the delta causes fish to be drawn into diversion pumps. Millions of juvenile salmon and striped bass have been lost each year due to pump entrainment. Organisms in the bay appear to be significantly impacted by reduced freshwater inflow, as demonstrated by the period of extremely low freshwater flow into the bay during the summer of 1977. At this time, the discharge of the Sacramento-San Joaquin river system dropped below 100 m/s, and phytoplankton biomass in the upper estuary decreased by 80% from normal levels. The abundance of zooplankton also declined markedly. Other notable effects ascribed to the diminished freshwater discharge included the absence of summer phytoplankton blooms in northern San Francisco Bay, low striped bass recruitment, and a general suppression of the pelagic food web.

The volume of river flow regulates stratification in the estuary which, in turn, affects phytoplankton production.[13] Peak phytoplankton production and biomass historically occur during spring in South San Francisco Bay and during summer in the northern reaches.[13] Primary production is usually higher in South Bay. Increased zooplankton and benthic grazing crop a major fraction of the phytoplankton biomass.[2] Suppression of the pelagic food web during periods of drought and extremely low riverine inflow can significantly lower the production of the entire estuarine system.

B. CHEMICAL CONTAMINANTS

San Francisco Bay, a highly modified urbanized system, receives chemical contaminant inputs from a wide array of municipal and industrial point sources, as well as nonpoint source runoff from urban and agricultural lands. It has been plagued for years by the introduction of domestic and industrial wastes from dozens of point sources, such as major municipal waste treatment facilities and industrial outfalls, and numerous smaller dischargers. However, as these point sources of contamination have become more tightly regulated, nonpoint source runoff from agricultural and urban areas has assumed greater importance. Despite improvements in waste disposal and treatment facilities, chemical contamination persists in bay sediments and organisms.

1. Nutrient Enrichment and Organic Loading

More efficient sewage treatment operations in bay watersheds have reduced nutrient inputs, concentrations of pathogens, and oxygen-consuming organic matter in the estuary. The effects of waste-derived nutrients are greater in South San Francisco Bay than elsewhere in the estuarine system because storm drains and waste treatment plants are the primary sources of freshwater inflow in the southern reach.[7] Highest nutrient levels in the estuarine system usually occur in South San Francisco Bay during the summer–fall period when runoff is low and dilution and mixing with Central San Francisco Bay water diminishes.[1] Hence, despite improvements of waste treatment facilities along the perimeter of South San Francisco Bay, waste-derived nutrients are more evident in this part of the estuary.

Today, San Francisco Bay does not exhibit symptoms of eutrophication — nuisance phytoplankton blooms and associated hypoxia or anoxia. The water column is consistently well oxygenated and typically unstratified with respect to dissolved oxygen.[14] While summer depletion of dissolved oxygen was common in South San Francisco Bay 20–30 years ago, it no longer occurs there.[15] However, localized drift macroalgal blooms have developed in some years, indicating that nutrient concentrations may be periodically high in some areas of the system. Sewage treatment facilities are critically important in preventing eutrophication of the estuary, especially South San Francisco Bay which is susceptible to nutrient accumulation in the summer months. They have reduced the concentrations of oxygen-consuming organic matter and ammonia, as well as the amount of enteric bacteria in the bay.[1]

Organic inputs to South Bay differ markedly from those to the northern bays. About 90% of the organic inputs to South Bay originate from autochthonous sources, notably phytoplankton and benthic algal production. Organic deposition is greatest in the spring. Allochthonous organic material predominates in the northern bays, with the Sacramento and San Joaquin Rivers and delta delivering as much as 75% of the organic input to these areas.[16]

2. Hydrocarbons

Among the many toxic organic compounds found in San Francisco Bay, polycyclic aromatic hydrocarbon compounds (PAHs) are particularly notable because of their potential carcinogenicity, mutagenicity, and teratogenicity to estuarine and marine organisms. PAHs vary considerably in toxicity. The low-molecular-weight PAH compounds (LMWpah) tend to be acutely toxic but noncarcinogenic to aquatic organisms. The high-molecular-weight PAH compounds (HMWpah) are less toxic; however, they have greater carcinogenic potential.[17] Although not all PAHs are potent carcinogens and mutagens, some of them become carcinogenic and mutagenic after metabolic activation.[18] Metabolized carcinogenic and mutagenic PAHs are a serious human health concern.

A major fraction of PAHs in San Francisco Bay, as in most urbanized estuaries, derives from several anthropogenic sources (e.g., oil, municipal and industrial waste-waters, combustion of fossil fuels, and urban runoff). Of secondary importance are natural PAH sources (e.g., biosynthesis and sediment diagenesis) which yield small concentrations of the compounds relative to anthropogenic inputs. Owing to their relative insolubility in seawater and strong adsorption to particulate matter, PAHs tend to concentrate in bottom sediments (Table 2). While some benthos take up sediment-bound PAHs and metabolize them rather rapidly, others (e.g., algae and mollusks) cannot metabolize the contaminants and tend to accumulate them. The bioconcentration of PAHs is highly variable in marine and estuarine organisms.

The types of PAH compounds accumulating in estuarine sediments and organisms reflect their source. For instance, the LMWpah compounds generally originate from relatively fresh, unburned petroleum. The HMWpah compounds derive principally from fossil fuel combustion. Pereira et al.[5] indicated that PAH compounds in sediments and clams (*Potamocorbula amurensis*) of Suisun Bay accumulated mainly from combustion sources. The National Status and Trends Program ranked

TABLE 2
Concentrations of PAHs in Sediments from Selected Estuaries in the United States

Estuary	Total PAHs
Casco Bay, ME	7320.00
Merrimack River, MA	1730.00
Salem Harbor, MA	10,220.00
Boston Harbor, MA	26,440.00
Buzzard's Bay, MA	1710.00
Narragansett Bay, RI	2350.00
East Long Island Sound, NY	48,560.00
West Long Island Sound, NY	8430.00
Raritan Bay, NJ	5010.00
Delaware Bay, DE	330.00
Lower Chesapeake Bay, VA	410.00
Pamlico Sound, NC	219.25
Charleston Harbor, SC	802.98
Sapelo Sound, GA	22.28
St. Johns River, FL	1926.91
Charlotte Harbor, FL	26.51
Tampa Bay, FL	27.10
Apalachicola Bay, FL	200.25
Mobile Bay, AL	96.79
Round Islands, MS	52.36
Mississippi River Delta, LA	603.41
Barataria Bay, LA	106.08
Galveston Bay, TX	68.05
San Antonio Bay, TX	8.98
Corpus Christi Bay, TX	28.17
Lower Laguna Madre, TX	0.00
San Diego Harbor, CA	5000.00
San Diego Bay, CA	0.00
Dana Point, CA	22.87
Seal Beach, CA	257.96
San Pedro Canyon, CA	527.00
Santa Monica Bay, CA	68.25
San Francisco Bay, CA	5976.03
Bodega Bay, CA	11.00
Coos Bay, OR	234.67
Columbia River Mouth, OR/WA	145.03
Nisqually Reach, WA	0.00
Commencement Bay, WA	1200.00
Elliott Bay, WA	4700.00
Lutak Inlet, AK	0.00
Nahku Bay, AK	100.00

Note: Relatively high concentrations in San Francisco Bay, parts per billion.

From NOAA, *National Status and Trends Program for Marine Environmental Quality; Progress Report and Preliminary Assessments of Findings of the Benthos Surveillance Project — 1984*, National Oceanic and Atmospheric Administration Office of Ocean Resources Conservation and Assessment, Rockville, MD, 1987.

San Francisco Bay among the highest coastal systems in the country in terms of total PAH concentrations in mussels (Table 3).

Petroleum hydrocarbons have a patchy distribution in the estuary. In some organisms, the concentrations of petroleum hydrocarbons reach levels comparable to those in highly contaminated estuaries.[1] For example, Kockelman et al.[15] reported similar concentrations of petroleum hydrocarbons in mussels from Central San Francisco Bay and those from Los Angeles and San Diego Harbors which are exposed to frequent petroleum hydrocarbon contamination. Jung et al.[19] ascertained that the levels of petroleum hydrocarbons in striped bass from the estuary are as high as those which adversely affect the hatching success of eggs in the laboratory.

3. Organochlorine Contaminants

More than 20,000 metric tons (mt) of pesticides (about 500 different varieties), representing 10% of the total pesticides used each year in U.S. agriculture, are applied annually to farmlands in the Central Valley.[20] The widespread use of herbicides, pesticides, and fertilizers in Central Valley farmlands results in the runoff of considerable quantities of chlorinated hydrocarbon contaminants into influent systems of San Francisco Bay, particularly the San Joaquin River. More than 20% of the total San Joaquin River flow consists of agricultural wastewater returned to the river in subsurface pipe drainage.[1] This wastewater contains salts and contaminants leached from the soil. The Sacramento River also delivers chlorinated hydrocarbon contaminants from farmlands into San Francisco Bay. The Colusa Basin Drain, an artificial channel that discharges to the Sacramento River at Knights Landing, directs runoff and agricultural return-flows from more than 4×10^5 ha of farmland. DDT, chlordane, Dacthal, nonachlor, and PCBs have been documented in fish, invertebrates, and sediments from the Colusa Basin Drain, and particle-bound fractions of these organochlorine compounds eventually accumulate in sediments and biota of the estuary. This contaminant pool is augmented by PCBs and DDT originating from point sources located on the shore of the San Francisco Bay and delta, as well as from atmospheric fallout.[21] As a consequence, San Francisco Bay remains a target estuary for continued monitoring of chlorinated hydrocarbon contamination in both biotic and abiotic compartments.

Investigations of chlorinated hydrocarbons in biota of San Francisco Bay have focused on PCBs and DDT. Analysis of historical datasets on organochlorine contamination in San Francisco Bay reveals several general trends. Surveys conducted during the 1960s showed a well-developed 100-fold gradient of pesticide concentration in bay water, decreasing from the San Joaquin and Sacramento Rivers through the northern bays to the mouth at the Golden Gate. Highest PCB levels occurred in fish and shellfish in South San Francisco Bay and lowest levels in San Pablo Bay (Figure 2). There is some evidence of increasing PCB concentrations in fish liver samples collected after 1976 (Table 4) and decreasing PCB concentrations in bivalve samples collected after 1981. The total PCB concentrations in fish liver samples collected from South San Francisco Bay during the mid–1980s by the National Status and Trends Program are high relative to other estuarine and coastal marine systems (Table 4).

TABLE 3
National Status and Trends Sites Ranking Among the Highest 20 in 1986, 1987, and 1988 and Overall for Total PAHs in Mussels (*Mytilus edulis, Mytilus californianus*) and Oysters (*Crassostrea virginica, Ostrea sandivicensis*)

Code	Location	State	Species	1986	1987	1988	Overall	Significant Difference?	Trend
SAWB	St. Andrew Bay	FL	cv	1	17t	v	3	no	no
EBFR	Elliott Bay	WA	me	2	13t	1	1	yes	no
HRUB	Hudson/Raritan Estuary	NY	me	3	2	v	4	yes	d
CBRP	Coos Bay	OR	me	4	v	v	9	yes	no
BHDB	Boston Harbor	MA	me	5	8	6	6	no	no
LITN	Long Island Sound	NY	me	6	10	15t	8	no	no
SGSG	Pt. St. George	OR	mc	7	v	v	16	yes	d
BHDI	Boston Harbor	MA	me	8t	4	3	5	yes	no
CBCH	Coos Bay	OR	mc	8t	v	v	13t	yes	no
BHHB	Boston Harbor	MA	me	10t	v	v	15	yes	no
PLLH	Pt. Loma	CA	mc	10t	v	v	v	yes	no
SIWP	Sinclair Inlet	WA	me	10t	—	v	19	yes	no
SCFP	Santa Cruz Island	CA	mc	13t	v	v	v	yes	no
SSSS	San Simeon Point	CA	mc	13t	v	v	v	yes	d
SCBR	S. Catalina Island	CA	mc	15	v	v	v	yes	no
BPBP	Barber's Point	HI	os	16	v	v	v	no	no
HRJB	Hudson/Raritan Estuary	NY	me	17	3	4	7	no	no
NYSH	New York Bight	NJ	me	18	v	v	v	yes	no
NYLB	New York Bight	NJ	me	19t	15	v	v	yes	d
CFBI	Cape Fear	NC	cv	19t	v	v	v	no	no
CHSF	Charleston Harbor	SC	cv	19t	v	v	v	no	no'
SDHI	San Diego Bay	CA	me	v	1	v	10	yes	no
MSBB	Mississippi Sound	MS	cv	v	5t	v	18	no	no
HHKL	Honolulu Harbor	HI	os	v	5t	v	v	yes	no

SFEM	San Francisco Bay	CA	me	—	7	13	11t	—	—
BBMB	Barataria Bay	LA	cv	v	9	v	v	no	no
CBSR	Choctawhatchee Bay	FL	cv	v	11	v	v	yes	no
BHBI	Boston Harbor	MA	me	v	12	5	13t	no	no
NYSR	New York Bight	NJ	me	v	13t	v	v	no	no
CBTP	Commencement Bay	WA	me	v	16	9t	v	no	no
BBPC	Biscayne Bay	FL	cv	—	17t	—	v	no	no
BBAR	Buzzards Bay	MA	me	v	19	17	v	yes	i
RSJC	Roanoke Sound	VA	cv	v	20	v	v	no	no
PCMP	Panama City	FL	cv	—	—	2	2	no	no
NMML	North Miami	FL	cv	—	—	7t	11t	no	no
APDB	Apalachicola Bay	FL	cv	v	v	7t	v	no	no
CBCI	Chincoteague Bay	VA	cv	v	v	9t	v	no	no
LICR	Long Island Sound	CT	me	v	v	9t	v	no	i
BBSM	Bellingham Bay	WA	me	v	v	9t	v	yes	i
DBAP	Delaware Bay	DE	cv	v	v	14	v	no	no
IRSR	Indian River	FL	cv	—	—	15t	17	—	—
UISB	Unakwit Inlet	AK	me	v	v	18	v	yes	i
LINH	Long Island Sound	CT	me	v	v	19	v	yes	i
SSBI	South Puget Sound	WA	me	v	v	20	v	yes	i
AIAC	Absecon Inlet	NJ	me	—	—	v	20	—	—

Note: cv = *Crassostrea virginica*; me = *Mytilus edulis*; mc = *Mytilus californianus*; os = *Ostrea sandivicensis*; t = two or more concentrations were equal; d = decreasing trends; i = increasing trends. Note high ranking for San Francisco Bay relative to other systems.

From NOAA, *A Summary of Data on Tissue Contamination from the First Three Years (1986–1988) of the Mussel Watch Project*, NOAA Tech. Mem. NOS OMA 49, National Oceanic and Atmospheric Administration, Rockville, MD, 1989.

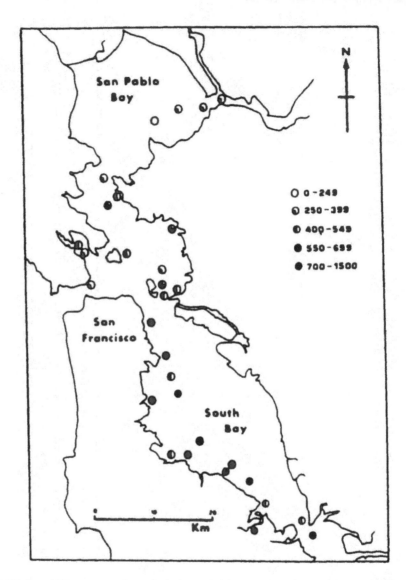

FIGURE 2   PCB concentrations (ng/g dry weight) in whole soft parts of native bay mussels (*Mytilis edulis*) in San Francisco Bay. Highest concentrations occur in South Bay and lowest concentrations in San Pablo Bay. (From Risebrough, R. W., Chapman, J. W., Okazaki, R. K., and Schmidt, T. T., Toxicants in San Francisco Bay and Estuary, Report of the Association of Bay Area Governments, Berkeley, CA, 1978.)

DDT and its metabolites have been released slowly from Central Valley soils, accounting for the persistence of moderate levels of contamination in some biota and sediments.[22] DDT contamination in fish and shellfish declined markedly throughout the bay between 1969 and 1977 and then leveled off. Mean total DDT concentrations recorded in native mussels (*Mytilus edulis*) and transplanted mussels (*M. californianus*) in the bay between 1979 and 1986 were >100 ng/g dry weight.

TABLE 4
Comparison of Average PCB Concentrations in Livers of Nearshore and Estuarine Fish from San Francisco Bay and Eight Other Areas Occupied by National Surveys 1976–1977 and in 1984

Area	tPCB (ppm wet weight)	
	1976–1977	1984
Western Long Island Sound	0.62	0.81
Lower Chesapeake Bay, VA	0.62	0.28
Duwamish River/Elliott Bay, WA	26.7	4.23
Nisqually Reach, WA	0.31	0.49
Columbia River, OR	0.24	0.20
Coos Bay, OR	(less than) 20.0	0.15
Southern San Francisco Bay, CA	0.22	1.23–2.30
Palos Verdes/San Pedro Canyon, CA	18.63	2.27
Dana Point, CA	0.07	0.38
Median	0.31	0.49
Range		
Minimum	(less than) 0.20	0.15
Maximum	26.7	4.23

Note: See original source for particular studies.

From Mearns, A. J., Matta, M. B., Simecek-Beatty, D., Buchman, M. F., Shigenaka, G., and Wert, W. A., *PCB and Chlorinated Pesticide Contamination in U.S. Fish and Shellfish: A Historical Assessment Report,* NOAA Tech. Mem. NOS OMA 39, National Oceanic and Atmospheric Administration, Seattle, WA, 1988.

At some locations, DDT concentrations in the mussels exceeded 1000 ng/g dry weight.[23-25] The highest levels of DDT occurred in samples from the Sante Fe Channel in Richmond Harbor which lies in close proximity to the former site of a pesticide formulation and packaging plant on the banks of the Lauritzen Canal. Pereira et al.,[21] assessing chlorinated hydrocarbon contamination in striped bass (livers) collected in 1992, reported total DDT concentrations as high as 396 ng/g wet weight and total chlordane as high as 42 ng/g dry weight. Dieldrin contamination has been minimal, with concentrations of the pesticide in San Francisco Bay fish and shellfish never exceeding 0.1 µg/g wet weight.[26]

Risebrough et al.[27] chronicled baywide PCB contamination in native mussels of San Francisco Bay, with lowest concentrations observed in samples from San Pablo Bay (mainly 0–400 ng/g dry weight) and highest concentrations in samples from South Bay (typically 500–1500 ng/g dry weight). Other surveys of native mussels (*Mytilus edulis*) and transplanted mussels (*M. californianus*) revealed similar PCB concentrations of 150–1000 µg/g dry weight.[20] From 1979–1986, PCB concentrations in mussels ranged from approximately 50–1800 ng/g dry weight.[23-25]

PCBs have persisted in fish populations of San Francisco Bay despite being banned from use for many years. During the mid–1980s, NOAA[28] registered high PCB levels in fish (livers) from San Pablo Bay, Southampton Shoal, and Hunter's

Point totalling 1191, 3734, and 6990 ng/g dry weight, respectively. Much lower values (total PCBs <270 ng/g wet weight) were observed by Pereira et al.[21] in striped bass (livers) in 1992. Chlorinated hydrocarbons tend to accumulate in the hepato-biliary tract and other lipid-rich tissues of striped bass, where they may generate hepatoxic effects that can be lethal. Another principal concern regarding PCB and DDT contamination of fish, in general, is the potential inhibition of the ATPase system responsible for the regulation of ion balance. At sufficiently high levels, these chlorinated organic compounds may also hinder fish adaptation to changes in salinity. Such effects may contribute substantially to the decline of populations, especially in "hot spot" locations.

Waterfowl that overwinter in the bay catchment tend to accumulate significant concentrations of DDE and PCBs. Species of note include canvasbacks (*Aythya valisineria*), lesser scaups (*Aythya affinis*), northern pintails (*Anas acuta*), and north-ern shovelers (*Anas clypeata*).[29] The eggs of some waterfowl (i.e., black-crowned night-herons, Caspian terns, Forster's terns, and snowy egrets) also accumulate DDE.[30-32] In 1982, for example, DDE concentrations in the eggs of these species ranged from 1.92–6.93 µg/g. Concentrations of PCBs in black-crowned night-heron eggs collected from the Bair Island National Wildlife Refuge at South Bay were as high as 52 µg/g. Caspian tern eggs collected from San Francisco Bay exhibited significantly higher PCB levels (4.85 µg/g) than those collected from San Diego Bay (1.70 µg/g) and Monterey Bay (1.83 µg/g).

Relatively low concentrations of total DDT (DDT + DDD = DDE) and PCBs occur in bottom sediments of the bay. The National Status and Trends Program of the National Oceanic and Atmospheric Administration (NOAA)[33] recorded concen-trations of total DDT in bottom sediments amounting to 1–6 ng/g dry weight; the U.S. Environmental Protection Agency (U.S. EPA)[22] registered values generally <10 ng/g dry weight. Similar values (0.8–9.0 ng/g dry weight) were reported in bay sediments by Pereira et al.,[21] who consider runoff from the Sacramento and San Joaquin Rivers and atmospheric deposition as the primary sources of the contami-nants. In regard to PCBs in bottom sediments, NOAA[33] documented concentrations ranging from 9–61 ng/g dry weight at four bay locations, and Pereira et al.[21] values ranging from 1.3–8.1 ng/g dry weight at 17 sites.

Tributaries to the bay appear to have higher concentrations of organochlorine compounds. For example, Law and Goerlitz[34] calculated total DDT concentrations >300 ng/g dry weight in bottom sediments of Belmont Creek. PCB concentrations in bottom sediments of San Francisquito Creek and San Rafael Creek amounted to 430 and 350 ng/g dry weight, respectively. Chlordane residues were found in 92% of the sediment samples examined from bay tributaries. Streams draining into South Bay had the highest levels of contamination, although several streams flowing into the western perimeter of San Pablo and Central San Francisco Bays also contained considerable quantities of chlorinated hydrocarbon residues in bottom sediments.

The highest concentrations of PCBs have been observed in sediments of the delta area. At Mormon Channel, Mormon Slough, and the Port of Stockton turning basin, for instance, the sum of Arochlor® 1242, 1254, and 1260 ranged from 7100 to 17,000 ng/g dry weight. Sediments are more heavily contaminated by organochlo-rine compounds along the perimeter of the estuary — particularly in areas of restricted water circulation abutting the shoreline — than in open waters. PCBs appear to be the most ubiquitous chlorinated hydrocarbon contaminants in bottom sediments of the estuary.[22]

4. Heavy Metals

San Francisco Bay receives heavy metals from multiple sources. While major influent systems deliver the largest fraction of heavy metals to the bay, direct discharges from wastewater treatment plants and industrial facilities as well as runoff from mine spoils also release considerable quantities. Upstream enrichment of heavy metals caused by runoff from mine spoils has occasionally resulted in fish kills (e.g., in the Sacramento River), and elevated levels of some heavy metals have been found in fish inhabiting streams and rivers that once received mine wastes.[9] Table 5 shows the relative importance of the different heavy metal sources in the bay. Riverine inflow is the most important source of copper (203 kg/d) and zinc (288 kg/d) loadings. Urban runoff accounts for the largest fraction of lead (250 kg/d) and a substantial amount of zinc (250 kg/d). Among the array of metals entering the system from point sources of contamination, nickel (29 kg/d), copper (31 kg/d), and zinc (74 kg/d) predominate. Approximately 6×10^6 m^3/yr of metal-enriched sediments are dredged from harbors and marinas and dumped at three dredged-spoil locations in the Central Bay and the northern reach of the estuary. In addition, numerous

---

TABLE 5

Sources of Trace Metals in San Francisco Bay, Including Municipal and Industrial Point Sources, Urban Runoff, and the Sacramento-San Joaquin River System[a]

Metal	Point Source	Urban Runoff	Rivers	Point Source Riverine[b]	Anthropogenic Riverine[c]
Ag	7.5	?	26.0	0.28	<0.28
As	5.7	9.0	37.0	0.15	0.39
Cd	4.0	3.0	27.0	0.15	0.26
Cr	14.0	15.0	92.0	0.15	0.32
Cu	31.0	59.0	203.0	0.15	0.44
Hg	0.8	0.2	3.0	0.26	0.33
Ni	29.0	?	82.0	0.35	>0.35
Pb	17.0	250.0	66.0	0.26	4.00
Se	2.5	?	7.4	0.33	>0.33
Zn	74.0	268.0	288.0	0.25	1.19

[a] Concentrations in kg/dry wt.

[b] Ratios of point source inputs and total anthropogenic input (sum of point source and urban runoff) relative to riverine input. Data from Walters, R. A. and Gartner, J. W., *Est. Coastal Shelf Sci.*, 21, 17, 1985.

[c] Data from Gunther, A. J., Davis, J. A., and Phillips, D. J. H., An Assessment of the Loading of Toxic Contaminants to the San Francisco Bay-Delta. Tech. Rep., Aquatic Habitat Institute, Richmond, CA, 1987.

---

hazardous waste disposal sites occur along the estuarine shore and may be leaching metals into the bay. Smaller local sources of heavy metals (e.g., marinas, naval bases) augment the major contaminant sources.

Certain heavy metal contamination in the San Francisco Bay is coupled to specific human activities. For example, selenium derives from discharges of oil refineries (primarily selenite) in the northern reach of the bay, as well as from the leaching of seleniferous soils on agricultural land in the San Joaquin catchment. Industrial wastewater discharges and municipal treatment plant effluents containing industrial-derived metals deliver most of the silver found in the bay. Point source loadings of silver into Suisun, San Pablo, Central, and South bays amount to 0.3–1.1 kg/d, 2.3–3.8 kg/d, 1.0–2.4 kg/d, and 0.2–7.0 kg/d, respectively.[35] Historic placer gold mines represent the main sources of mercury in the estuary, with natural cinnabar deposits in the Coast Range of the Central Valley constituting secondary sources. The Sacramento and San Joaquin Rivers are the principal influent systems for mercury.

Heavy metals in San Francisco Bay are characterized by large temporal and spatial variability. A patchy mosaic of heavy metal contamination typifies the system.[10] Seasonal differences in river inflow have dramatic effects on the magnitude of heavy metal inputs to the estuary. During the period of high flow in winter, river discharges and suspended sediment loads commonly are more than 10-fold greater than during the period of low flow in summer. The dominance of heavy metal concentrations by riverine influx occurs during high flow periods, and the dominance by local anthropogenic influences on the estuary takes place during low flow periods.[9] Within a year, heavy metal concentrations in surface sediments of intertidal and shallow subtidal areas can fluctuate as much as 10-fold.[10]

Localized anthropogenic inputs and variable physical and geochemical processes result in heterogeneous spatial distribution of heavy metals in the estuary as reflected by variable concentrations in the water column, sediments, and biota. While broad expanses of the estuary do not have high levels of heavy metals, localized areas of extreme enrichment are characteristic. For instance, patches of heavy metal contamination often lie in close proximity to municipal and industrial outfalls or marinas, such as observed in South Bay (Table 6). Where municipal and industrial discharges are elevated relative to natural river inflow (e.g., South Bay), some metals are more enriched than in other areas (e.g., the northern reach). The concentrations of many heavy metals increase at the head of the estuary in the urbanized/industrialized portion of the system. In the Central and South Bays, highest concentrations appear to occur in shallow waters along the perimeter, although less data have been collected in deeper, mid-bay areas. Bottom sediments, particularly in South Bay, may serve as a long-term source of heavy metal contamination.

As expected, the greatest impact of heavy metals on biota of the estuary is evident among benthic populations inhabiting extremely contaminated sediments. Some benthic populations have heavy metal concentrations comparable to those of benthos from the most contaminated systems known. In several reaches of San Francisco Bay, patches of very high metal concentrations have been observed in benthic biota (e.g., copper in clams, mussels, and snails) where sediments are only moderately enriched, suggesting enhanced bioavailability in these areas.[9,10,36]

TABLE 6
Range of Heavy Metal Concentrations in South Bay Compared to California Coastal and Other Estuarine Waters

Location, Date	Copper (nM)	Lead (pM)	Cadmium (pM)	Cobalt (nM)	Nickel (nM)	Silver (pM)
San Diego Bay, June 1989	14–44	120–184	541–1736	0.5–2.4	6–16	66–307
South Bay,						
Apr 1989	22–68	86–288	495–614	1.2–2.5	21–45	24–49
Aug 1989	32–73	52–308	1210–1470	0.7–5.3	34–71	70–244
Dec 1989	23–43	75–191	820–1390	0.4–1.7	27–48	25–133
California coastal waters	1.6–1.7	18–63	30–170	0.10–0.12	4.6–5.5	3–11
Estuarine waters	5.0–60	20–900	20–850	0.20–10	5–85	6–244

From Flegal, A. R. and Sanudo-Wilhelmy, S. A., *Environ. Sci. Technol.*, 27, 1934, 1993. With permission.

However, the patchy nature of heavy metal distributions in the estuary is often manifested in localized enrichment of contaminants in the benthos.

Heavy metals of greatest concern in San Francisco Bay based on the frequency or severity of contamination of biota, sediments, and the water column, as well as the magnitude of loadings from anthropogenic sources, are copper, cadmium, mercury, selenium, and silver.[9,37,38] Luoma et al.[10] uncovered chronic copper, cadmium, and chromium contamination in the bivalve *Corbicula* sp. at the mouth of the Sacramento and San Joaquin Rivers in northern San Francisco Bay. Silver is strongly bioaccumulated by benthic invertebrates. Although the concentrations of silver are typically low in benthic organisms in the northern reach, they are much higher in benthos in the South and Central Bays. Elevated silver concentrations in bottom sediments often is associated with metal-induced stress in benthic organisms of South Bay.[9,35] Between 1981 and 1987, silver enrichment was discerned in *Mytilus californianus* and *M. senhousia* in the Central Bay and lower South Bay. Extremely high concentrations of silver were found in the tellinid clam *Macoma balthica* (~200 μg/g dry weight) and the mussel *M. californianus* at an intertidal site near Palo Alto. Silver concentrations in *M. balthica* generally range from 7–117 μg/g dry weight.[9] During the 1980s, silver concentrations in *M. californianus* transplanted to the South and Central Bays ranked among the highest 15% of 358 determinations by the California State Mussel Watch Program.[39]

Silver contamination in the estuary is primarily ascribable to wastewater industrial discharges. In South Bay, silver concentrations in bottom sediments are as high as 5 μg/g dry weight and in benthic organisms at least one order of magnitude above the baseline concentrations of estuarine organisms.[10,35,40-42] Because silver is extremely toxic to many estuarine and marine organisms (e.g., microorganisms, green plants, and benthic invertebrates), high levels of contamination in benthos of South Bay are a major potential problem.

In South Bay, the concentration of biologically available heavy metals may be coupled to river-driven circulation. Winters of high freshwater inflow correspond to periods of low silver accumulation in estuarine organisms. Copper concentrations in benthos exposed to wastewater discharges in South Bay also decrease subsequent to periods of heavy precipitation and flood conditions.[12] The probability of death in estuarine organisms due to metal stress and toxicity increases substantially during periods of low river inflow in summer.

Selenium is another heavy metal responsible for serious biological impacts at high concentrations. Although an essential element for animals and some algae, selenium is toxic at elevated concentrations and may be teratogenetic and carcinogenetic to higher animals.[43] It also causes reproductive failure in animals at excessively high levels.[44] Selenium contamination of biota in the San Francisco Bay system is associated with pollution of subsurface agricultural drainage waters. Adverse effects of selenium were first noted in 1983 at the Kesterson National Wildlife Refuge in the San Joaquin Valley, where a mean selenium concentration of 73 mg/kg dry weight was recorded in submerged aquatic vegetation that served as forage for ducks on agricultural drainwater ponds. Levels of selenium in aquatic insects averaged 100 mg/kg dry weight in 1983 and more than 60 mg/kg in 1984.[45] In addition, the amount of selenium in invertebrate prey animals of aquatic birds, wildfowl, and waders was high, ranging from 20–218 mg/kg dry weight compared to a range of 1–3 mg/kg dry weight in control animals.[46] Biological and physiological effects of selenium contamination in wild birds included deformities, impaired reproduction, histopathological lesions, teratogenesis, and increased mortality.[45] Aside from selenium contamination of aquatic insects, invertebrates, and birds, elevated concentrations of selenium were found to be nearly 100 times higher in the Kesterson National Wildlife Refuge area than at a nearby control site.[47] Selenium contamination in fish may also be localized. For example, Saiki and Palawski[48] revealed that selenium levels in striped bass (*Morone saxatilis*) from the estuary proper were only one-fourth to one-half the levels measured in the most contaminated fish from the San Joaquin River.

III. OTHER ANTHROPOGENIC IMPACTS

A. INTRODUCED SPECIES

Some significant changes in biotic communities of San Francisco Bay are unrelated to chemical contamination but to other human activities, such as the introduction of plant and animal species which commenced more than an century ago. A number of East Coast invertebrates became established in the estuary when unintentionally shipped with American oysters (*Crassostrea virginica*) to the bay area during the latter half of the 19th Century. Recreationally and commercially important shellfish introduced at this time included the soft-shelled clam *Mya arenaria* and the Japanese little-neck clam *Tapes japonica*. Less desirable invertebrates that became established in the bay were the oyster drill *Urosalpinx cinerea* and the shipworm *Teredo navalis*. Damage to wooden structures, boats, bridges, and piers by shipworms soon developed into a major crises around the estuary. In total, more than 100 invertebrate species were introduced into San Francisco Bay, and many of them have become remarkably successful to the point of upsetting the ecological balance of the system. Today, most of the common macroinvertebrates along the inner shallows of the bay are introduced species. In addition, some planktonic species and algae are also introduced forms.

Numerous fish species have likewise been introduced into San Francisco Bay, the most notable being the striped bass (*Morone saxatilis*). Introduced from the East Coast of the United States in 1879, the striped bass now constitutes the primary sports fishery of the bay.[7] However, many other introduced species have dramatically altered the species composition of fish communities in different regions of the bay. For example, 20 of the 42 fish species inhabiting the sloughs of Suisun Bay marshes are introduced forms. The modern biological communities in both the northern and southern reaches of the bay consist of a large fraction of hardy, adaptable, temporally variable, and spatially patchy introduced populations.[9,49] It is necessary to understand the biology of these populations if human impacts on the communities are to be properly assessed.

B. LAND RECLAMATION

Nichols et al.[1] discussed the problem of land reclamation along San Francisco Bay. Prior to 1850, tidal marsh covered a 2200 km² area — 800 km² of saltwater marsh fringing the shoreline of the bay and 1400 km² of freshwater marsh surrounding the confluence of the Sacramento and San Joaquin Rivers. To create new farmland (especially in the delta area), tidal marshes were increasingly diked. Reclamation of marshlands for residential and industrial development accelerated rapidly during the late 1800s and early 1900s. While reclamation of saltwater marshes continued into the early 1970s, filling of freshwater marshlands in the delta was essentially complete by the early 1920s. Only 125 km² of undiked marsh still exist along the perimeter of the estuary, and hence, there is considerable speculation regarding the impact of marshland elimination on the overall ecology of the bay. Tidal marshlands are ecologically and economically important to San Francisco Bay because they serve as valuable nursery grounds for recreationally important fish. In addition, birds and wildlife populations utilize the marsh habitat. For instance, about one-half of the migratory birds on the Pacific Flyway overwinter on or near San Francisco Bay.[50] The loss of marsh as food and habitat can have a detrimental impact on waterfowl and other populations residing along the margins of the bay. Since tidal marshes also play an important role as a source or sink of nutrients and organic matter, their losses may have a significant effect on productivity of the system.

C. URBAN RUNOFF

Much research on the San Francisco Bay ecosystem has focused on point-source pollution. For example, elevated concentrations of copper and nickel in saline waters of South Bay have been positively correlated to wastewater discharges from municipal (2.4×10^{11} l/yr) and industrial (6.9×10^9 l/yr) facilities.[51] Less work has been conducted on nonpoint-source pollution. However, urban runoff can cause significant receiving-water impacts on aquatic life. The effects are usually very site specific.

Pitt[52] examined the receiving-water impacts of urban runoff in Coyote Creek, near San Jose. The Coyote Creek watershed is approximately 70 km long and 15 km wide, with about 80,000 ha of land area (Figure 3). Coyote Creek flows northwesterly along the western edge of the watershed, intersects with Silver Creek, and ultimately drains into the south terminus of San Francisco Bay. Developed urban areas concentrated in the northwest sector cover nearly 15% of the watershed. There are considerable differences in the water quality of Coyote Creek in the urban vs. nonurban areas (Table 7). Pitt[52] has indicated that stormwater is responsible for the high levels of many toxic substances in the receiving waters and sediments along developed areas of the creek. An average of 0.6–3.0 storm-drain outfalls per kilometer occur along the urban reach of the creek that he studied.

Pitt[52] found distinct differences in the taxonomic composition and relative abundance of aquatic biota in Coyote Creek. In the urban sections of the creek, the biotic community was dominated by pollution-tolerant species, such as tubificid worms and mosquitofish. In addition, low species diversity was evident in the community.

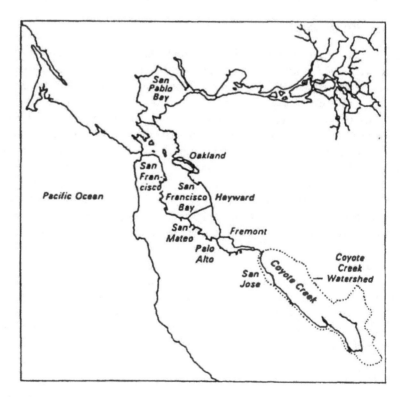

FIGURE 3   Location of the Coyote Creek and Coyote Creek watershed. (From Pitt, R. E., Effects of urban runoff on aquatic biota, in *Handbook of Ecotoxicology*, Hoffman, D. J., Rattner, B. A., Burton, G. A., Jr., and Cairns, J., Jr., Eds., Lewis Publishers, Boca Raton, FL, 1995, 609. With permission.)

TABLE 7
Water Quality Conditions in Urban and Nonurban Areas of Coyote Creek[a]

	Urban Area		Nonurban Area	
	Wet Weather	Dry Weather	Wet Weather	Dry Weather
Common parameters and major ions				
pH	7	8	—[b]	8
Temperature °(C)	16	17	—	16
Calcium	20[c]	100	40	100
Magnesium	6	70	20	60
Sodium	0.01	—	—	20
Potassium	2	4	2	2
Bicarbonate	50	150	—	200
Sulfate	20	60	—	40
Chloride	10	60	—	20
Total hardness	70	500	200	600
Total alkalinity	50	300	150	300
Residuals				
Total solids	350	1000	600	1000
Total dissolved solids	150	1000	300	1000
Suspended solids	300	4	600	20
Volatile suspended solids	60	2	90	10
Turbidity (NTU)	50	15	—	20
Specific conductance (μmhos/cm)	200	500	—	400
Organics and oxygen demand materials				
Dissolved oxygen	8	7	—	9.0
Biochemical oxygen demand (5–day)	25	—	5	—
Chemical oxygen demand	100	40	90	30.0
Total organic carbon	110	—	—	0.6
Nutrients				
Total Kjeldahl nitrogen	7.0	0.5	2.0	<0.3
Nitrate (as N)	0.7	0.8	—	1.2
Nitrite (as N)	—	0.02	—	<0.002
Ammonia (as N)	0.1	0.8	0.1	0.3
Orthophosphate	0.2	0.5	0.1	0.4
Heavy metals				
Lead (μg/l)	2000	40	200	2
Zinc (μg/l)	400	30	200	20
Copper (μg/l)	20	10	50	5
Chromium (μg/l)	20	10	5	5
Cadmium (μg/l)	5	<1	5	<1
Mercury (μg/l)	1	0.2	1	0.2
Arsenic (μg/l)	4	3	5	2
Iron (μg/l)	10,000	1000	20,000	2000
Nickel (μg/l)	40	<1	80	<1

[a] Most of this information was based on Coyote Creek monitoring results from Pitt, R. and Bozeman, M., *Sources of Urban Runoff Pollution and Its Effects on an Urban Creek*, EPA-600/52-82-090, U.S. Environmental Protection Agency, Cincinnati, 1979.

[b] Blanks signify no data were available.

[c] mg/l unless otherwise noted.

In contrast, a more diverse assemblage of organisms occurred in the nonurban portions of the creek, which were characterized by numerous benthic macroinvertebrate taxa and an abundance of native fish. Long-term effects of urban runoff, such as the inability of organisms to adjust to repeated toxic chemical exposure, the deposition and accumulation of toxic sediments, and habitat degradation, are probably much more important than short-term effects on communities associated with episodic events.

IV. BIOTIC COMMUNITIES

A. BENTHIC MACROFAUNAL COMMUNITIES

Multi-year investigations of the soft-bottom macrobenthic invertebrate community of San Francisco Bay reveal large fluctuations in abundance of populations that reflect: (1) within-year periodicity of reproduction, recruitment, and mortality that is not necessarily coincident with seasonal changes of the environment; and (2) aperiodic density changes following random perturbations of the environment.[53] It is clear that major restructuring of the benthic community can result from anomalous changes of the benthic habitat. For example, dredging directly alters the benthic habitat and dramatically impacts the benthos. Aside from habitat disturbances caused by anthropogenic activities, natural variations in environmental conditions such as freshwater inflow, tidal mixing, currents, and sediment erosion/deposition contribute substantially to the abundance and distribution patterns observed in the benthos. Random disturbance events (e.g., aperiodic major storms) also appear to play a role in the organization of benthic communities in the bay.

Nichols and Thompson[53] provide the most detailed information on the structure and temporal dynamics of the benthic macroinvertebrate community of San Francisco Bay. Their work shows that a homogeneous benthic macroinvertebrate community exists in the estuary, with small short-lived species comprising the most abundant populations. The density of these small, short-lived forms tends to peak between spring and fall each year. Surveys conducted since the 1950s indicate a general consistency in species composition and their relative abundances within each region of the bay.[54] The benthic community of Suisun Bay is typified by low diversity and numerical dominance by a few species. Fewer than 10 macrobenthic species occur in Suisun Bay, including the annelids *Limnodrilus hoffmeisteri* and *Nereis succinea*; the amphipods *Corophium spinicorne* and *C. stimpsoni*; and the bivalve mollusks *Corbicula fluminea*, *Macoma balthica*, and *Mya arenaria*. Much greater numbers of species compose the benthic community of San Pablo Bay. The following species are commonly found there: the amphipods *Ampelisca abdita*, *Corophium* spp., and *Grandidierella japonica*; the polychaetes *Glycinde* sp., *Heteromastus filiformis*, and *Streblospio benedicti*; and the mollusks *Gemma gemma*, *Ilyanassa obsoleta*, *Musculus senhousia*, and *Tapes japonica*. Among the numerical dominants in the shallower reaches of South Bay are *Ampelisca abdita*, *Gemma gemma*, and *Streblospio benedicti*. The same species commonly found in San Pablo Bay (see above) also occur in the deeper mud habitats of South Bay, together with the large, tube-dwelling polychaete *Asychis elongata*. More optimum marine conditions in the

Central Bay promote the development of a benthic community that is much more diverse and composed of species commonly found along the California coast.[49]

Historically, benthic communities in the northern bays have differed markedly from those in South San Francisco Bay.[2] According to Nichols and Pamatmat,[55] macrofaunal biomass in the northern bays is low, ranging from 60–630 g wet wt/m². Surface-dwelling bivalves (*Potamocorbula amurensis*) with a biomass of 200 g wet wt/m² predominate in the channel of the northern bays.[56] In South Bay, macrofaunal abundance and biomass values are much higher than in the northern bays.[55] This is particularly true for deposit feeders which are prevalent in South Bay. The polychaete *Asychis elongata* attains high numbers in South Bay, where it can occur in very dense patches.[2,49]

Introduced species have greatly affected the composition and structure of benthic communities in the estuary. Of the species listed above, all are introduced forms except for the native polychaete *Glycinde* sp. and the bivalve mollusk *Macoma balthica*. In the last 150 years, more than 100 benthic invertebrate species have been accidentally or intentionally introduced in the estuary, and some have altered the ecological balance.[57] For example, the Asian clam *Potamocorbula amurensis* successfully proliferated in South San Francisco Bay after its invasion, essentially displacing other members of the benthic community.[56]

Benthic macrofaunal species in San Francisco Bay exhibit large within- and between-year fluctuations in abundance. For many species, recruitment does not fall within a repeated annual cycle. Hence, periodic increases resulting from recruitment, which underlie within-year changes in abundance of benthic invertebrate species populations, often do not occur on a seasonally predictable basis. Nichols and Thompson,[53] investigating three species (i.e., *Ampelisca abdita*, *Gemma gemma*, and *Streblospio benedicti*) on an intertidal mudflat in South San Francisco Bay, determined that only *A. abdita* displayed statistically significant year-to-year consistency in the timing of abundance fluctuations. These investigations also demonstrated that abundance peaks of *G. gemma* and *S. benedicti* were highly variable year to year over a ten-year period (Figure 4). Such observations are commonly reported in the literature for estuarine benthic communities.[58]

Variations in benthic invertebrate species abundance and distribution may also arise from environmental perturbations. Both physical and biological disturbances greatly influence the abundance, distribution, and diversity of estuarine benthic communities. While physical disturbances of the benthos are caused by major storms, bottom currents, wave activity, sediment deposition/erosion, and pulsing from freshets, biological disturbances are incurred principally from grazing, predation, and competition.[59-64] Changes in seafloor chemistry may also play a paramount role.[60] In addition to natural processes, anthropogenic alteration of benthic habitats (e.g., by dredging and dredged material disposal, sewage sludge dumping, deposition of drilling muds and cuttings, and municipal and industrial effluents) can cause major reorganization of the resident benthos. The successional pattern of a benthic community, in turn, is closely coupled to the frequency and nature of these disturbances. Repeated benthic habitat impacts, for example, produce spatial mosaic patterns in which parts of the community exist at different levels of succession.[65]

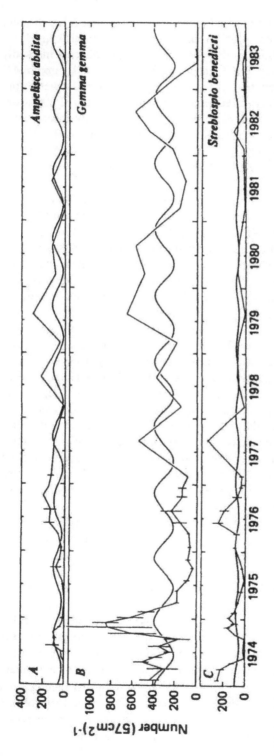

FIGURE 4 Abundance of the three numerically dominant benthic macroinvertebrate species at a South San Francisco Bay mudflat study site over a 10-year period (straight lines), and computed average annual abundance cycles from least-squares regressions of the data (curvilinear lines). Variation (± 1 std. dev.) among replicates shown for the first three years when samples were collected monthly or bimonthly. (From Nichols, F. H. and Thompson, J. K., Time scales of change in the San Francisco Bay benthos, *Hydrobiologia*, 129, 121, 1985. With permission.)

Estuarine and coastal marine environments subjected to continual physical disturbance are characterized by benthic communities with nonequalibrium structure.[66] Opportunistic macrobenthic species frequently dominate these environments.[59,60] The number of nonopportunistic taxa tends to decrease in such highly unstable systems.[67]

Stochastic events or random perturbations of the benthic habitat associated with physical disturbances commonly are of overwhelming importance in the restructuring of estuarine benthic communities. The significance of biological disturbances in modulating benthic community organization can be easily overlooked due to confounding physical factors. For instance, in some systems predators regulate the abundance of prey directly by consuming larvae, juveniles, and adults. They indirectly influence survivorship of their prey by burrowing through sediments, disturbing the sediment surface, and reducing larval settlement.[68] Because observed community patterns may result from responses of organisms to spatial and temporal changes in the environment, to competition, to predation, to anomalous events, or to the synergistic effect of interacting processes, detection of the forces largely responsible for shaping the community structure is often obscured.[69]

In San Francisco Bay, variations in freshwater inflow strongly affect the abundance and distribution of benthic macrofauna. For example, conditions of prolonged drought and rising salinity during 1976/77 prompted the mass migration of estuarine species to normally depauperate brackish/freshwater areas upestuary. This migration accounted for an increase in both abundance and diversity of the benthic community in Suisun Bay. Species that settled in the normally brackish Suisun Bay included *Ampelisca abdita*, *Corophium ascherusicum*, *Mya arenaria*, and *Streblospio benedicti*. In contrast, an extremely wet period during 1982/83 eliminated freshwater-sensitive species (e.g., *Ampelisca abdita*) from extensive regions of the estuary. Following that unusually wet period, this species did not become re-established as a numerical dominant on the mudflats of South Bay for at least two years. *Ampelisca abdita* apparently moves seasonally in the estuary in response to changing river runoff.[1] In winter when freshwater inflow is high, the amphipod moves from upestuary to downestuary locations, or from shallow to deep water sites. The organism returns to shallower locations and upestuary sites during dryer summer months when freshwater inflow declines.

Sediment erosion and smothering by decaying algae have an acute short-term impact on benthic community structure along intertidal mudflats.[1] These phenomena disturb the sediment surface and may reduce settlement and recruitment. Although such aperiodic perturbations severely depress abundance of the benthos, their effects are typically localized. However, it is unknown how widespread these events are throughout the estuary.

Anthropogenic impacts also appear to contribute substantially to variability of benthic communities in San Francisco Bay. The diversion of freshwater for farm irrigation and other purposes alters river inflow and consequently affects the distribution of estuarine species as noted above. Long-term inputs of halogenated hydrocarbons, heavy metals, and other chemical contaminants cannot be overlooked. Their occurrence can significantly influence reproduction, growth, and mortality of the benthos. However, the greatest impact of chemical inputs on the benthos of San

Francisco Bay has been instances of localized acute contamination, which have eliminated some species from affected areas. The effects of low-level chronic chemical contamination on the structure of benthic communities in the estuary ·have yet to be adequately addressed. More investigations must be conducted on chemical contamination in bottom sediments to better understand benthic community dynamics in the estuary.

B. Fish Communities

More than 100 species of fish have been recorded in San Francisco Bay. Among the most important forms in the estuary are several resident species (e.g., shiner perch *Cymatogaster aggregata*, starry flounder *Platichthys stellatus*, splittail *Pogonichthys macrolepidotus*, and leopard shark *Triakis semifasciata*) and a number of anadromous forms (e.g., America shad *Alosa sapidissima*, striped bass *Morone saxatilis*, and chinook salmon *Oncorhynchus tshawytscha*). Fishes of the bay can be divided into two groups based on their seasonal trends: (1) species having seasonal cycles of occurrence each year (e.g., northern anchovy *Engraulis mordax*, Pacific herring *Clupea harengus pallasi*, and striped bass *M. saxatilis*); and (2) species having no consistent seasonal trend (e.g., white surgeon *Acipenser transmontanus*, Pacific tomcod *Microgadus proximus*, and brown smoothhound shark *Mustelus henlei*).[4] Species which exhibit seasonal cycles of occurrence in the estuary also display seasonal variations in abundance. For example, the abundance of northern anchovy *E. mordax*, Pacific herring *C. harengus pallasi*, and striped bass *M. saxatilis* may be 2- to 10-fold greater in spring and summer than in winter and early spring. The summer peaks in abundance typically consist of young-of-the-year fish which migrate into the bay and spawn there and fish which move into the estuary after reaching capturable size. Spawning and migration are strongly influenced by environmental conditions in the bay (e.g., salinity, temperature, and food availability).

Freshwater inflow is a major factor affecting the abundance and distribution of fish populations in the estuary. Armor and Herrgesell[4] delineated three time scales of altered fish abundance and distribution related to variations of freshwater inflow: short-term (3–4 weeks), seasonal, and interannual. They also subdivided fish populations into wet-response, dry-response, and mixed-response species, depending upon when the fish reached peak abundance. A population is considered to be a dry-response species if it attains peak baywide abundance during years of low freshwater inflow. If peak abundance occurs during wet years when high freshwater inflow predominates, the fish is categorized as a wet-response species. A mixed-response species, in turn, is one which may reach high abundance during either dry or wet years. Analysis of trawl catches in past years indicates that most fish are wet-response and mixed-response species (Table 8).

Wet years appear to enhance the abundance of many species in the estuary. For example, 24 species of fish, including most of the estuarine forms, attained highest abundance during wet years when freshwater inflow increased. This infers that some factors associated with freshwater inflow contribute to increased abundance of fish in the estuary. The amount of freshwater inflow clearly affects other environmental factors, such as salinity, temperature, currents, food supply, and turbidity, and these

TABLE 8
Species Response to Water Year Type in the San Francisco Bay Estuary

	Wet response	Dry response	Mixed Response
Fresh	Threadfin shad	Sacramento squawfish	Inland silverside
	Carp	Tuleperch	Splittail
	Prickly sculpin		White catfish
Anadromous	White sturgeon	Pacific lamprey	American shad (14)
	Green sturgeon		King salmon
	Steelhead		Striped bass (5)
Estuarine	Threespine stickleback		Delta smelt
	Yellowfin goby (10)		Bay goby (11)
	Longfin smelt (2)		
	Staghorn sculpin (8)		
	Starry flounder (12)		
Marine-estuarine	Pacific herring (3)	Arrow goby	White croaker (9)
	Cheekspot goby	Walleye surfperch	Northern anchovy (1)
			Plainfin midshipman (15)
			Shiner perch (4)
Marine	Leopard shark	Bay ray	Night smelt
	Pile perch	White seaperch	Bay pipefish
	Speckled sanddab (7)	Jacksmelt (13)	Barred surfperch
	Diamond turbot	Black perch	Brown rockfish
	Sand sole	Rubberlip seaperch	Lingcod
	California tonguefish	Pacific butterfish	England sole (6)
	Brown smoothhound	Bonyhead sculpin	Dwarf perch
	Spiny dogfish		Big skate
	Pacific tomcod		Surf smelt
	Topsmelt		Curlfin turbot
	Snowy snailfish		

Note: Number in parenthesis is that fish's rank in the 15 most abundant species.

From Armor, C. and Herrgesell, P. L., *Hydrobiologia*, 129, 211, 1985. With permission.

factors often have a profound effect on fish abundance. Greater freshwater flow into the bay can result in higher food production either directly by the transport of organic matter into the system from allochthonous sources or indirectly by the nutrient enhancement of primary production within the system. More food in the estuary favors greater recruitment of some marine species. Higher food levels are particularly advantageous to species utilizing the estuary as a nursery area.

The volume of freshwater inflow also affects the distribution of fish populations. During wet years and high freshwater inflow, abundance of some species is consistently greater downestuary. This general pattern is evident among freshwater and estuarine species, as well as some anadromous forms. The exact nature of the relationship between freshwater inflow and the distribution of fish populations in the bay has not been determined.

Considering the coupling of freshwater inflow and the observed abundance and distribution patterns of fish in the estuary, the impact of freshwater diversions on land takes on added significance. Apart from freshwater diversions, insidious pollution of the estuary appears to have accelerated the rapid decline of certain fishery resources. Legislation restricting the harvest of anadromous fishes (e.g., salmon, striped bass, and sturgeon) substantially reduced commercial catches. At one time the American shad *A. sapidissima*, chinook salmon *O. tshawytscha*, striped bass *M. saxatilis*, green sturgeon *A. medirostris*, and white sturgeon *A. transmontanus* supported commercial fisheries in the bay, but these fisheries were terminated in 1956, 1956, 1935, 1917, and 1917, respectively. Clam and oyster fisheries, likewise once bountiful, have disappeared from the bay.[70] The Pacific herring *C. harengus pallais* remains the only major commercial fishery in the estuary. The American shad *A. sapidissima*, chinook salmon *O. tshawytscha*, California halibut *Paralichthys californicus*, striped bass *M. saxatilis*, starry flounder *P. stellatus*, green sturgeon *A. medirostris*, white sturgeon *A. transmontanus*, jacksmelt *Atherinopsis californiensis*, brown rockfish *Sebastes auriculatus*, staghorn sculpin *Leptocottus armatus*, white catfish *Ictalurus cactus*, brown smoothhound shark *M. henlei*, leopard shark *T. semifasciata*, and various surf perch support large recreational fisheries.[4] Thus, a gradual shift in the character of the fisheries has occurred over the years from primarily commercial to principally recreational.

V. CONCLUSIONS

Human-induced changes in San Francisco Bay and surrounding land areas have resulted in numerous impacts on benthic macroinvertebrate and fish communities. Chemical wastes derived from agricultural and industrial activities, while locally detrimental to many plant and animal species in the estuary, have not been adequately assessed. Chronic regional effects of chemical contamination have also been noted, but effects of this contamination on aquatic communities in the estuary are largely inferential. Arguably the greatest influence of human activity on the biological communities in the bay is attributable to the alteration of riverine inflow due to freshwater diversions. Decreased freshwater inflow can alter delicate balances among plant and animal communities, leading to marked changes in abundance, biomass, and distribution of invertebrate and fish populations.[1] Such changes greatly influence the dynamics of biotic communities in the estuary.

REFERENCES

1. Nichols, F. H., Cloern, J. E., Luoma, S. N., and Peterson, D. H., The modification of an estuary, *Science*, 231, 567, 1986.
2. Caffrey, J. M., Spatial and seasonal patterns in sediment nitrogen remineralization and ammonium concentrations in San Francisco Bay, California, *Estuaries*, 18, 219, 1995.
3. Smith, G. J. and Flegal, A. R., Silver in San Francisco Bay estuarine waters, *Estuaries*, 16, 547, 1993.

4. Armor, C. and Herrgesell, P. L., Distribution and abundance of fishes in the San Francisco Bay estuary between 1980 and 1982, *Hydrobiologia*, 129, 211, 1985.

5. Pereira, W. E., Hostettler, F. D., and Rapp, J. B., Bioaccumulation of hydrocarbons derived from terrestrial and anthropogenic sources in the Asian clam, *Potamocorbula amurensis*, in San Francisco Bay estuary, *Mar. Pollut. Bull.*, 24, 103, 1992.

6. Walters, R. A. and Gartner, J. W., Subtidal sea level and current variations in the northern reach of San Francisco Bay, *Est. Coastal Shelf Sci.*, 21, 17, 1985.

7. Conomos, T. J., Ed., *San Francisco Bay: The Urbanized Estuary*, American Association for the Advancement of Science, San Francisco, CA, 1979.

8. Long, E., Macdonald, D., Matta, M. B., Van Ness, K., Buchman, M., and Harris, H., Status and Trends in Concentrations of Contaminants and Measures of Biological Stress in San Francisco Bay, NOAA Tech. Mem. NOS OMA 41, National Oceanic and Atmospheric Administration, Seattle, WA, 1988.

9. Luoma, S. N. and Phillips, D. J. H., Distribution, variability, and impacts of trace elements in San Francisco Bay, *Mar. Pollut. Bull.*, 19, 413, 1988.

10. Luoma, S. N., Dagovitz, R., and Axtmann, E., Temporally intensive study of trace metals in sediments and bivalves from a large river-estuarine system: Suisun Bay/Delta in San Francisco Bay, *Sci. Total Environ.*, 97/98, 685, 1990.

11. Cross, R. and Williams, D., Eds., Proceedings of the National Symposium on Freshwater Inflow to Estuaries, Tech. Rept. FWS/OBS-81/04, U.S. Fish and Wildlife Service, Washington, D.C., 1981.

12. Cloern, J. E. and Nichols, F. H., Eds., *Temporal Dynamics of an Estuary: San Francisco Bay*, Junk, Dordrecht, Netherlands, 1985.

13. Cloern, J. E., Annual variations in river flow and primary production in the South San Francisco Bay estuary, in *Estuaries and Coasts: Spatial and Temporal Intercomparisons*, Elliott, M. and Ducroty, J. P., Eds., Olsen and Olsen, Fredensborg, Denmark, 1991.

14. Kuwabara, J. S. and Luther, G. W., III, Dissolved sulfides in the oxic water column of San Francisco Bay, California, *Estuaries*, 16, 567, 1993.

15. Kockelman, W. J., Conomos, T. J., and Leviton, A. E., Eds., *San Francisco Bay: Use and Protection*, American Association for the Advancement of Science, San Francisco, CA, 1982.

16. Jassby, A., Cloern, J. E., and Powell, T. M., Organic carbon sources and sinks in San Francisco Bay: freshwater flow induced variability, *Mar. Ecol. Prog. Ser.*, 93, 39, 1993.

17. Eisler, R., Polycyclic Aromatic Hydrocarbon Hazards to Fish, Wildlife, and Invertebrates: A Synoptic Review, Biol. Rep. 85 (1.11), U.S. Fish and Wildlife Service, Washington, D.C., 1987.

18. Kennish, M. J., *Ecology of Estuaries: Anthropogenic Effects*, CRC Press, Boca Raton, FL, 1992.

19. Jung, M., Whipple, J. A., and Moser, M., Summary Report of the Cooperative Striped Bass Study, Institute of Aquatic Resources, Santa Cruz, CA, 1984.

20. Wright, D. A. and Phillips, D. J. H., Chesapeake and San Francisco Bays: a study in contrasts and parallels, *Mar. Pollut. Bull.*, 19, 405, 1988.

21. Pereira, W. E., Hostettler, F. D., Cashman, J. R., and Nishioka, R. S., Occurrence and distribution of organochlorine compounds in sediment and livers of striped bass (*Morone saxatilis*) from the San Francisco Bay-Delta estuary, *Mar. Pollut. Bull.*, 28, 434, 1994.

22. Phillips, D. J. H. and Spies, R. B., Chlorinated hydrocarbons in the San Francisco estuarine ecosystem, *Mar. Pollut. Bull.*, 19, 445, 1988.

23. Hayes, S. P., Phillips, P. T., Martin, M., Stephenson, M., Smith, D., and Linfield, J., California State Mussel Watch, Marine Water Quality Monitoring Program 1983–84, Water Quality Monitoring Report No. 85-2WQ, California Water Research Control Board, Sacramento, 1985.

24. Hayes, S. P. and Phillips, P. T., California State Mussel Watch Marine Water Quality Monitoring Program 1984 to 1985, Water Quality Monitoring Rept., No. 86-3WQ, California Water Research Control Board, Sacramento, 1986.

25. Stephenson, M., Smith, D., Ichikawa, G., Goetzl, J., and Martin, M., State Mussel Watch Program Preliminary Data Report 1985–86, Tech. Rept., State Water Resources Control Board, California Department of Fish and Game, Monterey, 1986.

26. Mearns, A. J., Matta, M. B., Simecek-Beatty, D., Buchman, M. F., Shigenaka, G., and Wert, W. A., PCB and Chlorinated Pesticide Contamination in U.S. Fish and Shellfish: A Historical Assessment Report, NOAA Technical Memorandum NOS OMA 39, National Oceanic and Atmospheric Administration, Seattle, WA, 1988.

27. Risebrough, R. W., Chapman, J. W., Okazaki, R. K., and Schmidt, T. T., Toxicants in San Francisco Bay and Estuary, Report of the Association of Bay Area Governments, Berkeley, CA, 1978.

28. National Oceanic and Atmospheric Administration, National Status and Trends Program for Marine Environmental Quality: Progress Report — A Summary of Selected Data on Chemical Contaminants in Tissues Collected During 1984, 1985, 1986, Tech. Rept., OAD/OMA/NOS/NOAA, Department of Commerce, Rockville, MD, 1987.

29. Ohlendorf, H. M. and Miller, M. R., Organochlorine contaminants in California waterfowl, J. Wildl. Manage., 48, 867, 1984.

30. Ohlendorf, H. M and Fleming, W. J., Birds and environmental contaminants in San Francisco and Chesapeake Bays, Mar. Pollut. Bull., 19, 487, 1988.

31. Custer, T. W., Hensler, G. L., and Kaiser, T. E., Clutch size, reproductive success, and organochlorine contaminants in Atlantic Coast black-crowned night- herons, Auk, 100, 699, 1983.

32. Henny, C. J., Blus, L. J., Krynitsky, A. J., and Bunck, C. M., Current impact of DDE on black-crowned night-herons on the Intermountain West, J. Wildl. Manage., 48, 1, 1984.

33. National Oceanic and Atmospheric Administration, National Status and Trends Program for Marine Environmental Quality: Progress Report and Preliminary Assessments of Findings of the Benthic Surveillance Project — 1984, Tech. Rept., OAD/NOS/NOAA, Department of Commerce, Rockville, MD, 1987.

34. Law, L. M. and Goerlitz, D. F., Selected chlorinated hydrocarbons in bottom material from streams tributary to San Francisco Bay, Pestic. Monit. J., 8, 33, 1974.

35. Smith, G. J. and Flegal, A. R., Silver in San Francisco Bay estuarine waters, Estuaries, 16, 547, 1993.

36. Luoma, S. N. and Cloern, J. E., The impact of wastewater discharge on biological communities in San Francisco, in San Francisco Bay: Use and Protection, Kockelman, W., Conomos, T. J., and Leviton, A. E., Eds., American Association for the Advancement of Science, San Francisco, CA, 1982, 137.

37. Flegal, A. R., Smith, G. J., Gill, G. A., Sanudo-Wilhelmy, S., and Anderson, L. C. D., Dissolved trace element cycles in San Francisco Bay, Mar. Chem., 36, 329, 1991.

38. Van Geen, A. and Luoma, S. N., Trace metals (Cd, Cu, Ni, and Zn) and nutrients in coastal waters adjacent to San Francisco Bay, California, Estuaries, 16, 559, 1993.

39. Hayes, S. P. and Phillips, P. T., California State Mussel Watch, Marine Water Quality Monitoring Program 1985–86, Water Quality Monitoring Report No. 87-2WQ, California Water Research Control Board, Sacramento, 1987.

40. Cain, D. J. and Luoma, S. N., Copper and silver accumulation in transplanted and resident clams (*Macoma balthica*) in South San Francisco Bay, *Mar. Environ. Res.*, 15, 115, 1985.

41. Cain, D. J. and Luoma, S. N., Influence of seasonal growth, age, and environmental exposure on Cu and Ag in a bivalve indicator, *Macoma balthica* in San Francisco Bay, *Mar. Ecol. Prog. Ser.*, 60, 45, 1990.

42. Luoma, S. N., Cain, D., and Johansson, C., Temporal fluctuations of silver, copper, and zinc in the bivalve *Macoma balthica* at five stations in South San Francisco Bay, *Hydrobiologia*, 129, 109, 1985.

43. Liu, D. L., Yang, Y. P., and Hu, M. H., Selenium content of marine food chain organisms from the coast of China, *Mar. Environ. Res.*, 22, 151, 1987.

44. Heinz, G. H., Pendleton, G. W., Krynitsky, A. J., and Gold, L. G., Selenium accumulation and elimination in mallards, *Arch. Environ. Contam. Toxicol.*, 19, 374, 1990.

45. Hoffman, D. J., Sanderson, C. J., LeCaptain, L. J., Comartie, E., and Pendleton, G. W., Interactive effects of arsenate, selenium, and dietary protein on survival, growth, and physiology in mallard ducklings, *Arch. Environ. Contam. Toxicol.*, 22, 55, 1992.

46. Goede, A. A., Wolterbeek, H. T., and Koese, M. J., Selenium concentrations in the marine invertebrates *Macoma balthica*, *Mytilus edulis*, and *Nereis diversicolor*, *Arch. Environ. Contam. Toxicol.*, 25, 85, 1993.

47. Hamilton, S. J. and Buhl, K. J., Acute toxicity of boron, molybdenum, and selenium to fry of chinook salmon and coho salmon, *Arch. Environ. Contam. Toxicol.*, 19, 366, 1990.

48. Saiki, M. K. and Palawski, D. U., Selenium and other elements in juvenile striped bass from the San Joaquin Valley and San Francisco estuary, California, *Arch. Environ. Contam. Toxicol.*, 19, 717, 1990.

49. Nichols, F. H., Natural and anthropogenic influences on benthic community structure in San Francisco Bay, in *San Francisco Bay: The Urbanized Estuary*, Conomos, T. J., Ed., American Association for the Advancement of Science, San Francisco, CA, 1979, 231.

50. National Oceanic and Atmospheric Administration, National Estuary Study, U.S. Fish and Wildlife Service, Washington, D.C., 1970, Volume 5.

51. Flegal, A. R. and Sanudo-Wilhelmy, S. A., Comparable levels of trace metal contamination in two semi-enclosed embayments: San Diego Bay and South San Francisco Bay, *Environ. Sci. Technol.*, 27, 1934, 1993.

52. Pitt, R. E., Effects of urban runoff on aquatic biota, in *Handbook of Ecotoxicology*, Hoffman, D. J., Rattner, B. A., Burton, G. A., Jr., and Cairns, J., Jr., Eds., Lewis Publishers, Boca Raton, FL, 1995, 609.

53. Nichols, F. H. and Thompson, J. K., Time scales of change in the San Francisco Bay benthos, *Hydrobiologia*, 129, 121, 1985.

54. California Department of Water Resources, Sacramento-San Joaquin Delta Water Quality Surveillance Program 1981 — Monitoring Results Pursuant to Conditions Set Forth in Delta Water Rights Decision 1485, State Department, Sacramento, CA, 1982, Vol. 3.

55. Nichols, F. H. and Pamatmat, M. M., The Ecology of the Soft-bottom Benthos of San Francisco Bay: A Community Profile, U. S. Fish and Wildlife Service, Biological Report 85(7.19), Washington, D.C., 1988.

56. Nichols, F. H., Thompson, J. K., and Schemel, L. E., Remarkable invasion of San Francisco Bay (California, USA) by the Asian clam *Potamocorbula amurensis*. II. Displacement of a former community, *Mar. Ecol. Prog. Ser.*, 66, 95, 1990.

57. Carlton, J. T., Introduced invertebrates of San Francisco Bay, in *San Francisco Bay: The Urbanized Estuary*, Conomos, T. J., Ed., American Association for the Advancement of Science, San Francisco, CA, 1979, 427.

58. Kennish, M. J., *Ecology of Estuaries: Biological Aspects*, CRC Press, Boca Raton, FL, 1990.

59. Rhoads, D. C. and Boyer, L. F., The effects of marine benthos on physical properties of sediments: a successional perspective, in *Animal-Sediment Relations: The Biogenic Alteration of Sediments*, McCall, P. L. and Tevesz, M. J. S., Eds., Plenum Press, New York, 1982, 3.

60. Rhoads, D. C. and Germano, J. D., Interpreting long-term changes in benthic community structure: a new protocol, *Hydrobiologia*, 142, 2911, 1986.

61. Dayton, P. K., Competition, disturbance, and community organization: the provision and subsequent utilization of space in a rocky intertidal community, *Ecol. Monogr.*, 41, 351, 1971.

62. Rhoads, D. C., McCall, P. L., and Yingst, J. Y., Disturbance and production on the estuarine seafloor, *Am. Sci.*, 66, 577, 1978.

63. Germano, J. D., Infaunal Succession in Long Island Sound: Animal-Sediment Interactions and the Effects of Predation, Ph.D. thesis, Yale University, New Haven, CT, 1983.

64. Whitman, J. D., Refuges, biological disturbance, and rocky subtidal community structure in New England, *Ecol. Monogr.*, 55, 421, 1985.

65. Johnson, R. G., Conceptual models of benthic marine communities, in *Models in Paleobiology*, Schopf, T. J. M., Ed., Freeman, Cooper, & Company, San Francisco, CA, 1972, 148.

66. Reice, S. R., Nonequilibrium determinants of biological community structure, *Am. Sci.*, 82, 424, 1994.

67. Hall, S. J., Raffaelli, D., and Thrush, S. F., Patchiness and disturbance in shallow water benthic assemblages, in *Aquatic Ecology-Scale Pattern and Process*, Giller, S., Hildrew, A. G., and Raffaelli, D. G., Eds., Blackwell Scientific Publications, Boston, 1994, 333.

68. Ambrose, W. G., Jr., Role of predatory infauna in structuring marine soft-bottom communities, *Mar. Ecol. Prog. Ser.*, 17, 109, 1984.

69. Flint, R. W. and Kalke, R. D., Biological enhancement of estuarine benthic community structure, *Mar. Ecol. Prog. Ser.*, 31, 23, 1986.

70. Smith, S. E. and Kato, S., The fisheries of San Francisco Bay: past, present, and future, in *San Francisco Bay: The Urbanized Estuary*, Conomos, T. J., Ed., American Association for the Advancement of Science, San Francisco, CA, 1979, 445.

6 Case Study 5: Puget Sound

I. INTRODUCTION

One of the most heavily contaminated inland marine systems is Puget Sound, which consists of a fjord-like basin and adjacent embayments. For more than a century, Puget Sound has been a major repository of various types of wastes derived from municipal and industrial wastewater discharges, dumping operations, spills, urban and agricultural runoff, and other sources.[1] These wastes have delivered significant concentrations of inorganic and organic chemical contaminants, such as polycyclic aromatic hydrocarbons (PAHs), chlorinated hydrocarbons, and heavy metals. The highest concentrations of these contaminants occur in bottom sediments of urbanized bays including Elliott Bay near Seattle, Commencement Bay near Tacoma, Budd Inlet near Olympia, and Sinclair Inlet near Bremerton.[2-4] Several contaminant groups found in these areas have a high potential for bioaccumulation in Puget Sound organisms, and some of the contaminants are a potential hazard to human health.[5]

Industrial activities in adjacent land areas are responsible for a wide range of toxic chemicals that locally enter urbanized bays. For example, chemical manufacturing plants, oil refineries, pulp mills, marinas, and other facilities have released many chemical contaminants to embayments. Effects of industrial operations, as well as other anthropogenic sources, are perhaps manifested most conspicuously in Commencement Bay, a system designated by the U.S. Environmental Protection Agency (U.S. EPA) in 1981 as one of the 10 highest priority hazardous waste disposal sites in the United States for remedial investigation under the Superfund Program. Here, higher concentrations of many synthetic organic compounds have been detected in water, sediment, and biota than in other urban embayments.[6] Chlorinated hydrocarbon compounds, such as DDT, polychlorinated biphenyls (PCBs), and chlorinated butadienes (CBDs), reached peak levels in the 1950s and 1960s and may be a principal cause of increased frequencies of neoplasms, necroses, lesions, and other disorders observed in fishes and invertebrates.[1,2,7] Historically, sediments dredged from waterways of the sound were periodically dumped at two disposal sites in Commencement Bay. Since many chemical contaminants tend to be sequestered by particulate matter and accumulate in fine grained sediments of the sound, these disposal sites retain elevated contaminant levels. High concentrations of toxic chemicals in bottom sediments are a major concern because the food web of Puget Sound is detritus-based. Historical accumulations of chemical contaminants in bottom sediments may be slowly leaching into the water column and contributing to present-day chemical toxic impacts.[8] Hence, there has been an ongoing effort to analyze these sediments and the benthic organisms inhabiting them.

This chapter provides an overview of chemical contamination in Puget Sound. It assesses the sources and fates of contaminants in the system. It also reviews available data on pollution-related biological impacts, especially those occurring in benthic infauna, epibenthic fauna, and demersal fish populations. In addition, the various histopathological conditions found in resident fauna are assessed. Finally, the ecological health of Puget Sound and potential detrimental effects of food web contamination on human health is examined.

II. PHYSICAL DESCRIPTION

Located in northwestern Washington State, Puget Sound is a glacially scoured basin with depths to ~300 m. The fjord-like estuary extends roughly north–south for ~130 km from Admiralty Inlet to Olympia, but is only ~10–40 km in width. Because the sound is divided into a number of branches and large islands rise above the sea surface at various locations, the shoreline configuration in highly irregular (Figure 1).

Several major rivers (i.e., Skagit, Stillaguamish, Snohomish, Duwamish, Puyallup, and Nisqually rivers) flow into Puget Sound along the eastern shoreline. Chief among these influent systems are the Skagit, Stillaguamish, and Snohomish rivers which account for more than 75% of the freshwater input to the sound. Other freshwater sources include precipitation, surface runoff, and groundwater inflow. While salinities in surface waters of Puget Sound are reduced near the mouth of these larger rivers, they typically average >30‰ in the basin. In high energy channel and nearshore environments, bottom sediments consist of sand and cobbles. Clayey silt and clayey fine sand predominate in deeper water areas. Surface deposits are generally oxic.[9]

The Puget Sound basin has an area of 2600 km^2 and a volume of 169 km^3.[10] Seawater enters Puget Sound at depth through the Strait of Juan de Fuca, and there is a net surface outflow. Large oscillating tidal currents, superimposed on this general circulation pattern, largely drive horizontal and vertical mixing in the basin. The mean residence time for water in the central basin is about 120–140 days, but is much longer in isolated inlets and in restricted, deep basins.

III. CHEMICAL CONTAMINATION

Although extensive deep portions of Puget Sound and its remote, rural bays are relatively contaminant-free, parts of those bays bordering urban, industrialized centers contain high concentrations of toxic chemicals.[11] In these local impacted areas, substantial volumes of domestic and industrial wastes have released high concentrations of many different chemical compounds to bottom sediments. For example, Malins et al.[4] reported more than 900 individual organic compounds in bottom sediments of Commencement Bay, notably hundreds of aromatic and chlorinated hydrocarbons, as well as various bromine-, sulfur-, nitrogen-, and oxygen-containing compounds. In nearly every sediment sample analyzed, they detected PCBs, CBDs, hexachlorobenzene (HCB), and pesticides. Large numbers of toxic chemicals have also been recorded in Elliott Bay, Bellingham Bay, and other urban embayments.

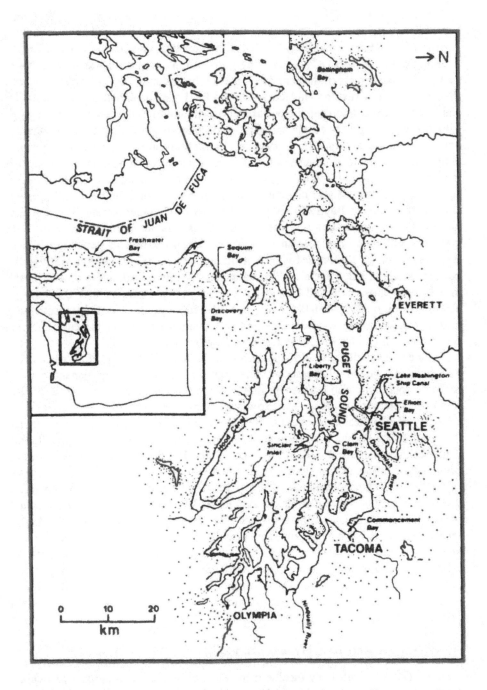

FIGURE 1 Map of Puget Sound, Washington (see inset), showing embayments and water-ways. (From Ginn, T. C. and Barrick, R. C., in *Oceanic Processes in Marine Pollution*, Vol. 5, *Urban Wastes in Coastal Marine Environments*, Wolfe, D. A. and O'Connor, T. P., Eds., Robert E. Krieger Publishing Company, Malabar, FL, 1988, 157. With permission.)

Sediment bioassays suggest that the three most toxic areas include Elliott Bay, the Duwamish River, and the Commencement Bay waterways.[8]

Barrick and Prahl[12] and Barrick[13] found highest total combustion-derived PAHs ranging from 8–73 μg/g OC (16–2400 ng/g dry sediment) in surface sediments (0–2 cm) near urban centers of the sound (Figure 2). In Commencement Bay sediments, Becker et al.[14] reported low-molecular-weight PAH (i.e., naphthalene, acenaphthylene, acenaphthene, fluorene, phenanthrene, and anthracene) concentrations ranging from 40–20,000 ng/g dry weight and high-molecular-weight PAH (i.e., fluoranthene, pyrene, benzo(a)anthracene, chrysene, benzofluoranthenes, benzo(a)pyrene, indeno(1,2,3-c,d)pyrene, dibenzo(a,h)anthracene, and benzo(g,h,i)perylene) concentrations ranging from 77–34,000 ng/g dry weight. Long[7] recorded the highest mean concentrations (dry weight) of total PCBs in sediments of Elliott Bay (380 ng/g) and Commencement Bay (270 ng/g). Sediments of Commencement Bay had the highest mean concentration (dry weight) of CBDs, amounting to 16,000 ng/g. The mean concentration of total aromatic hydrocarbons in sediments peaked in Elliott Bay (13,000 ng/g). Bottom sediments of the urbanized bays also contained highest concentrations of heavy metals. While arsenic, cadmium, lead, and mercury levels in sediment samples collected throughout the sound ranged from 1600–470,000 ng/g, 390–18,000 ng/g, 2300–630,000 ng/g, and 20–1900 ng/g dry weight, respectively, highest concentrations were documented in Commencement Bay (arsenic), Elliott Bay (cadmium and lead), and Bellingham Bay (mercury).

Acutely toxic chemical effluents prior to the late 1970s were often associated with pulp and paper mill operations. The high biochemical oxygen demand caused by these operations also resulted in large kills of commercially and recreationally important fish, such as salmonids. However, many other industries also discharged a variety of chemical contaminants into the sound during this earlier period that impacted biota. Pollution abatement has significantly reduced chemical pollution from peak levels in the 1940s, 1950s, and 1960s. The dumping of chemical waste, principally from industries along the Hylebos Waterway, accounted for high levels of CBDs in Commencement Bay during the 1960s. Aromatic hydrocarbon (AH) levels increased in the early 1900s and peaked in 1950. Heavy metals reached maximum concentrations in the 1960s. A lead-copper smelting facility of the American Smelting and Refining Company near Tacoma contributed large amounts of arsenic and lead to Commencement Bay.[8] Both acute lethal and sublethal effects related to multiple contaminant inputs were evident in organisms inhabiting these urbanized areas of the sound. These effects were manifested as histopathological abnormalities, reproductive failures, and increased mortality of invertebrate and fish populations.

A. POLYCYCLIC AROMATIC HYDROCARBONS

Because PAHs are relatively insoluble in seawater and sorb strongly to particulate matter, they tend to accumulate in bottom sediments which represent the primary source of contaminant exposure for estuarine and marine organisms. The distributions and concentrations of PAHs in bottom sediments are important because of the

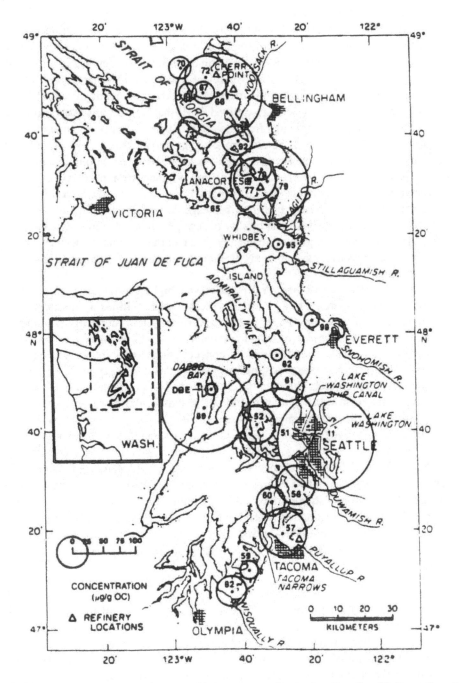

FIGURE 2 Total combustion-derived PAHs at sampling stations (numbers) in Puget Sound. The radius of the circle about each station is proportional to the total concentration (µg/g OC) of the PAHs. (From Barrick, R. C. and Prahl, F. G., *Est. Coastal Shelf Sci.*, 25, 175, 1987. With permission.)

effects these contaminants have on biological systems. Thus, investigations of PAHs in environmental media of Puget Sound have focused on the benthic regime.

Analysis of sediment cores from the central basin of Puget Sound showed that the maximum concentrations of PAHs and combustion PAHs occurred in sediments deposited between 1940 and 1950.[1] Municipal wastewater discharges and atmospheric deposition have accounted for most of the PAH compounds in central Puget Sound sediments. Most of the sewage-derived PAHs that have accumulated in bottom sediments over the years originated from Seattle's METRO sewage treatment facility at West Point.[13] Effluent PAHs containing ≥4 rings derive mainly from stormwater input, incorporating atmospheric dust, road-wear particles, and automotive oils and greases, all of which yield PAH compounds that originate from pyrolysis and combustion processes. The interaction of street-dust accumulation and precipitation events is responsible for a large fraction of sewage-derived PAHs ultimately released to the sound. The improved effluent quality of sewage treatment plants, which reduce solids emissions, and the greater efficiency of air pollution control devices, which lower particle emissions, have probably limited the concentrations of particle-sorbed PAHs entering the sound.

More recently, Barrick and Prahl[12] observed highest PAH concentrations in modern sediments within a few kilometers of industrial facilities in northern Puget Sound, urban areas in central Puget Sound, and river systems draining coal-bearing strata. While combustion-derived PAH concentrations peak in surface sediments at isolated regions of the sound, minimal concentrations exist in regions lying in between. This pattern indicates that atmospheric and water transport do not widely disperse PAH compounds through the system, and a substantial fraction of the compounds are removed to bottom sediments in close proximity to major sources. Thus, apart from the main municipal sources of combustion-derived PAHs in central Puget Sound, other regional sources of these contaminants also appear to play a significant role.

Since the mid-1950s, the concentrations of combustion-derived PAHs have declined in bottom sediments. Barrick and Prahl[12] documented a median surface concentration of combustion-derived PAHs of about 30 µg/g OC in Puget Sound. Long and Chapman[11] reported elevated levels of combustion-derived PAHs in urbanized bays: (1) Elliott Bay (3880–46,500 ng/g dry weight); (2) Commencement Bay (500–38,830 ng/g dry weight); and (3) Sinclair Inlet (1960–2690 ng/g dry weight).

B. Chlorinated Hydrocarbons

As noted previously, chlorinated hydrocarbon compounds tend to concentrate in bottom sediments of urbanized bays, particularly Commencement and Elliott Bays. For example, high levels of CBDs and hexachlorobenzene (HCB) occur in the Hylebos Waterway of Commencement Bay and elevated levels of PCBs in the Duwamish Waterway of Elliott Bay. Malins et al.[4] compared the concentrations of chlorinated hydrocarbon compounds in bottom sediments of urbanized embayments (i.e., Commencement Bay, Elliott Bay, Sinclair Inlet, and eastern Port Gardner) to those in nonurbanized embayments. They chronicled the following mean concentrations of PCBs, CBDs, and HCB, respectively.

1. In Commencement Bay: 0.27, 1.6, and 0.07 μg/g;
2. In Elliott Bay: 0.38, 0.01, and 0.0002 μg/g;
3. In Sinclair Inlet: 0.13, 0.06, and 0.0003 μg/g;
4. In eastern Port Gardner: 0.08, 0.02, and 0.0004 μg/g;
5. In nonurban embayments: 0.005, 0.003, and 0.0001 μg/g.

In Commencement Bay, the concentrations of CBDs and HCB were much higher than in the other embayments. The concentrations of PCBs in the urban embayments greatly exceeded those in the nonurban embayments. All of the urban embayments exhibited marked variation in organochlorine contaminant concentrations.

Pesticide concentrations were less variable than the aforementioned organochlorine contaminants. The concentrations of lindane, heptaclor, aldrin, *alpha*-chlordane and *trans*-nonachlor were generally <0.002 μg/g in bottom sediments of urban embayments and <0.001 μg/g in nonurban areas. The levels of DDT and its derivatives in urban embayments were <0.01 μg/g.

Long and Chapman[11] documented PCB levels of 383–1329 ng/g dry weight in sediments of Commencement Bay and 338–2823 ng/g dry weight in sediments of Elliott Bay. Lower PCB concentrations (176–218 ng/g dry sediment) were registered in Sinclair Inlet sediments. Lowest PCB values (<100 ng/g dry weight) were recorded in sediments of Case Inlet and Samish Bay. PCB concentrations in urban embayments of Puget Sound are compared to those in other U.S. estuarine and coastal marine systems (Table 1).

Chlorinated hydrocarbon concentrations vary considerably in Commencement Bay sediments.[14] CBDs had the highest contaminant levels, with mean and range values of 2900 ng/g dry weight and 2.0–66,000 ng/g dry weight, respectively. The concentrations of PCBs averaged 140 ng/g dry weight and ranged from 3.0–2000 ng/g dry weight. Those of chlorinated benzenes were somewhat lower, with a mean of 130 ng/g dry weight and a range of 2.5–1300 ng/g dry weight.

Stein et al.[15] obtained highest PCB measurements (570 ng/g wet weight) in sediments of the Duwamish Waterway. Moderate concentrations of PCBs were delineated in sediments of Hylebos Waterway (230 ng/g wet weight) and Everett Harbor (10 ng/g wet weight). Lowest PCB levels were discerned in Pilot Point sediments (21 ng/g wet weight) and Polnell Point sediments (21 ng/g wet weight).

C. HEAVY METALS

High concentrations of heavy metals such as arsenic, copper, mercury, and lead accumulate in waters near the major port areas of Puget Sound (i.e., Port of Seattle and Port of Tacoma), as well as in other areas affected by municipal or industrial discharges. Anthropogenic sources of heavy metals to Puget Sound include sewage effluents, industrial wastewater discharges, shipping, land runoff, automobile emissions, and atmospheric deposition. Natural sources of heavy metals to the sound are dissolved and particulate phases in seawater entering from the Strait of Juan de Fuca to the north and in freshwater entering from rivers draining surrounding lands.

Historically significant anthropogenic sources of heavy metals to Puget Sound are sewage treatment plants (particularly the METRO plant at West Point), a

TABLE 1
PCBs in Sediments from Selected Estuarine and Coastal Marine Systems in the United States, 1984[a]

Estuary	Total PCBs
Casco Bay, ME	95.28
Merrimack River, MA	52.97
Salem Harbor, MA	533.58
Boston Harbor, MA	17,104.86
Buzzard's Bay, MA	308.46
Narragansett Bay, RI	159.96
East Long Island Sound, NY	10.00
West Long Island Sound, NY	234.43
Raritan Bay, NJ	443.89
Delaware Bay, DE	2.50
Lower Chesapeake Bay, VA	51.00
Pamlico Sound, NC	ND[b]
Charleston Harbor, SC	9.10
Sapelo Sound, GA	ND
St. Johns River, FL	140.00
Charlotte Harbor, FL	ND
Tampa Bay, FL	ND
Apalachicola Bay, FL	12.00
Mobile Bay, AL	ND
Round Islands, MS	ND
Mississippi River Delta, LA	34.00
Barataria Bay, LA	ND
Galveston Bay, TX	ND
San Antonio Bay, TX	ND
Corpus Christi Bay, TX	ND
Lower Laguna Madre, TX	ND
San Diego Harbor, CA	422.10
San Diego Bay, CA	6.74
Dana Point, CA	7.06
Seal Beach, CA	46.71
San Pedro Canyon, CA	159.56
Santa Monica Bay, CA	14.00
San Francisco Bay, CA	123.46
Bodega Bay, CA	4.18
Coos Bay, OR	3.19
Columbia River Mouth, OR/WA	8.77
Nisqually Reach, WA	4.23
Commencement Bay, WA	20.60
Elliott Bay, WA	329.87
Lutak Inlet, AK	5.50
Nahku Bay, AK	6.60

Note: PCBs = polychlorinated biphenyls.

[a] Parts per billion.
[b] No Data.

From NOAA, *National Status and Trends Program for Marine Environmental Quality: Progress Report and Preliminary Assessments of Findings of the Benthos Surveillance Project — 1984*, National Oceanic and Atmospheric Administration Office of Ocean Resources Conservation and Assessment, Rockville, MD, 1987.

chlor-alkali industrial plant in Bellingham, and the American Smelting and Refining Company's (ASARCO) lead-copper smelting facility near Tacoma. Commencing operations in 1889, the ASARCO smelting facility released significant quantities of arsenic, antimony, copper, lead, and zinc to the southern part of Puget Sound. Crecelius et al.[16] conveyed that the smelter discharged arsenic and antimony into the surrounding environment via three pathways: (1) as stack into the air (2×10^5 kg/yr of As_2O_3 and 2×10^4 kg/yr of antimony oxides); (2) as dissolved arsenic and antimony species in liquid effluent discharged directly into Puget Sound ($2-7 \times 10^4$ kg As/yr and 2×10^3 kg Sb/yr); and (3) as crystalline slag particles dumped directly into the sound (1.5×10^6 kg As/yr and 1.5×10^6 kg Sb/yr). Smokestack emissions from the smelter and subsequent atmospheric fallout were deemed to be an important route of heavy metal entry to estuarine waters. Thus, Schell and Nevissi[10] calculated large inputs of copper (~408 mt/yr), lead (~2477 mt/yr), and zinc (~744 mt/yr) to Puget Sound via this process.

Between 1965 and 1970, a mercury cell chlor-alkali industrial plant in Bellingham discharged an estimated 4.5–9 kg Hg/d to Bellingham Bay. High discharges of mercury were terminated by the plant in 1970, when the discharge rate dropped to about 0.1 kg Hg/d. At this time, peak mercury concentrations in bottom sediments amounted to about 100 μg/g within 100 m of the plant outfall and decreased to background levels at distances greater than 9 km. In areas away from industrial activities, such as the Strait of Juan de Fuca and Skagit Bay, mercury concentrations in bottom sediments varied between 0.01 and 0.1 μg/g.[17]

Sewage treatment plants have been a source of arsenic, antimony, mercury, and other heavy metals to Puget Sound. Of all sewage treatment facilities, the METRO plant at West Point in northwest Seattle discharged most of the heavy metal contaminants over the years. Prior to 1973, this plant annually discharged into Puget Sound 800 kg of arsenic, 420 kg of mercury, and 250 kg of antimony and may have contributed significantly to sediment contamination near the plant outfall.[17] Schell and Nevissi[10] suggested that the dumping of dredged spoils or sludge from sewage treatment plants probably increased heavy metal concentrations at some locations. However, the total input of heavy metals from the sewage treatment plants compared to all other sources appears to have been small. For example, Schell and Nevissi[10] estimated that <1.5% of the total concentration of Cu, Pb, and Zn entering the sound from all sources was derived from the METRO sewage treatment plant. They attributed most of the input of these metals throughout Puget Sound to natural rather than man-made sources. Nevertheless, anthropogenic input of the contaminants was responsible for locally elevated concentrations, particularly near point sources in urban embayments.

Crecelius and Bloom[1] examined temporal trends of cadmium, copper, lead, mercury, and silver in sediment cores from central Puget Sound. Profiles for all of these metals, except cadmium, revealed increased concentrations commencing during the 1880–1890 period, which reached maximum levels between 1950 and 1960. While cadmium displayed no significant change in concentration over the entire length of the cores, lead, mercury, and silver declined substantially after 1955. For example, the mean concentration of lead decreased 12% between 1955 and 1981, and that of mercury dropped 20% between the 1950s and 1970s. The maximum decrease in silver was only 8%. Copper showed no significant change between the mid-1950s and mid-1980s. Although cadmium concentrations in the central basin did not change significantly over a century of time, there is evidence of cadmium contamination — together with the other heavy metals — in relatively shallow areas near contaminant sources.[18]

Crecelius et al.[16] presented data on the concentrations of heavy metals in surface sediments at various locations in Puget Sound. Antimony, arsenic, and mercury concentrations in Bellingham Bay sediments ranged from 0.83–1.43 μg/g dry weight, 9.7–15.4 μg/g dry weight, and 0.10–100 μg/g dry weight, respectively. Antimony and arsenic levels were locally much higher in the Tacoma area, reaching maximum values of 10,000 μg/g dry weight and 12,500 μg/g dry weight, respectively. Mercury concentrations in this area ranged from 0.101–0.511 μg/g dry weight. In southern Puget Sound, the levels of arsenic (2.9–39 μg/g dry weight) exceeded those of antimony (0.59–12 μg/g dry weight). Mercury concentrations (0.015–0.234 μg/g dry weight) were typically higher in sediments of Bellingham Bay and the Tacoma area. Table 2 shows concentrations of heavy metals in Commencement Bay, Elliott Bay, and Nisqually Reach to other U.S. estuarine systems.

Long and Chapman[11] reviewed copper, lead, and zinc concentrations in sediments of Elliott Bay, Commencement Bay, Sinclair Inlet, Samish Bay, and Case Inlet. In Elliott Bay, the concentrations of copper, lead, and zinc in bottom sediments ranged from 34–206 μg/g dry weight, 49–2109 μg/g dry weight, and 86–413 μg/g dry weight, respectively. Elevated levels of these metals were also recorded in Commencement Bay (copper = 72–581 μg/g dry weight; lead = 28–791 μg/g dry weight; zinc = 57–1190 μg/g dry weight). Sinclair Inlet sediments had higher concentrations of zinc (156–238 μg/g dry weight) than copper (151–184 μg/g dry weight) and lead (98–136 μg/g dry weight). Concentrations of the metals were all <100 μg/g dry weight in sediments of Samish Bay and Case Inlet.

Becker et al.[14] compared the concentrations of selected heavy metals in sediments of Commencement Bay with those of Carr Inlet. Mean concentrations of arsenic, cadmium, copper, lead, and mercury in Commencement Bay sediments amounted to 270, 5.1, 370, 220, and 1.0 μg/g dry weight, respectively. Much lower mean concentrations of these metals were found in Carr Inlet sediments (arsenic = 3.4 μg/g dry weight; cadmium = 0.95 μg/g dry weight; copper = 6.4 μg/g dry weight; lead = 9.2 μg/g dry weight; and mercury = 0.04 μg/g dry weight). The much higher levels of the heavy metals in Commencement Bay reflect the greater anthropogenic inputs in this area of Puget Sound.

IV. BIOTIC IMPACTS

Puget Sound contains rich biological resources. Aside from extensive stocks of salmon, a variety of other species (e.g., cod and rockfish) support major commercial and recreational fisheries. Shellfish (e.g., clams, oysters, and crabs) also provide valuable commercial and recreational harvests. A significant fraction of the finfish and shellfish harvests occurs in industrialized areas of Puget Sound, where the biota may be affected by municipal and industrial wastewater discharges and other chemical contaminant inputs. A major concern involves the bioaccumulation of the contaminants and the potential health hazards they pose to humans who ingest contaminated seafood. Hence, investigations of chemical contaminant impacts have focused on biota collected in urban embayments of the sound.

The principal contaminants identified in the tissues of organisms inhabiting industrialized areas of Puget Sound include AHs, chlorinated hydrocarbons (e.g., DDT, PCBs, CBDs, and chlorinated benzenes), phthalate esters, and heavy metals. Of these contaminants, DDT (and its metabolites, principally p,p'-DDE) and PCBs have been particularly troublesome. Benthic invertebrates, as well as demersal and semi-demersal fish, collected in urban areas have typically displayed higher concentrations of PCBs and DDT compared with biota of nonurban reference sites.[2-5,13,15]

An array of biological impacts appears to be pollution related in Puget Sound (Table 3). Disease is an important consequence of chemical pollution of estuarine and marine waters. Skeletal and genetic abnormalities, physiological malfunctions, metabolic disorders, and some forms of cancer in estuarine and marine organisms have been linked to chemical pollution.[19] However, because of insufficient data, it is often difficult to state conclusively that a cause-and-effect relationship exists between chemical contamination and specific diseases in these organisms. In Puget Sound, numerous histopathological conditions purportedly occur in crabs (e.g., *Cancer gracilis* and *Cancer magister*), shrimp (e.g., *Crangon alaskensis*), demersal fishes (e.g., English sole *Pleuronectus vetulus* and starry flounder *Platichthys stellatus*) and other organisms inhabiting urban embayments, including necrotic and abnormal conditions in crab and shrimp hepatopancreas, antennal gland, bladder and midgut; necroses and melanized nodules and granulomas in shrimp gills; neoplasms (tumors), necroses, lesions, and other cellular disorders in fish livers; lesions and hyperplasia in fish gills; and lesions in fish kidneys, skin, fin, heart, gastrointestinal tract, spleen, gonad, and gall bladder. While some of these conditions are idiopathic with unknown etiology, others are likely caused by exposure to chemical contaminants.[7]

Gardner[20] specified six types of lesions in crabs and shrimp that may be related to chemical pollution of Puget Sound sediments based on thorough histological examination. These are: (1) tubular metaplasia; (2) vesiculate hepatopancreatic cells; (3) necrosis of the hepatopancreas; (4) necrosis of the bladder; (5) melanized nodules in gill stems of crabs; and (6) necrosis of digestive tubular epithelium of clams. Lesions appear less commonly in organs of reproduction and neurological control. Malins[21] ascertained that pathologic pigmentation (melanotic pigmentation of the carapace) in crabs and shrimp of Puget Sound corresponds to pollutant exposure.

The role of chemical contaminants in the induction of various neoplasm types in fish is the subject of ongoing investigations.[22-24] A strong correlation has been demonstrated between prevalences of hepatic neoplasms and neoplasia-related lesions in bottom-feeding fish (e.g., English sole *Pleuronectus vetulus*) and exposure to chemical contaminants in Puget Sound sediments.[4,24-27] The liver, in particular, has been a target organ of histopathological studies to assess the effects of xenobiotic chemical toxicants and carcinogens. Idiopathic liver lesions, including neoplasms, have been described in four demersal fish species from chemically contaminated waterways and embayments of Puget Sound: the English sole *Pleuronectus vetulus*, rock sole *Lepidopsetta bilineata*, staghorn sculpin *Leptocottus armatus*, and starry

TABLE 2
Trace Metals in Sediments from Selected Estuarine and Coastal Marine Systems in the United States (ppm)

Estuary	Chromium	Copper	Lead	Zinc	Cadmium	Silver	Mercury
Casco Bay, ME	92.10	16.97	29.13	76.27	0.15	0.09	0.12
Merrimack River, MA	41.15	6.47	23.25	35.75	0.07	0.05	0.08
Salem Harbor, MA	2296.67	95.07	186.33	238.00	5.87	0.88	1.19
Boston Harbor, MA	223.67	148.00	123.97	291.67	1.61	2.64	1.05
Buzzard's Bay, MA	73.66	25.02	30.72	97.72	0.23	0.37	0.12
Narragansett Bay, RI	93.60	78.95	60.25	144.43	0.35	0.56	0.00
East Long Island Sound, NY	37.63	11.26	22.13	58.83	0.11	0.15	0.09
West Long Island Sound, NY	131.50	111.00	69.75	243.00	0.73	0.68	0.48
Raritan Bay, NJ	181.00	181.00	181.00	433.75	2.74	2.06	2.34
Delaware Bay, DE	27.76	8.34	15.04	49.66	0.24	0.11	0.09
Lower Chesapeake Bay, VA	58.50	11.32	15.70	66.23	0.38	0.08	0.10
Pamlico Sound, NC	79.67	14.13	30.67	102.67	0.33	0.09	0.11
Sapelo Sound, GA	51.80	5.93	16.00	38.33	0.09	0.02	0.03
St. Johns River, FL	37.67	9.77	26.00	67.67	0.18	0.11	0.07
Charlotte Harbor, FL	26.47	1.17	4.33	7.20	0.08	0.01	0.02
Tampa Bay, FL	23.70	4.97	4.67	9.10	0.15	0.08	0.03
Apalachicola Bay, FL	69.17	16.93	30.67	111.67	0.05	0.06	0.06
Mobile Bay, AL	93.00	17.40	29.67	161.00	0.11	0.11	0.12
Mississippi River Delta, LA	72.27	19.40	22.67	90.00	0.47	0.17	0.06
Barataria Bay, LA	52.07	10.50	18.33	59.33	0.19	0.09	0.05
Galveston Bay, TX	41.13	8.03	18.33	33.97	0.05	0.09	0.03
San Antonio Bay, TX	39.43	5.57	11.33	32.00	0.07	0.09	0.02
Corpus Christi Bay, TX	31.43	6.63	13.00	56.00	0.19	0.07	0.04
Lower Laguna Madre, TX	24.53	5.83	11.33	36.00	0.09	0.07	0.03
San Diego Harbor, CA	178.00	218.67	50.97	327.67	0.99	0.76	1.04

San Diego Bay, CA	49.70	7.67	11.61	58.67	0.04	0.76	0.04
Dana Point, CA	39.80	10.03	18.80	53.67	0.22	0.80	0.13
Seal Beach, CA	108.33	26.00	27.37	125.00	0.17	1.27	0.59
San Pedro Canyon, CA	106.50	31.33	17.33	118.33	1.17	1.20	0.32
Santa Monica Bay, CA	53.53	10.53	33.37	46.67	0.18	0.51	0.01
San Francisco Bay, CA	1466.67	160.71	67.39	501.66	0.51	0.37	0.25
Bodega Bay, CA	246.33	0.06	2.17	38.33	0.18	1.74	0.14
Coos Bay, OR	110.30	1.47	4.65	32.00	0.62	0.31	0.11
Columbia River Mouth, OR/WA	29.53	17.00	15.90	107.67	0.86	2.14	0.25
Nisqually Reach, WA	118.07	13.33	24.57	105.33	0.68	2.62	0.32
Commencement Bay, WA	69.50	51.33	34.63	101.00	0.77	5.90	0.01
Elliott Bay, WA	114.37	96.00	20.23	166.00	0.84	1.18	0.11
Lutak Inlet, AK	58.27	26.67	15.90	180.33	0.96	0.09	0.24
Nahku Bay, AK	23.27	9.80	43.30	191.33	1.09	4.37	0.23
Charleston, SC	86.33	16.03	27.33	72.67	—	—	—

From Young, D. and Means, J., in *National Status and Trends Program for Marine Environmental Quality, Progress Report on Preliminary Assessment of Findings of the Benthic Surveillance Project, 1984*, National Oceanic and Atmospheric Administration, Rockville, MD, 1987; U.S. Geological Survey, National Water Summary, 1986.

TABLE 3
Types of Biological Impacts That May Be Pollution Related in Puget Sound[a]

Effects	Species	References(s)	Comments
Acute lethality	Fish and invertebrates	Quinlan et al., 1985	Mass mortalities were common prior to effluent controls on pulp mills and other discharges
		Strickland, 1983	Mortalities occur in areas where oxygen deficiencies occur (i.e., Budd Inlet)
	Phoxocephalid amphipods, esp. *Rhepoxynius abronius*	Quinlan et al., 1985	Absent from contaminated areas (i.e., Commencement Bay, Denny Way CSO) possible due to sediment lethality
	Oyster larvae, *Crassostrea gigas*	Cardwell et al., 1977, 1979	Mortalities occur due to dinoflagellate blooms, which may be pollution-related
Histopathological abnormalities			
Lesions	Clams, *Macoma carlottensis*	Malins et al., 1980	Limited sampling
	Shrimp, *Pandalus danae*	Malins et al., 1980	Limited sampling
	Crab, *Cancer magister*	Malins et al., 1980	Limited sampling
	Crab, *Cancer productus*	Malins et al., 1980	Limited sampling
	Crab, *Cancer gracilis*	Malins et al., 1980	Limited sampling
	Sculpin, *Leptocottus armatus*	Malins et al., 1980, 1982a	Broad-scale sampling
	Sole, *Parophrys vetulus*	Malins et al., 1980, 1982a	Broad-scale sampling
	Sole, *Microstomus pacificus*	Malins et al., 1980, 1982a	Broad-scale sampling
	Tomcod, *Microgadus proximus*	Malins et al., 1980	Limited sampling
Fin rot	Sole, *Parophrys vetulus*	Harper-Owes, 1982; Malins et al., 1982b	Broad-scale sampling
	Flounder, *Platichthys stellatus*	Harper-Owes, 1982; Malins et al., 1982b	Broad-scale sampling
Community changes	Phytoplankton	Quinlan et al., 1985	An increased prevalence of red tides and dinoflagellate blooms may be occurring in Puget Sound

TABLE 3 (CONTINUED)
Types of Biological Impacts That May Be Pollution Related in Puget Sound[a]

Effects	Species	References(s)	Comments
	Benthic fauna	Quinlan et al., 1985	Differences in benthic community species composition have been noted in chemically contaminated areas compared to reference areas
Avoidance response	Polychaetes: *Mediomastus californiensis, Laonice cirrata, Nephthys ferrunginea, Tauperia oculata, Acontorhina cyclia, Euphilomedes producta*	Comiskey et al., 1984	"Avoid" the area of the Denny Way CSO
Possible reproductive effects	Marine mammals: Seal, *Phoca vitulina richardsii* Porpoise, *Phocena phocena*	Calambokidis et al., 1984	High tissue PCB levels, possible reproductive failures in 1960–1970s
	Waterfowl: Guillemot, *Cephus columba*	Riley et al., 1984	High tissue PCB levels, addled eggs

Note: See original source for reference(s) listed.

[a] Does not include laboratory sediment bioassay data; refers to *in situ* effects observed over large spatial areas.

From Chapman, P. M., in *Oceanic Processes in Marine Pollution*, Vol. 5, *Urban Wastes in Coastal Marine Environments*, Wolfe, D. A. and O'Connor, T. P., Eds., Robert E. Krieger Publishing Company, Malabar, FL, 1988, 170. With permission.

flounder *Platichthys stellatus*.[4,13,28-32] Table 4 summarizes histomorphological characteristics of hepatic lesions in English sole. Continual exposure of flatfish to hepatocarcinogens and hepatotoxins in sediments of urban embayments seems to play an integral role in the development of multiple types of lesions, such as hepatocellular and biliary neoplasms, non-neoplastic proliferative conditions, and foci of hepatocellular alteration (putative preneoplastic lesions).[31]

Although lesions in Puget Sound bottom fish are often detected in the liver, other organs commonly affected by these disease conditions include the kidney and gills. Pathological changes in filtering organs commonly develop in fish inhabiting chemically polluted environments. However, some of the statistically most defensible associations of pollution and disease in fish inhabiting these degraded environments can be found in integumentary lesions that predominate in individuals continuously in contact with chemical contaminants in bottom sediments.[33]

TABLE 4
Histomorphological Characteristics of Hepatic Lesions Found in English Sole from Eagle Harbor, Washington

Liver Lesion	Tissue Architecture	Compression of Adjacent Tissue	Presence of Associated Degeneration or Necrosis	Presence of Associated Regenerative Tissue	Presence of Associated Fibroplasia
Nonspecific necrosis	Usually non-nodular, normal muralial architecture; includes coagulation and liquefaction necrosis and hyalinization	No	—	Frequently	Occasionally
Cystic parenchymal degeneration	Loss of parenchyma in cystic pattern without cellular debris; cystic area filled with fine flocculent material; little or no cellular inflammatory response	No	Rarely	No	No
Nuclear pleomorphism	Nonhyperplastic; non-nodular; normal muralial architecture	No	Rarely	Occasionally	Rarely
Megalocytic hepatosis	Nonhyperplastic, non-nodular normal muralial architecture, distribution from focal to diffuse	Occasionally, minor; due to cell hypertrophy	Yes	Yes	Yes
Hepatic steatosis	Nonhyperplastic; normal muralial architecture; usually non-nodular	No	Occasionally	No	No
Hemosiderosis	Nonhyperplastic; non-nodular; normal muralial architecture	No	Frequently	Occasionally	Occasionally
Eosinophilic focus	Hyperplastic; normal muralial architecture; nodular, well-defined focus; muralia of focus blend into surrounding parenchyma	No or minor; sinusoidal compression within focus	No	No	No
Basophilic focus	Hyperplastic; normal muralial architecture; often non-nodular focus with irregular margins; muralia of focus blend into surrounding parenchyma	No or minor; occasional sinusoidal compression within focus	No	No	No
Clear cell focus	Hyperplastic; normal muralial architecture; nodular, well-defined focus; muralia of focus blend into surrounding parenchyma	No or minor; sinusoidal compression within focus	No	No	No

Hyperplastic regenerative focus	Hyperplastic; normal muralial architecture usually retained; non-nodular with poorly defined margins; muralia >2 cell layers thick	No	Yes	Yes	Occasionally
Liver cell adenoma	Hyperplastic; usually normal muralial structure retained; absence of biliary, pancreatic, and macrophagic components; occasional sinusoidal dilation or ectasia within tumor; usually nodular, well-defined area with clear separation from surrounding parenchyma	Yes; expansive but not invasive	Occasionally	No	Rarely, but sometimes encapsulated
Hepatocellular carcinoma	Hyperplastic; usually trabecular, occasional sheetlike or acinar morphology; absence of biliary, pancreatic, and macrophagic components; frequent sinusoidal dilation or ectasia within tumor; irregular muralial architecture; clear separation from surrounding parenchyma	Yes; usually also invasive; compression of sinusoids	Occasionally	No	Occasionally
Cholangioma	Retention of tubular architecture, but occasional regions of squamous to cuboidal epithelia; usually nodular, well-defined tumor	Yes	Occasionally	No	Yes; usually dense; occasionally encapsulated
Cholangiocellular carcinoma	Loss of tubular architecture; squamous to cuboidal epithelia in broad or papillary sheets with irregular borders	Yes; frequently invasive; metastases demonstrated	Occasionally	No	Yes; usually dense
Mixed hepatocholangio-cellular carcinoma	Hepatocellular component with architecture of hepatocellular carcinoma, often with complete loss of organization; biliary component with architecture of cholangiocellular carcinoma; extensive intermingling of components of both cell types; absence of macrophagic components, other hepatic elements	Yes; frequently invasive	Frequently	No	Yes; usually with biliary component
Nonhyperplastic regeneration	Non-nodular; normal muralial architecture	No	Yes	Yes	Occasionally
Cholangiofibrosis	Hyperplastic; tubular architecture retained, but tubules are dilated	No	No	No	Yes; usually dense; encapsulated

From Myers, M. S., Rhodes, L. D., and McCain, B. B., *J. Natl. Cancer Inst.*, 78, 334, 1987. With permission.

A. BENTHIC INFAUNAL COMMUNITIES

1. Community Structure

Variations in species abundance of benthic invertebrates of Puget Sound, as in other estuaries, can be dramatic. Interannual variations in species abundance often greatly exceed seasonal variations in estuaries, even though the species composition may remain reasonably stable over long periods of time.[33-37] Large changes in abundance of the macrofauna over an annual period frequently are ascribed to normal period-icities of reproduction, recruitment, and mortality. However, in many cases they have been attributed to random environmental and anthropogenic perturbations, where vagaries in physical or chemical conditions cause large aperiodic density changes. While physical and chemical factors clearly influence the success of benthic infauna in an area, biological factors (e.g., competition and predation) must also be considered since they can act as major limiting factors. Hence, the mere tolerance of a species to physical and chemical conditions commonly do not provide sufficient explanation for observed abundance and distribution patterns.

Benthic data collected over a 20-year period at 200-m depth in the main basin of Puget Sound with a 0.1-m^2 van Veen grab sampler indicated large shifts in numerical abundance of the common species in spite of low frequency of physical disturbance.[37] The bivalve *Macoma carlottensis* consistently dominated the benthic invertebrate community in most years, with an average density of 600/m^2 and, on two sampling dates, with a density exceeding 2000/m^2. Of the more than 120 benthic invertebrate species identified in the main basin of Puget Sound, many are equilibrium species with a longevity greater than 1 year. Nearly all of these species occur in densities much less than 100/m^2.

Nichols[37] tracked gradual changes in the numerical dominance of benthic invertebrates in Puget Sound from 1963 to 1983. Community dominance changed rapidly between 1964 and 1978. The collective abundance of *Macoma carlottensis*, *Pectinaria californiensis*, *Ampharete acutifrons*, and *Axinopsida sericata* — the four species contributing most to the total faunal abundance — progressively increased during this interval. Numerical shifts in dominance of the common species appeared at irregular multiyear frequencies. The clam *M. carlottensis* was the dominant species of the benthic community from 1964 to 1968. It declined sharply in abundance in 1969 and 1970, being replaced as the dominant species by the polychaete *Pectinaria californiensis* which reached a density of 1000/m^2. The polychaete *Ampharete acutifrons*, normally a minor component of the community, underwent pulsing to a mean abundance greater than 2000/m^2 in fall 1979. In 1982 and 1983, the bivalve *Axinopsida sericata* increased abruptly in abundance to nearly 100/m; the rise in abundance approached 1000/m^2 during the mid-1980s. Although *M. carlottensis* decreased acutely in abundance in the late 1960s, it began to return to numerical dominance in late 1970 and 1971, attaining a mean abundance above 2000/m^2 in 1977 and 1982.

It is unclear if pollution has played a significant role in the temporal variations observed in species abundance of benthic infauna in the central basin. The relative importance of natural vs anthropogenic factors in controlling the structure of benthic communities has not been unequivocally established for most areas of the sound. The high natural variability in population abundance and composition tends to confound interpretation of pollution data for the benthic species.

In Commencement Bay, there is some evidence that chemical pollution has impacted the benthic community which is composed mainly of bivalves and poly-chaetes. For example, Long[7] stated that infauna taxa richness was highest in reference areas (10–11 taxa) and generally lowest (2–3 taxa) in parts of Commencement Bay. Although in some areas of Commencement Bay the benthic community is charac-terized by low diversity, the few taxa present have high abundance.[38] These patterns typify stressed communities. Similar community structure is apparent in some of the waterways of Puget Sound, such as Hylebos Waterway[38] and Kellogg Island in the Duwamish River,[39] where contamination by PCBs, CBD, or other chemical compounds is high, and more sensitive organisms and life history stages may be excluded from impacted sites. At the Denny Way combined sewage overflow, the pollution-tolerant polychaete *Capitella capitata* exhibited great seasonal variation but was consistently the most common species found in this area during a five-year study period.[40] The persistent occurrence of *C. capitata* at this site reflects the long-term influence of pollutant loadings.

Long and Chapman[11] indicated that in waters impacted by chemical pollution in Puget Sound (i.e., Commencement Bay, Elliott Bay, Duwamish Waterway, and Sinclair Inlet), polychaetes and mollusks comprised the majority of the fauna (73.2–98.4%). At nonimpacted reference stations (i.e., Case Inlet and Samish Bay), abundance of polychaetes and mollusks decreased to 23.8–47.0% of the totals, with echinoderms and arthropods being more abundant than at the aforementioned impacted areas. Members of the pollution-sensitive amphipod family Phoxocepha-lidae, while abundant at the reference stations in Case Inlet and Samish Bay, were absent at the pollution impacted sites. In Elliott Bay, phoxocephalids comprised only 0.7% of the benthic community. Chapman et al.[41] corroborated these findings.

Swartz et al.[6] examined the toxicity of Commencement Bay sediments on marine infaunal amphipods (*Rhepoxynius abronius*). Survival of the amphipods was assessed by short-term (10-day) exposure to sediments collected from the offshore, deeper parts of the bay as well as major industrialized waterways (i.e., Blair, City, Hylebos, and Sitcum waterways). Results showed that survival of the organisms relative to controls was high after exposure to sediments from the deeper portion of Com-mencement Bay but low after exposure to sediments from some parts of the water-ways. A correlation also existed between amphipod distribution and sediment tox-icity, with higher amphipod density and species richness observed in the deeper areas of the bay than in the waterways. Phoxocephalid amphipods were absent from the waterways but ubiquitous in the deeper bay, reflecting the effects of sediment contamination.

Long[7] determined that lesions in crabs (i.e., *Cancer gracilis* and *C. magister*) and shrimp (i.e, *Crangon alaskensis*) were generally most common in Commence-ment and Elliott Bays. For example, gill nodules predominated in crabs from the lower Duwamish Waterway, midgut nodules in crabs from the lower Duwaminsh Waterway, and midgut necroses in crabs from the Seattle waterfront. Antennal gland

and gill necroses in shrimp occurred most frequently in specimens from the Elliott Bay and the upper Duwamish Waterway, respectively.

2. Contaminant Concentrations

Ginn and Barrick[5] compared PCB concentrations in invertebrate taxa from industrialized urban areas of the Duwamish River estuary (Elliott Bay) and Hylebos Waterway (Commencement Bay) to those from nonurban reference locations. PCB concentrations in tissues of polychaete, mussel, clam, and shrimp species from the Duwamish River estuary amounted to 250, 92–210, 50–180, and 480 ng/g wet weight, respectively, and those from the Hylebos Waterway, 66–260, 72, 54–120, and 800 ng/g wet weight, respectively. Much lower concentrations were recorded at reference sites for polychaete (45 ng/g wet weight), mussel (10–30 ng/g wet weight), clam (<2–30 ng/g wet weight), and shrimp (26–54 ng/g wet weight) species. The highest PCB levels were observed in the hepatopancreas of crabs from the Duwamish River estuary (9600 ng/g wet weight) and Hylebos Waterway (3600 ng/g wet weight). Again, much lower PCB concentrations occurred in the hepatopancreas of crabs from reference areas (130 ng/g wet weight).

Highest concentrations of heavy metals in invertebrates have also been documented in industrialized areas. For instance, the levels of chromium, lead, and zinc in mussels near major Seattle sewage outfalls were substantially higher than in mussels from control sites.[5,42] Similarly, the concentrations of chromium, copper, and lead in dungeness crabs at the City and Hylebos Waterways in Commencement Bay were 2–5 times higher than at control sites (e.g., Discovery Bay).[5]

Bivalve mollusks tend to bioaccumulate heavy metals, because they have little capability of regulating metal concentrations in their tissues. While blue mussels *Mytilus edulis* accumulated high concentrations of copper (75 mg/kg wet weight) and mercury (0.836 mg/kg wet weight) near the ASARCO refinery in Commencement Bay, the concentrations of copper (7.4 mg/kg wet weight) and mercury (0.109 mg/kg wet weight) in *M. edulis* removed from pollutant sources in the central basin of the sound were comparable to those of reference areas in Sequim Bay (copper = 5.8 mg/kg wet weight; mercury = 0.102 mg/kg wet weight).[5] These values reflect the overriding importance of local sources of heavy metal contamination in the urban embayments.

B. Fish

1. Fisheries Trends

Pollution-related effects on fish in Puget Sound are primarily manifested as diseases in demersal species. Various histopathological conditions identified in these bottom-dwelling forms, such as neoplasms, necroses, and lesions, have served as useful indicators of environmental degradation in certain areas of the sound. Aside from the conspicuous histopathological effects chronicled in demersal fish of Puget Sound,[3,4,13,15,26-32] there is often little evidence of pollution impacts on fish assemblages. For example, the richness and diversity of demersal fish species appear to be highest in chemically contaminated urban areas such as Commencement and

Elliott Bays.[3,7,43] Contaminated waterways in Commencement Bay are important rearing and feeding areas for many species. The mouth of the Hylebos Waterway contains high densities (>30,000/m^2) of gammarid amphipods and harpacticoid copepods. This area represents a major feeding area for salmon and other valuable species (e.g., flatfish).[38]

Puget Sound supports a diverse assemblage of fish populations. Long[7] delineated highest demersal fish abundance, species richness, and species diversity in Commencement and Elliott Bays, the two most contaminated embayments in Puget Sound. Pedersen and DiDonato[44] listed a total of 53 species of ground fishes which contribute to the recreational and commercial fisheries of the sound. However, fisheries catch statistics have not been stable over the years. Kimura and Millikan[45] noted a downward trend for Pacific hake harvests, as did Bargman[46] for lingcod harvests. Erratic recruitment and overfishing are possible causes for the declines.[43] Harper-Owes[47] uncovered significant decreases of chum salmon runs. Landahl and Johnson[48] revealed diminishing commercial bottom-trawl catches of English sole throughout the sound since 1967. Although declines in various fisheries of Puget Sound have developed over the years, data needed to substantiate the possibility of direct effects of toxicants on the fisheries are largely lacking. The causes of variable fisheries catches in Puget Sound are not completely understood, but at present the effects of chemical contaminants as a potentially vital factor cannot be discounted.

2. Contaminant Concentrations

Demersal fish collected at various locations in Puget Sound have been analyzed for concentrations of selected halogenated hydrocarbons and heavy metals. Malins et al.[4] contended that the concentrations of metabolically resistant organochlorine contaminants (e.g., PCBs, CBDs, and HCB) in the liver and muscle of English sole *Pleuronectes vetulus* were generally high relative to the concentrations in bottom sediments. However, the levels of the more metabolically labile compounds (e.g., AHs) and heavy metals were typically lower in the fish than in the sediments. The mean concentrations of AHs, PCBs, CBDs, and HCB in the liver of English sole from the Hylebos Waterway in Commencement Bay amounted to <0.05, 39, 1.7, and 1.1 μg/g dry weight, respectively, and in the liver of English sole from the Duwamish Waterway of Elliott Bay, <0.05, 47, <0.07, and 0.035 μg/g dry weight, respectively. Lower values were recorded in English sole from nonurban embayments (i.e., Port Madison and Case Inlet), where the mean concentrations of AHs, PCBs, CBDs, and HCB were <0.05, 2.3, <0.015, and 0.01 μg/g dry weight, respectively. In muscle tissue of English sole from the Hylebos Waterway, the mean concentrations of AHs (<0.05 μg/g dry weight), PCBs (3.4 μg/g dry weight), CBDs (0.26 μg/g dry weight), and HCB (0.17 μg/g dry weight) compared favorably with the mean concentrations of AHs (<0.05 μg/g dry weight), PCBs (4.8 μg/g dry weight), CBDs (<0.015 μg/g dry weight), and HCB (0.008 μg/g dry weight) in muscle tissue of English sole from the Duwamish Waterway. No values were reported for PCBs, CBDs, and HCB in muscle tissue of samples from Case Inlet and Port Madison.

Stein et al.[15] determined PCB concentrations in liver tissue of three species of benthic flatfish (English sole *Pleuronectes vetulus*, rock sole *Lepidopsetta bilineata*, and starry flounder *Platichthys stellatus*) from five locations in Puget Sound. Fish samples were collected from the highly contaminated Duwamish and Hylebos Waterways, the moderately contaminated Everett Harbor, and the lowly contaminated Pilot Point and Polnell Point areas. Highest mean concentrations of PCBs were found in Duwamish Waterway samples; values in English sole (3700 ng/g wet weight) and rock sole (3000 ng/g wet weight) far exceeded those in starry flounder (1900 ng/g wet weight). Lowest mean concentrations were generally observed in samples from the less contaminated sites of Everett Harbor, Pilot Point, and Polnell Point (Table 5).

TABLE 5
PCB Concentrations in Livers of Three Flatfish Species from Five Locations in Puget Sound[a,b]

	Location				
Species	Duwamish Waterway	Hylebos Waterway	Everett Harbor	Pilot Point	Polnell Point
English sole	3700 ± 370 (4)	2400 ± 800 (5)	260 ± 31 (4)	300 ± 63 (4)	400 ± 57 (4)
Rock sole	3000 ± 650 (4)	550 ± 44 (3)	730 ± 14 (2)	540 ± 68 (4)	340 ± 67 (4)
Starry flounder	1900 ± 140 (40)	—	—	—	280 ± 30 (4)

[a] Concentrations shown as mean ± SE.
[b] Number of samples in parentheses.

Compiled from Stein et al., *Mar. Environ. Res.*, 35, 95, 1993. With permission.

Figure 3 illustrates concentrations of chlorinated hydrocarbons (i.e., PCBs and DDT) in sediments and livers of English sole from Commencement Bay, Elliott Bay, and Case Inlet. By comparison, concentrations of chlorinated hydrocarbons in sediments and livers of three other species of fish are depicted for Santa Monica Bay in the Southern California Bight. High PCB concentrations are evident in sediments and fish livers from both Commencement and Elliott Bays.

Muscle tissues of English sole collected from both Commencement and Elliott Bays also exhibit elevated levels of PCBs, with mean values ranging from 190–900 µg/g wet weight in contaminated areas. By comparison, Discovery Bay reference concentrations were <13 ng/g wet weight. Similarly, the mean values of DDT in muscle tissues of English sole collected from both Commencement and Elliott Bays (7–20 ng/g wet weight) exceeded those from Discovery Bay (<5 ng/g wet weight).[5] Volatile chlorinated ethylenes have also been confirmed in muscle tissues of fish from Hylebos Waterway in Commencement Bay with concentrations of ~30–40 ng/g wet weight.[49,50]

PCB concentrations in fish livers from urban embayments in Puget Sound rank among the highest estuarine and coastal marine systems in the United States.

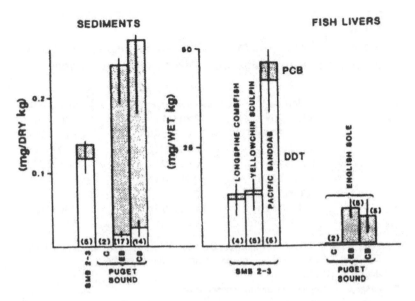

FIGURE 3 Comparison of concentrations of chlorinated hydrocarbons in sediment and fish livers at Puget Sound, Washington, with concentrations found at Santa Monica Bay, Southern California Bight. Sampling sites in Puget Sound: Commencement Bay (CB), Elliott Bay (EB), and Case Inlet (C); Case Inlet is the control. Sampling Sites in Santa Monica Bay (SMB 2-3). (Modified from Brown, D. A. et al., *Mar. Environ. Res.*, 18, 291, 1986. With permission.)

Systems with the highest mean PCB values in fish livers, based on results of the 1984 National Status and Trends Benthic Surveillance Project, included Elliott Bay (4.23 µg/g wet weight), Boston Harbor (2.62 µg/g wet weight), and Commencement Bay (2.30 µg/g wet weight). Data collected in the 1970s and 1980s do not indicate any substantial decrease in PCB contamination of estuarine and coastal marine fish livers on a national basis.[51] Since the mid-1980s, there has been no clear evidence of a largescale nationwide decrease of PCB contamination in estuarine and marine environments, although major declines have been recorded in proximity to some known industrial sources and other "hot spot" areas.[52]

Puget Sound fish have considerable ability to regulate heavy metal concentrations in their muscle tissues unlike benthic invertebrates. For example, analysis of heavy metals (arsenic, cadmium, chromium, copper, lead, mercury, and zinc) in the muscle of fish from Commencement and Elliott Bays revealed only neglible increases compared to fish from nonindustrialized Discovery Bay (Table 6). Hence, the concentrations of copper in fish muscle from Commencement Bay (0.62 mg/kg wet weight) and Elliott Bay (0.75 mg/kg wet weight) were only slightly higher than those from Discovery Bay (0.45 mg/kg wet weight). Similar consistent heavy metal values in muscle tissue of fish from Commencement, Elliott, and Discovery bays have been documented for cadmium (0.005, 0.006, and 0.005 mg/kg wet weight, respectively), chromium (0.18, 0.16, and 0.10 mg/kg wet weight, respectively), lead (1.18, 0.024, and 0.40 mg/kg wet weight, respectively), and zinc (7.4, 4.4, and 5.5 mg/kg wet weight, respectively). These data signify a lack of substantial heavy

TABLE 6
Comparison of Heavy Metal Concentrations (mg/kg wet weight)
in the Muscle Tissue of Fish from the Puget Sound Region[a]

Metal	Elliott Bay[b]	Pt. Defiance[c] (Commencement Bay)	Discovery Bay[c]
As	3.12 (6.10)[d]	7.80 (34.1)	2.40 (5.3)
Cd	0.006 (0.006)	0.005 (0.012)	0.005 (0.013)
Cr	0.16 (0.16)	0.18 (0.41)	0.10 (0.48)
Cu	0.75 (0.89)	0.62 (1.2)	0.45 (0.93)
Hg	1.130 (0.180) (0.08)	0.050 (0.11)	
Pb	0.024 (0.032)	1.18 (10.4)	0.40 (0.61)
Zn	4.4 (5.0)	7.4 (10.7)	5.5 (7.3)

[a] Species sampled include sole (three species), walleye pollock, cod, salmon, and sculpin.
[b] Galvin et al. (1984). (*Note:* See original source for this reference.)
[c] Gahler et al. (1982).
[d] Average concentration (maximum concentration).

From Ginn, T. C. and Barrick, R. C., in *Oceanic Processes in Marine Pollution*, Vol. 5, *Urban Wastes in Coastal Marine Environments*, Wolfe, D. A. and O'Connor, T. P., Eds., Robert E. Krieger Publishing Company, Malabar, FL, 1988, 157. With permission.

metals accumulating in Puget Sound fish muscle, even in highly contaminated areas. In contrast, some heavy metal concentrations (i.e., copper and lead) in liver tissues of English sole collected from waters near Seattle have been elevated relative to controls.

3. Diseases

There is a growing database linking accumulation of chemical contaminants in bottom sediments of Puget Sound with adverse biological effects in benthic fish. English sole exposed to toxic chemicals, in particular, display increased prevalences of neoplasms and other hepatic diseases, as well as reproductive impairment.[4,15,26-32,48,53-55] Puget Sound sediments are contaminated by high levels of AHs, and they contain multiple hepatocarcinogens.[31] PAH compounds in bottom sediments have been strongly implicated as causative agents in the etiology of neoplasms and preneoplastic lesions in the liver of this species. Other species appear to be less affected. For example, Collier et al.[56] asserted that in the Duwamish Waterway the prevalence of hepatic neoplasms (hepatocellular carcinoma, hepatocellular adenoma, and cholangiocellular carcinoma) in adult English sole typically ranges from 20–30% compared to <1% prevalence of hepatic neoplasms in adult starry flounder from the same area. Higher prevalences of several putatively preneoplastic liver lesions also occur in subadult English sole than in subadult starry flounder. The associations between chemical contaminants in bottom sediments and the genesis of lesions in fish tissues have been substantiated by field surveys, long-term laboratory studies, statistical analyses, and epizootiological models.[57]

Long[7] followed the trends of four idiopathic liver conditions in English sole and rock sole from Puget Sound. Nonspecific necroses (i.e., degenerative lesions encompassing several subcellular and cellular perturbations), although observed in fish throughout the sampling areas, occurred in highest percentages in fish collected in Commencement and Elliott Bays (e.g., 24% in English sole from the upper Duwamish Waterway). Specific degenerative/necrotic lesions (i.e., megalocytic hepatosis and nuclear pleomorphism) also were most common in fish from these two bays, being highest in English sole from the upper Duwamish Waterway (21.3%) and near the Seattle waterfront (19.9%). Hepatic neoplasms (principally of hepatocellular, cholangiocellular, and mesenchymal cell origin) were most frequently discerned in fish from the upper Duwamish Waterway (16.2% English sole; 5.5% rock sole). They were not seen in fish from reference sites. Preneoplastic conditions peaked in fish (21% in English sole) from the upper Duwamish Waterway. Thus, geographical trends in the frequency of liver disorders correlate well with high contaminant levels.

Sol et al.[58] investigated the impact of chemical contaminants in Puget Sound on reproductive function in English sole and rock sole. Fish exposure to and uptake of AHs appear to be closely correlated with the development of several types of reproductive dysfunctions, including inhibited gonadal development, depressed plasma concentrations of reproductive steroids, reduced spawning success, reduced egg size, and reduced egg and larval viability. Population models developed for English sole from polluted areas of Puget Sound suggest that contaminant-related decreases in reproductive output may potentially lower the growth rate of English sole subpopulations in urban areas.

Johnson et al.[59] inferred that chemical contaminant exposure may interfere with ovarian development in female English sole. They conveyed that female sole from heavily contaminated sites in Puget Sound are less likely than other female sole to undergo ovarian development. These fish were significantly less likely to experience gonadal recrudescence, and they had lower mean levels of plasma estradiol than females from less contaminated sites. Although the effects of these abnormalities on the abundance of English sole in contaminated embayments of Puget Sound are not known, the reproductive condition and long-term viability of flatfish stocks residing in these areas may be potentially threatened by continued anthropogenic contaminant accumulation in bottom sediments.

Stein et al.[60] recently evaluated the effects of contaminant exposure of juvenile chinook salmon (*Oncorhynchus tshawytscha*) in the Duwamish Waterway and Commencement Bay. They found significantly higher mean concentrations of PAHs and PCBs in stomach contents and PCBs in the liver of salmon from these two urban estuaries compared to fish from a nonurban system (Nisqually estuary). They determined that increased exposure of the salmon to toxic chemicals in the two urban embayments led to the induction of a hepatic enzyme (cytochrome P4501A) that plays a central role in the metabolism of carcinogenic PAHs. Mean concentrations of fluorescent aromatic compounds in bile, an estimate of exposure to PAHs, were significantly higher in fish from the two contaminated urban systems than from the nonurban estuary. Damage to hepatic DNA was also associated with fish exposure to the contaminants, as evidenced by increased levels of putative xenobiotic-DNA adducts. Altered immunocompetence in juvenile chinook salmon from the Duwamish Waterway suggest that the contaminants may adversely affect the physiological fitness of these organisms as they migrate through the waterway and Commencement Bay. It is unknown whether exposure to the contaminants and the resulting altered immunocompetence lowers survivorship of the fish.

C. MARINE MAMMALS

PCB contamination, while present in all biotic groups examined in Puget Sound, is pronounced in marine mammals due to biomagnification in the food chain. Calambokidis et al.[61] described PCB and DDT contamination in various marine mammals of Puget Sound, focusing on harbor and Dall's porpoise, sea lions, harbor seals, as well as minke, killer, and pygmy sperm whales. Based on analysis of ~100 marine mammals from the sound, the authors concluded that the animals bioconcentrate PCBs and DDT from the fish they consume. Harbor seals, in particular, have elevated levels of PCBs in their tissues, ranking them among the highest found anywhere in the world. Harbor seal populations with the highest pup mortalities bioaccumulated both PCBs and DDT.[8,62] Because of declining DDT levels in Puget Sound, however, the principal pollutant threat to the marine mammals appears to be PCBs. No other chemical contaminants measured in the tissues of this group of animals, including heavy metals, PAHs, and other organochlorine compounds, approach the concentrations great enough to pose a serious health threat.

D. MARINE BIRDS

The adipose tissues of great blue herons collected from Commencement and Elliott Bays had PCB levels ranging from 14–80 mg/kg wet weight. By comparison, PCB concentrations in liver tissue of these birds ranged from 0.19–3.2 mg/kg wet weight.[63] These values are greater than those of birds sampled in reference areas, and they likely reflect biomagnification effects.

Dames and Moore[38] recorded 76 species of waterfowl (both resident and migratory forms) in contaminated areas of Commencement Bay, as did the U.S. Army Corps of Engineers[64] along the Duwamish River. Riley et al.[63] observed only two nests of pigeon guillemot in the Duwamish West Waterway. Two live young and four addled eggs were found, indicating a hatching success rate of only 33%. One of the eggs exhibited high PCB concentrations. Elevated levels of PCBs and mercury were also present in tissues of waterfowl. Available data suggest, therefore, that some marine birds in Puget Sound may also be adversely affected by chemical contaminant exposure.[8]

REFERENCES

1. Crecelius, E. A. and Bloom, N., Temporal trends of contamination in Puget Sound, in *Oceanic Processes in Marine Pollution*, Vol. 5, *Urban Wastes in Coastal Marine Environments*, Wolfe, D. A. and O'Connor, T. P., Eds., Robert E. Krieger Publishing Company, Malabar, FL, 1988, 149.
2. Malins, D. C., McCain, B. B., Brown, D. W., Sparks, A. K., and Hodgins, H. O., Chemical Contaminants and Biological Abnormalities in Central and Southern Puget Sound, NOAA Tech. Memo. OMPA-2, U.S. National Oceanic and Atmospheric Administration, Boulder, CO, 1980.
3. Malins, D. C., McCain, B. B., Brown, D. W., Sparks, A. K., Hodgins, H. O., and Chain, S. L., Chemical Contaminants and Abnormalities in Fish and Invertebrates from Puget Sound, NOAA Tech. Memo. OMPA-19, U.S. National Oceanic and Atmospheric Administration, Boulder, CO, 1982.
4. Malins, D. C., McCain, B. B., Brown, D. W., Chan, S.-L., Myers, M. S., Landahl, J. T., Prohaska, P. G., Friedman, A. J., Rhodes, L. D., Burrows, D. G., Gronlund, W. D., and Hodgins, H. O., Chemical pollutants in sediments and diseases of bottom-dwelling fish in Puget Sound, Washington, *Environ. Sci. Technol.*, 18, 705, 1984.
5. Ginn, T. C. and Barrick, R. C., Bioaccumulation of toxic substances in Puget Sound organisms, in *Oceanic Processes in Marine Pollution*, Vol. 5, *Urban Wastes in Coastal Marine Environments*, Wolfe, D. A. and O'Connor, T. P., Eds., Robert E. Krieger Publishing Company, Malabar, FL, 1988, 57.
6. Swartz, R. C., Deben, W. A., Sercu, K. A., and Lamberson, J. O., Sediment toxicity and distribution of amphipods in Commencement Bay, Washington, USA, *Mar. Pollut. Bull.*, 13, 359, 1982.
7. Long, E. R., An assessment of marine pollution in Puget Sound, *Mar. Pollut. Bull.*, 13, 380, 182.
8. Chapman, P. M., Summary of biological effects in Puget Sound — past and present, in *Oceanic Processes in Marine Pollution*, Vol. 5, *Urban Wastes in Coastal Marine Environments*, Wolfe, D. A. and O'Connor, T. P., Eds., Robert E. Krieger Publishing Company, Malabar, 1988, 169.
9. Barrick, R. C., Hedges, J. I., and Peterson, M. L., Hydrocarbon geochemistry of the Puget Sound region. I. Sedimentary acyclic hydrocarbons, *Geochim. Cosmochim. Acta*, 44, 1349, 1980.
10. Schell, W. R. and Nevissi, A., Heavy metals from waste disposal in central Puget Sound, *Environ. Sci. Technol.*, 11, 887, 1977.
11. Long, E. R. and Chapman, P. M., A sediment quality triad: measures of sediment contamination, toxicity, and infaunal community composition in Puget Sound, *Mar. Pollut. Bull.*, 16, 405, 1985.
12. Barrick, R. C. and Prahl, F. G., Hydrocarbon geochemistry of the Puget Sound region. III. Polycyclic aromatic hydrocarbons in sediments, *Est. Coastal Shelf Sci.*, 25, 175, 1987.
13. Barrick, R. C., Flux of aliphatic and polycyclic aromatic hydrocarbons to central Puget Sound from Seattle (Westpoint) primary sewage effluent, *Environ. Sci. Technol.*, 16, 682, 1982.
14. Becker, D. S., Ginn, T. C., Landolt, M. L., and Powell, D. B., Hepatic lesions in English sole (*Parophrys vetulus*) from Commencement Bay, Washington (USA), *Mar. Environ. Res.*, 23, 153, 1987.

15. Stein, J. E., Collier, T. K., Reichert, W. L., Casillas, E., Hom, T., and Varanasi, U., Bioindicators of contaminant exposure and sublethal effects in benthic fish from Puget Sound, WA, USA, *Mar. Environ. Res.*, 35, 95, 1993.

16. Crecelius, E. A., Bothner, M. H., and Carpenter, R., Geochemistries of arsenic, antimony, mercury, and related elements in sediments of Puget Sound, *Environ. Sci. Technol.*, 9, 325, 1975.

17. Crecelius, E. A., Bothner, M. H., and Carpenter, R., Geochemistries of arsenic, antimony, mercury, and related elements in sediments of Puget Sound, *Environ. Sci. Technol.*, 9, 325, 1975.

18. Crecelius, E. A., Riley, R. G., Bloom, N. S., and Thomas, B. L., History of Contamination of Sediments in Commencement Bay, Tacoma, Washington, NOAA Tech. Memo. NOS OMA 14, U.S. National Oceanic and Atmospheric Administration, Rockville, MD, 1985.

19. Sindermann, C. J., *Ocean Pollution: Effects on Living Resources and Humans*, CRC Press, Boca Raton, FL, 1996.

20. Gardner, G. R., Chemically induced histopathology in aquatic invertebrates, in *Pathobiology of Marine and Estuarine Organisms*, Couch, J. A. and Fournie, J. W., Eds., CRC Press, Boca Raton, FL, 1993, 359.

21. Malins, D. C., Alterations in the cellular and subcellular structure of marine teleosts and invertebrates exposed to petroleum in the laboratory and field: a critical review, *Can. J. Fish. Aquat. Sci.*, 39, 877, 1982.

22. Mix, M. C., Cancerous diseases in aquatic animals and their association with environmental pollutants: a critical review, *Mar. Environ. Res.*, 20, 1, 1986.

23. Hinton, D. E., Toxicologic histopathology of fishes: a systematic approach and overview, in *Pathobiology of Marine and Estuarine Organisms*, Couch, J. A. and Fournie, J. W., Eds., CRC Press, Boca Raton, FL, 1993, 177.

24. Moore, M. J. and Myers, M. S., Pathobiology of chemical-associated neoplasia in fish, in *Aquatic Toxicology: Molecular, Biochemical, and Cellular Perspectives*, Malins, D. C. and Ostrander, G. K., Eds., Lewis Publishers, Boca Raton, FL, 1994, 327.

25. Landolt, M. L., Holmes, E. H., and Ostrander, G. K., Preneoplastic cellular changes associated with exposure to environmental contaminants in Puget Sound, Washington, *Mar. Environ. Res.*, 17, 334, 1985.

26. Malins, D. C., Krahn, M. M., Brown, D. W., Rhodes, L. D., Myers, M. S., McCain, B. B., and Chan, S.-L., Toxic chemicals in marine sediment and biota from Mukilteo, Washington: relationships with hepatic neoplasms and other hepatic lesions in English sole (*Parophrys vetulus*), *J. Natl. Cancer Inst.*, 74, 487, 1985.

27. Rhodes, L. D., Myers, M. S., Gronlund, W. D., and McCain, B. B., Epizootic characteristics of hepatic and renal lesions in English sole, *Parophrys vetulus*, from Puget Sound, *J. Fish Biol.*, 31, 395, 1987.

28. Pierce, K. V., McCain, B. B., and Wellings, S. R., Pathology of hepatomas and other liver abnormalities in English sole (*Parophrys vetulus*) from the Duwamish River estuary, Seattle, Washington, *J. Natl. Cancer Inst.*, 60, 1445, 1978.

29. Pierce, K. V., McCain, B. B., and Wellings, S. R., Histopathology of abnormal livers and other organs of starry flounder *Platichthys stellatus* (Pallas) from the estuary of the Duwamish River, Seattle, Washington, USA, *J. Fish Dis.*, 3, 81, 1980.

30. Krahn, M. M., Rhodes, L. D., Myers, M. S., Moore, L. K., MacLeod, W. D., and Malins, D. C., Associations betweem metabolites of aromatic compounds in bile and the occurrence of hepatic lesions in English sole (*Parophrys vetulus*) from Puget Sound, Washington, *Arch. Environ. Contam. Toxicol.*, 15, 61, 1986.

31. Myers, M. S., Rhodes, L. D., and McCain, B. B., Pathologic anatomy and patterns of occurrence of hepatic neoplasms, putative preneoplastic lesions, and other idiopathic hepatic conditions in English sole (*Parophrys vetulus*) from Puget Sound, Washington, *J. Natl. Cancer Inst.*, 78, 333, 1987.

32. Myers, M. S., Landahl, J. T., Krahn, M. M., and McCain, B. B., Relationships between hepatic neoplasms and related lesions and exposure to toxic chemicals in marine fish from the U.S. West Coast, *Environ. Health Perspect.*, 90, 7, 1991.

33. Flint, R. W. and Younk, J. A., Estuarine benthos: long-term community structure variations, Corpus Christi Bay, Texas, *Estuaries*, 6, 126, 1983.

34. Holland, A. F., Long-term variations of macrobenthos in a mesohaline region of Chesapeake Bay, *Estuaries*, 8, 93, 1985.

35. Holland, A. F., Shaughnessy, A. T., and Hiegel, M. H., Long-term variation in mesohaline Chesapeake Bay macrobenthos: spatial and temporal patterns, *Estuaries*, 10, 227, 1987.

36. Dauer, D. M. and Alden, R. W., III, Long-term trends in the macrobenthos and water quality of the lower Chesapeake Bay (1985–1991), *Mar. Pollut. Bull.*, 30, 840, 1995.

37. Nichols, F. H., Abundance fluctuations among benthic invertebrates in two Pacific estuaries, *Estuaries*, 8, 136, 1985.

38. Dames & Moore, Baseline Studies and Evaluations for Commencement Bay Study/Environmental Impact Assessment, Vols. I–VII, Series of Reports Submitted to the Seattle District, U.S. Army Corps of Engineers, Seattle, WA, 1981.

39. Leon, H., Benthic Community Impact Study for Terminal 107 (Kellogg Island) and Vicinity, Technical Report by Pacific Rim Planners Submitted to the Port of Seattle, Seattle, WA, 1980.

40. Comiskey, C. A., Farmer, T. A., and Brandt, C. C., Dynamics and Biological Impacts of Toxicants in the Main Basin of Puget Sound and Lake Washington, Vol. IIA, Technical Report Prepared by Science Applications International Corporation, Submitted to the Municipality of Metropolitan Seattle, Seattle, WA, 1984.

41. Chapman, P. M., Dexter, R. N., Kathman, R. D., and Erickson, G. A., Survey of Biological Effects of Toxicants upon Puget Sound Biota. IV. Inter-relationships of Infauna, Sediment Bioassay and Sediment Chemistry Data, NOAA Tech. Memo. NOS OMA-9, U.S. National Oceanic and Atmospheric Administration, Rockville, MD, 1985.

42. Schell, W. R., Nevissi, A., Piper, D., Christian, G., Murray, J., Spyradakis, D., Olsen, S., Huntamer, D., Knudsen, E., and Zafiropoulos, D., Heavy Metals near the West Point Outfall and in the Central Basin of Puget Sound, Final Report Prepared for the Municipality of Metropolitan Seattle by the College of Fisheries and the Departments of Oceanography, Chemistry, and Civil Engineering, University of Washington, Seattle, WA, 1977.

43. Chapman, P. M., Summary of biological effects in Puget Sound — past and present, in *Oceanic Processes in Marine Pollution*, Vol. 5, *Urban Wastes in Coastal Marine Environments*, Wolfe, D. A. and O'Connor, T. P., Eds., Robert E. Krieger Publishing Company, Malabar, FL, 1988, 169.

44. Pedersen, M. G. and DiDonato, G., Groundfish Management Plan for Washington's Inside Waters, Progress Report 170, Washington Department of Fisheries, Olympia, WA, 1982.

45. Kimura, D. K. and Millikan, A. R., Assessment of the Population of Pacific Hake (*Merluccius productus*) in Puget Sound, Washington, Technical Report No. 35, Washington State Department of Fisheries, Seattle, WA, 1977.

46. Bargman, G. G., The Biology of Fisheries for Ling Cod (*Ophiodon elongatus*) in Puget Sound, Technical Report No. 66, Washington State Department of Fisheries, Seattle, WA, 1982.

47. Harper-Owes, Water Quality Assessment of the Duwamish Estuary, Technical Report, Submitted to the Municipality of Metropolitan Seattle, WA, 1982.

48. Landahl, J. T. and Johnson, L. L., Contaminant exposure and population growth of English sole in Puget Sound: the need for better early life-history data, *Am. Fish. Soc. Symp.*, 14, 117, 1993.

49. Gahler, A. R., Cummins, J. M., Blazevich, J. N., Rieck, R. H., Arp, R. L., Gangmark, C. E., Pope, S. V. W., and Filip, S., Chemical Contaminations in Edible Nonsalmonid Fish and Crabs from Commencement Bay, Washington, Report No. EPA 910/9-82-093, U.S. Environmental Protection Agency, Seattle, WA, 1982.

50. Tetra Tech, Commencement Bay Nearshore/Tideflats Remedial Investigations, Vol. I–IV, Report No. EPA-910/9-85-134b, Prepared by Tetra Tech for the Washington Department of Ecology and U.S. Environmental Protection Agency, Seattle, 1985.

51. Mearns, A. J., Matta, M. B., Simecek-Beatty, D., Buchman, M. F., Shigenaka, G., and Wert, W. A., PCB and Chlorinated Pesticide Contamination in U.S. Fish and Shellfish: A Historical Assessment Report, NOAA Tech. Memo. NOS OMA 39, Seattle, WA, 1988.

52. Kennish, M. J., Ed., *Practical Handbook of Marine Science*, Vol. 2, CRC Press, Boca Raton, FL, 1994.

53. Collier, T. K. and Varanasi, U., Hepatic activities of xenobiotic metabolizing enzymes and biliary levels of xenobiotics in English sole (*Parophrys vetulus*) exposed to environmental contaminants, *Arch. Environ. Contam. Toxicol.*, 20, 462, 1991.

54. Collier, T. K., Stein, J. E., Sanborn, H. R., Hom, T., Myers, M. S., and Varanasi, U., A field study of the relationship between bioindicators of maternal contaminant exposure and egg and larval viability of English sole (*Parophrys vetulus*), *Mar. Environ. Res.*, 35, 171, 1993.

55. Arkoosh, M. R., Clemons, E., Huffman, P., Sanborn, H. R., Casillas, E., and Stein, J. E., Leukoproliferative response of splenocytes from English sole (*Pleuronectes vetulus*) exposed to chemical contaminants, *Environ. Toxicol. Chem.*, 15, 1154, 1996.

56. Collier, T. K., Singh, S. V., Awasthi, Y. C., and Varanasi, U., Hepatic xenobiotic metabolizing enzymes in two species of benthic fish showing different prevalences of contaminant-associated liver neoplasms, *Toxicol. Appl. Pharmacol.*, 113, 319, 1992.

57. Johnson, L. L., Stehr, C. M., Olson, O. P., Myers, M. S., Pierce, S. M., McCain, B. B., and Varanasi, U., National Benthic Surveillance Project: Northeast Coast, Fish Histopathology and Relationships Between Lesions and Chemical Contaminants (1987–89), NOAA Tech. Memo. NMFS-NWFSC-4, U.S. National Oceanic and Atmospheric Administration, Seattle, WA, 1992.

58. Sol, S. Y., Johnson, L. L., Collier, T. K., Krahn, M. M., and Varanasi, U., Contaminant Effects on Reproductive Output in North Pacific Flatfish, Proceedings of the International Symposium on North Pacific Flatfish, Alaska Sea Grant College Program, Fairbanks, AK, 1995.

59. Johnson, L. L., Casillas, E., Collier, T. K., McCain, B. B., and Varanasi, U., Contaminant effects on ovarian development in English sole (*Parophrys vetulus*) from Puget Sound, Washington, *Can. J. Fish. Aquat. Sci.*, 45, 2133, 1988.

60. Stein, J. E., Hom, T., Collier, T. K., Brown, D. W., and Varanasi, U., Contaminant exposure and biochemical effects in outmigrant juvenile chinook salmon from urban and nonurban estuaries of Puget Sound, Washington, *Environ. Toxicol. Chem.*, 14, 1019, 1995.

61. Calambokidis, J., Peard, T., Steiger, G. H., and Cubbage, J. C., Chemical Contaminants in Marine Mammals from Washington State, NOAA Tech. Memo. NOS OMS 6, U.S. National Oceanic and Atmospheric Administration, Seattle, WA, 1984.

62. Calambokidis, J., Bowman, K., Carter, S., Cubbage, J., Dawson, P., Fleischner, T., Schuett-Hames, J., Skidmore, J., and Taylor, B., Chlorinated Hydrocarbon Concentrations and the Ecology and Behavior of Harbor Seals in Washington State Waters, Tech. Rep., Evergreen State College, Olympia, WA, 1978.

63. Riley, R., Crecelius, E. A., Fitzner, R. E., Thomas, B. L., Gurtisen, J. M., and Bloom, N. S., Organic and Inorganic Toxicants in Sediments and Birds from Puget Sound, NOAA Tech. Memo. NOS OMS-1, U.S. National Oceanic and Atmospheric Administration, Rockville, MD, 1984.

64. U.S. Army Corps of Engineers, East, West, and Duwamish Waterways Navigation Improvement Study, Draft Feasibility Report and Draft Environmental Impact Statement, U.S. Army Corps of Engineers, Seattle District, Seattle, WA, 1982.

7 Case Study 6: Firth of Clyde and the Tees River Estuary

I. INTRODUCTION

Estuarine and coastal marine waters of England are affected by the same suite of organic and inorganic chemical pollutants that plague many U.S. coastal systems. Wastewater discharges, sewage sludge dumping, dredged-spoil disposal, and other sources of contaminants have been responsible for substantial concentrations of nutrients, halogenated hydrocarbons, polycyclic aromatic hydrocarbons, heavy metals, and radioactive substances in coastal waters. In systems subjected to long-term sewage sludge disposal (e.g., Garroch Head, Firth of Clyde, Scotland), enrichment of nutrients and organic carbon loading have significantly impacted biotic communities. The most acute effects are manifested among the benthos. Other systems strongly affected by contaminant inputs from chemical industry wastes and sewage discharges (e.g., Tees estuary) have also exhibited degraded water quality and impacted biotic communities. Heavily industrialized estuaries, such as the Tees estuary, are influenced by a wide range of chemical contaminants. Here, once thriving fisheries and bottom-dwelling communities have been degraded. The following discussion provides a detailed description of pollution impacts on biotic communities at Garroch Head in the Firth of Clyde, Scotland, and the Tees River estuary in northeast England.

II. FIRTH OF CLYDE, SCOTLAND

A. PHYSICAL DESCRIPTION

The Clyde Sea on the west coast of Scotland covers an area of approximately 450 km² and a volume of about 100 km³. It consists of the Clyde River estuary and a complex of long narrow sea locks to the north and deep basins located on each side of the island Arran to the south. A prominent sill lies south of Arran at a depth of ~50 m, and it influences circulation in the upper Clyde estuary. A critically important part of the Clyde Sea is the Firth of Clyde which forms a deep basin (the Arran Basin) between Arran and the Ayrshire coast (Figure 1). It supports valuable commercial and recreational fisheries and provides a deep waterway for passage of large vessels (500,000 mt or more) to load or unload cargo at highly developed urban and industrial sites. For more than a century, however, the Firth of Clyde has also received substantial amounts of anthropogenic wastes that threaten biotic communities and habitats.

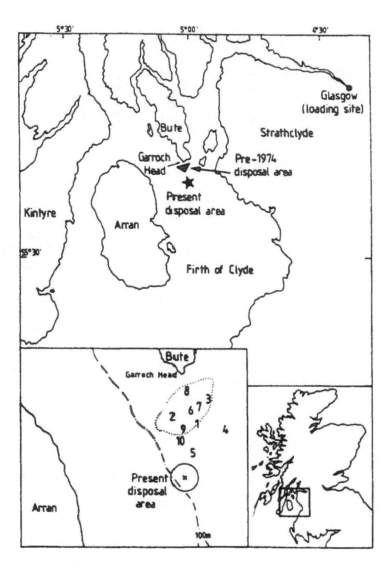

FIGURE 1 Map showing location of the pre-1974 sewage sludge disposal area (dumpsite 1) and the present sewage sludge disposal area in the Firth of Clyde, Scotland. (From Rodger, G. K., Davies, I. M., and Moore, D. C., *Mar. Ecol. Progr. Ser.*, 75, 293, 1991. With permission.)

The Firth of Clyde is characterized by extremely weak tidal currents and low freshwater inflow. Although the volume of the Firth is 10^{11} m³, freshwater input only averages 10^2 m³/s, mainly from the Clyde and Leven Rivers. Tidal currents are <10 cm/s, but rise to >100 cm/s south of the sill in the North Channel. The tidally well-mixed waters of North Channel contrasts markedly with the tidally weak and normally stratified waters of the Firth. A sharp front separates these two bodies of water.[1]

Current measurements in the Firth of Clyde indicate nearly negligable estuarine circulation patterns.[2] A compensating estuarine inflow occurs along the eastern shore of the Firth. Thus, Dooley[1] recorded lower salinities (<32.5‰) in surface waters of the Arran Basin and higher salinities (>33‰) in deeper waters adjacent to the Ayrshire coast. A well-developed pycnocline occurs at a depth of ~15 m. Above the pycnocline, waters flow slowly southward. Below the pycnocline to a depth of ~100 m, the water column has a layered structure owing to successive intrusions of water of different temperature and salinity over the sill.[2]

The residence time of water in the Firth of Clyde is relatively short, despite the occurrence of weak currents. Except in the deep northern basin and in some coastal embayments, water in the Firth is usually replaced within a period of 4 weeks or less. When wind-driven currents predominate, the residence time may be reduced to only a few days. Dooley[1] calculated a residence time ranging from 2–3 days during periods of active inflow over the sill to more than 3 months in isolated deep pockets where circulation is extremely poor. The forcing of large volumes of water over the sill during storms can rapidly increase water exchange in the Firth. This exchange is important because of an array of pollutants entering the system from a large population (>2 million people) within the catchment area of the Firth.

B. Sewage Sludge Dumpsites

1. Site Description

Two sewage sludge dumpsites occur immediately south of Garroch Head on the Isle of Bute in the Firth of Clyde. The first dumpsite, which lies ~2 km south of Garroch Head, received sewage sludge virtually continuously from 1904 to 1974.[3,4] The second dumpsite, which began receiving sewage waste in 1974, lies ~6.3 km south of Garroch Head (Figure 1).[5] The water depth 2 km south of Garroch Head is ~70–80 m, and current flow <10 cm/s. Because tidal currents are weak and water velocities low (~0.5 m/s) in the dumpsite areas, sludge dispersal has been slight. Consequently, when dumped, the sludge sinks fairly rapidly to the seafloor, where it remains locally concentrated with its associated load of trace metals and other chemical contaminants. Low near-bottom turbulence and bottom shear stresses also result in the retention of most of the nutrients at the dumpsite which tend to leach away slowly, thereby stimulating only moderately enhanced phytoplankton production. Similarly, toxic organic chemicals associated with the sewage sludge largely remain confined to the immediate dumping ground. A central "hot spot" area has been delineated with a decreasing gradient of contaminant concentrations falling off to background levels within a few kilometers.

The sediments at dumpsite 1 are rich in muds, with the amount of organic carbon ranging from 3–8% compared to 0.3–2.0% in clean muddy areas nearby.[6] In the 1970s, the dumpsite was enriched in a wide range of heavy metals. Highest metal and organic carbon concentrations occurred in a relatively restricted area (15–20 km²). Organochlorine compounds (e.g., DDT and PCBs) also accumulated at the dumpsite. Sediment contamination extended to a depth of ~40–50 cm.[5]

2. Other Pollution Sources

Numerous domestic, industrial, and agricultural sources of pollution have affected the Firth of Clyde for more than a century. Substantial amounts of untreated sewage have entered the Firth for years from inland populated sites as well as coastal towns. Since the early 1900s, sewage sludge has been dumped in the Garroch Head area, with various biotic impacts resulting from the disposal operations. Industrial facilities have released halogenated hydrocarbons, heavy metals, and other contaminants at various points along the coast. Agricultural runoff has delivered nutrients and pesticides. Petroleum hydrocarbons have reached the sea from accidental spills, routine shipping activities, fixed installations, municipal wastes, urban runoff, and effluents from nonpetroleum industries. The operation of electric generating stations, notably on the north coast of Ayrshire, has caused other impacts on biological communities due to entrainment of organism and in-plant passage, impingement of organisms on intake screens, thermal discharges, and biocidal releases.

The following discussion focuses on the effects of long-term sewage sludge disposal on biological communities in the Firth of Clyde. The dumping of large volumes of sewage sludge with an industrial waste component near Garroch Head has been a major concern because of the valuable recreational and commercial fisheries of the area. Several comprehensive investigations have been conducted on the benthic fauna and habitat of the disposal sites.

The amount of sewage sludge disposed of annually at dumpsite 1 increased gradually through time. In 1904, about 2×10^5 metric tons (mt) of sludge were dumped at the site. By 1925, this number had increased to nearly 4×10^5 mt and by 1938, it had risen to nearly 7×10^5 mt. By the late 1960s, more than a 1×10^6 mt of sewage sludge were dumped every year at this location. Over the 70-year period of waste disposal (1904–1974), dumpsite 1 received more than 30×10^6 mt of sewage sludge.

Dumpsite 2, a 6 km² designated disposal area, received nearly 15×10^6 mt (wet) of sewage sludge between 1974 and 1984. Although sludge dumping continued unabated at the site after 1984, the area of the seafloor covered by the organic waste was much smaller than at dumpsite 1. The reduced area of sludge coverage reflects the greater accuracy of waste discharge from the disposal vessels. Relatively weak currents (5–15 cm/s) also limited subsequent dispersal of the waste from the dumpsite.

3. Chemical Contaminants

Several studies in the late 1960s and 1970s examined chemical contaminants at dumpsite 1.[7-10] Results of these studies were later compared to investigations focusing on the recovery phase of the site following termination of sewage sludge disposal.[4,5,11,12] Water and sediment samples collected at the dumpsite during the early surveys concentrated on organochlorine compounds, particularly PCBs, and heavy metals. PCBs were identified as a major contaminant of the sludge, and some heavy metals (e.g., copper, lead, and zinc) also occurred at elevated concentrations. The PCBs were derived from accidental seepage into groundwater that subsequently entered the sewage disposal system. In addition to heavy PCB contamination of sediments in a confined area ascribable to sewage dumping, a more widespread low-level contamination of surface sediments was caused by the riverine transport of PCBs in association with fine organic particulate material. Halcrow et al.,[10] examining PCBs in surface sediments and benthic fauna at dumpsite 1, ascertained a marked increase in contaminant concentrations during the 1960s. The concentrations of PCBs in surface sediments of the study area ranged from 10–2890 ng/g dry weight.

The benthic fauna in the area was correspondingly contaminated. Surveys of benthic fauna inhabiting dumpsite 1 during the early 1970s revealed significant PCB contamination in some species, reflecting the buildup of the toxic chemicals in the fine organic particulate fraction that settled to the seafloor.[11] PCB levels recorded in the tissues of three species of molluscs from the Garroch Head area at this time were as follows: *Buccinum undatum* (38–376 ng/g wet weight), *Chlamys septemradiata* (55–471 ng/g wet weight), and *Nucula sulcata* (39–217 ng/g wet weight).[10]

PCBs may have contributed to reduced diversity of the benthos at the center of the disposal site.[5] However, with the elimination of PCB sources in the Clyde catchment area as well as restrictions on the manufacture and use of the contaminants during the 1970s, PCB levels at the Garroch Head dumping ground decreased substantially. Mackay[5] reported a 66% reduction of PCB concentrations in the sewage sludge during the 1970s.

In regard to heavy metals, Halcrow et al.[10] observed considerable enrichment of copper, lead, and zinc in a relatively small area within 2 km of the center of dumpsite 1. Earlier studies indicated that certain epibenthic fauna, such as the potentially economic *Buccinum* spp., *Crangon* spp., and *Pandalus* spp., accumulated some heavy metals.[7,8] Peak concentrations of copper, lead, and zinc in surface deposits at the dumping ground amounted to 300 μg/g dry weight (Cu), 403 μg/g dry weight (Pb), and 681 μg/g dry weight (Zn) between September 1971 and January 1972. Silver, not present in determinable quantities in the uncontaminated sediments of the area, existed in dumpsite deposits at concentrations of 2–6 μg/g dry weight. Highest levels of mercury were >2000 ng/g dry weight. While silver, copper, lead, and zinc were enriched by factors of 20, 16, 9, and 7 times, respectively, compared to control sites, other metals were also enriched at the dumpsite (Table 1).

Rodger et al.[12] chronicled changes in metal concentrations in both surface and subsurface sediments which had occurred at dumpsite 1 by 1985. Sediment cores obtained from 10 sites at the dumpsite in 1985 (Figure 1) showed a marked reduction in heavy metal concentrations in surface sediments at the dumping ground compared to concentrations reported while sludge disposal was in progress prior to 1974. Metal concentrations in 1985 were highest in surface sediments at four sites (2, 6, 8, and 9; Figure 1) in the western part of the study area, with values of cadmium, copper, lead, mercury, and zinc ranging from <0.2 to 0.7 mg/kg dry weight, 81.5–87.5 mg/kg dry weight, 133–163 mg/kg dry weight, 0.63–0.97 mg/kg dry weight, and 272–294 mg/kg dry weight, respectively. Subsurface metal concentrations at these four sites peaked at a depth of about 16–25 cm (e.g., ~3.0 mg/kg dry weight [Cd], 220 mg/kg dry weight [Cu], 315 mg/kg dry weight [Pb], and 650 mg/kg dry weight [Zn]). The lower metal concentrations in surface sediments collected in 1985 may

TABLE 1
Range and Mean Values of Organic Carbon and Heavy Metals in Bottom Sediments from Sewage Sludge Dumpsite 1 and Similar Sediments from Control Areas

Source of Sample	(μg/g)					Hg (ng/g)	Trace Elements (μg/g)								Fe$_2$O$_3$ (%)	Cd (%)	Organic carbon (%)
	Ag	As	Cd	Cr	Cu		Mn	Mo	Ni	Pb	Se	Sn	V	Zn			
Disposal area	4	21	6.4	122	269	2187[a]	407	7.0	77	361	<0.5	9	110	631	5.3	0.3	7.46
Clyde control	2–6	16–26	4–8	87–175	250–300	40:149[a]	300–430	2.7–10.5	58–87	269–403	7.5–11		80–120	437–681	3.0–6.8	0.004–0.80	6.66–7.93
	<0.2	7.0	1.6	33	16		355	0.33	30	42	<0.5	4	63	85	2.0	1.68	0.9
Solway control	6.0–8.0	N.A.[b]	1.0–3.0	10–65	9–20	N.A.[b]	283–432	0.3–0.4	14–50	24–67	3–5		35–100	60–130	1.53–6.09	0.38–8.45	0.4–1.7
	<0.2		<1	40	10		341	N.A.[b]	N.A.[b]	36	N.A.[b]	N.A.[b]	N.A.[b]	63	N.A.[b]	N.A.[b]	N.A.[b]
	N.A.[b]			15–62	5–16		240–700			12–66				36–105			

[a] Data from 1 and 2 samples, respectively.

[b] N.A. = Not Available.

From Halcrow, W., Mackay, D. W., and Thornton, I., *J. Mar. Biol. Assoc. U. K.*, 53, 721, 1973. With permission.

be due to two principal factors. First, metals accumulating in the 1960s and early 1970s at the dumpsite were buried by post-1974 sedimentation; hence, the concentrations of metals in surface sediments in 1985 were substantially reduced. Alternatively, higher rates of bioturbation and physical processes (e.g., storms, bottom currents) may have occurred during the mid-1980s, resulting in greater release of metals from the sediments. The marked subsurface increase in heavy metal concentrations in the cores suggests that the former is the more likely cause.

Mackay[5] discerned similar concentrations of organic carbon, copper, lead, and zinc in surface sediments at dumpsites 1 and 2. The maximum amount of organic carbon at both locations was ~8%. Peak concentrations of copper, lead, and zinc at each dumping ground was 200 μg/g dry weight, 300 μg/g dry weight, and 400 μg/g dry weight, respectively.

4. Biotic Impacts

With the implementation of the Dumping at Sea Act of 1974 (superseded by the Food and Environmental Protection Act, Part II, 1987), annual monitoring of the sludge disposal areas and constituent fauna was conducted. The Scottish Marine Biological Association Laboratory at Dunstaffnage carried out the earliest monitoring program of sewage sludge dumping effects at Garroch Head. Other investigators have also studied sewage sludge impacts at dumpsite 2, which was licensed by the Department of Agriculture and Fisheries of Scotland in 1974. Because the greatest effect of sewage sludge dumping has been the introduction of organic carbon and associated contaminants into bottom sediments, the principal focus of biological surveys has been on benthic community impacts.

a. Upper Clyde Estuary Macrobenthic Communities

Henderson[13] provided an overview of the macrobenthos of the upper Clyde estuary. She analyzed the long-term changes in species composition, density, and dominance patterns of benthic macroinvertebrates of the estuary. Because of organic pollution, the macrobenthic community is dominated by brackish, pollution-tolerant oligochaetes and polychaetes. However, both natural and anthropogenic factors are responsible for the temporal fluctuations in the constituent populations of the community.

The benthic monitoring program conducted by Henderson[13] was subdivided into two time periods spanning from 1974 to 1976 and from 1977 to 1980. The dominant members of the benthic macroinvertebrate community of the upper Clyde estuary between 1974 and 1976 included the Tubificid oligochaetes *Tubificoides benedeni*, *T. costatus*, *T. pseudogaster*, and *Monopylepohorus rubroniveus*; the Enchytraeid *Lumbricillus lineatus*; and the Naidids *Nais elinguis* and *Paranais littoralis*. These species primarily inhabited the midsegment of the upper estuary. Between 1977 and 1980 inclusive, additional marine representatives were more common (i.e., polychaetes — *Manayunkia aestuarina*, *Polydora ciliata*, and *Streblospio shrubsolii*). Some species (i.e., *Carcinus maenas*, *Corophium volutator*, and *Hydrobia ulvae*) appeared for the first time.

The abundance of the benthic fauna increased downestuary. Highest densities occurred at three stations: (1) the Cart confluence (<100–68,000 ind./m²);

(2) Dalmuir (<500–230,000 ind./m^2); and (3) Erskine (24,000–800,000 ind./m^2). Upestuary of the Cart confluence, densities dropped substantially, often falling to 0 ind./m^2 between 1974 and 1976. The densities generally increased at upestuary stations after 1976. They also remained high through 1980 at the Cart confluence and Dalmuir locations. At Erskine, however, a gradual diminution in density was evident, and by late 1980, it had fallen to only 1000 ind./m^2.

Of the Tubificidae at Erskine, *Tubicoides benedeni*, *T. costatus*, and *T. pseudogaster* attained peak densities of ≤75,000, >100,000, and 20,000 ind./m^2, respectively, between 1974 and 1976. Subsequent to 1976, however, the abundance of these forms decreased, and by 1980, the densities of *T. benedeni*, *T. costatus*, and *T. pseudogaster* were 26,000, 3000, and <200 ind./m^2, respectively. Also at Erskine during the years of 1974 and 1976, *Lumbricillus lineatus* and *Monopylephorus rubroniveus* reached highest densities, with *L. lineatus* having maximum counts up to 5000 ind./m^2 and *M. rubroniveus* >10,000 ind./m^2. Other species found in high densities during the entire survey were *Paranais littoralis*, exceeding 500,000 ind./m^2 at Dalmuir and Erskine, and *Capitella capitata* occasionally surpassing 100,000 ind./m^2 between 1974 and the end of 1977.

The dominance patterns of benthic macroinvertebrates changed temporally at the sampling sites. In some cases, a well-defined inverse relationship of dominance was evident among two species. For example, *Paranais littoralis* and *T. benedeni* always showed inverse dominance patterns despite highly erratic fluctuations in dominance between 1974 and 1976. A seasonal dominance pattern of these species emerged in 1976 and 1977, with *T. bendeni* dominating in summer and *P. littoralis* dominating in winter.

b. Benthic Fauna: Sewage Sludge Dumpsite 1

Halcrow et al.[9] examined the benthic infauna and epifauna at sewage sludge dumpsite 1 in 1971. They showed that polychaetes dominated the benthic community in the center of the disposal area, principally *Capitella capitata* (Fabricus) and species of Cirratulidae. *Scololepis fuliginosa* also occurred in the disposal area. A single oligochaete species, *Peloscolex* sp., was present in large numbers. The diversity of the infauna progressively decreased toward the center of the dumping ground. While uncontaminated reference sites nearby exhibited mixed fauna of mollusks, crustaceans, echinoderms, and polychaetes,[6] the only infaunal mollusk inhabiting the center area of the dumpsite was *Phacoides borealis* (L.). In the most heavily affected zone, important species of the upper Clyde such as *Abra* spp., *Nucula* spp., and *Pectinaria* spp., were totally absent. Figures 2–4 illustrate the distribution of species of Cirratulidae, *Nucula* spp., and capitellid worms in the disposal area.

Some polychaetes, notably the capitellids *Capitata capitata*, *Dasybranchus caducus* (Grube), and *Notomastus latericeus* M. Sars, exhibited distinct habitat preferences (Figure 4). For example, *C. capitata* occupied the dumpsite in large numbers, whereas *D. caducus* inhabited neighboring clean sediments and could not be found in the disposal area. Another polychaete *Pygospio elegans* Claparede flourished on the periphery of the dumping ground, as did the cnidarian *Cerianthus lloydi* Gosse.

Trawl samples collected in the study area revealed higher abundances of some epifaunal species (e.g., *Crangon allmani* Kinahan, *Buccinum undatum* (L.), and

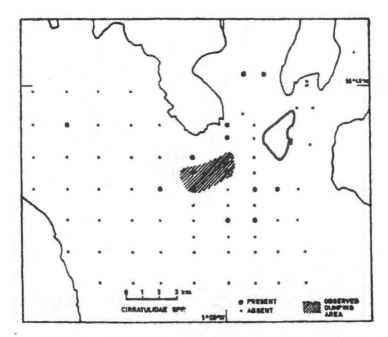

FIGURE 2   Distribution of Cirratulidae species at Garroch Head sewage sludge dumpsite 1 in 1971. *Chaetozone setosa* not included in distribution data. (From Halcrow, W., Mackay, D. W., and Thornton, I., *J. Mar. Biol. Assoc. U. K.*, 53, 721, 1973. With permission.)

Pandalus montagui Leach) in and around the disposal area (Table 2). However, there was no evidence of changes in species composition of epifauna at the dumpsite. Certain epifaunal species collected from the middle of the dumpsite displayed elevated levels of some heavy metals (Table 3). Most conspicuous in this regard was *B. undatum* which accumulated appreciable levels of zinc and to a lesser extent cadmium, lead, and nickel at the dumpsite relative to control areas. For most of the epifauna, however, the differences in metal concentrations were relatively small and not significant.

It is important to note that, prior to the initiation of sewage sludge disposal, the area off Garroch Head sustained the commercially important prawn *Nephrops norvegicus*. This species has continued to thrive in uncontaminated sediments away from the dumpsite. Unfortunately, the degraded habitat at the dumping ground caused considerable economic losses for shellfisheries.[6]

c. Benthic Community Recovery: Dumpsite 1

Sewage sludge disposal at dumpsite 1 was terminated in 1974, with the disposal operation being moved to a new licensed area 4 km farther south (Figure 1). Studies conducted in the early 1970s prior to closure of the dumpsite demonstrated that the benthic community experienced the classical effects of organic enrichment, including a decrease in species diversity and an increase in individual population abundance and biomass.[7-9,14] Effects of sludge dumping on benthic communities were compared with those reported for other systems.[15] Moore and Rodger[4] resurveyed the benthic

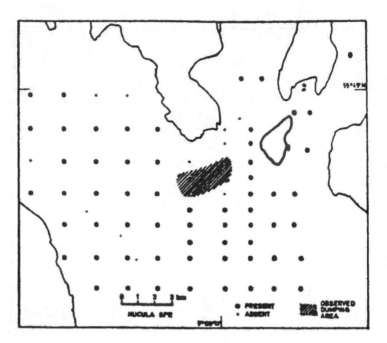

FIGURE 3   Distribution of *Nucula* spp. at Garroch Head sewage sludge dumpsite 1 in 1971. (From Halcrow, W., Mackay, D. W., and Thornton, I., *J. Mar. Biol. Assoc. U. K.*, 53, 721, 1973. With permission.)

community at dumpsite 1 in 1985, comparing results to those of Topping and McIntyre[14] who surveyed the benthos in 1971. The principal objective was to assess the degree of recovery of the benthic community following pollution abatement by resurveying the stations sampled by Topping and McIntyre[14] for benthic community analysis.

Analysis of the macrobenthos collected at 5 stations in 1971 revealed a total of 48 species/taxa, with densities ranging from 3290–52,500 ind./m^2 and biomasses from 30.14–162.71 g/m^2. The total number of species ranged from 13–35 at the sampling sites. In 1985, a total of 123 species/taxa and 4870 indivduals were collected. The mean density ranged from 910–18,145 ind./m^2, and the mean biomass, from 27.94–108.09 g/m^2. Tables 4 and 5 compare abundance and biomass values of the macrofauna for these surveys.

During both the 1971 and 1985 surveys, polychaetes dominated the benthic community at the dumpsite. However, the community in 1985 was much more diverse, being dominated less by opportunistic species. As is evident in Table 6, opportunistic polychaetes (Capitellidae and *Peloscolex* [*Tubificoides*]) accounted for most of the total faunal abundance in 1971. Among the dominant fauna, only two species were mollusks, *Nucula turgida* and *Thyasira flexuosa*. Crustaceans were not represented. With the exception of Nemertea, annelids constituted the remainder of the list.

Moore and Rodger[4] clearly demonstrated that the benthic community at dumpsite 1 had recovered substantially from pollution impacts 11 years after cessation of

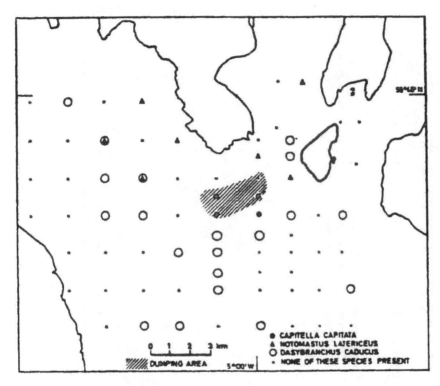

FIGURE 4 Distribution of three species of capitellids at Garroch Head sewage sludge dumpsite 1 in 1971. (From Halcrow, W., Mackay, D. W., and Thornton, I., *J. Mar. Biol. Assoc. U. K.*, 53, 721, 1973. With permission.)

dumping of sewage sludge at the accumulating disposal area. As stated by Moore and Rodger,[4] the benthic community "... now approaches what should be considered as normal for this slightly enriched area." Comparisons of faunal indices at the 5 sampling sites in 1971 to those in 1985 supports this conclusion. For example, diversity and evenness indices at 4 stations (1, 2, 3, and 5) in 1971 indicated very modified benthic communities. The benthic community at station 4 was the least modified by the disposal operation. By 1985, faunal indices at station 1 to 4 were similar to those of benthic communities at reference locations remote from the dumpsite. The benthic community at station 5, which had the lowest diversity and evenness indices in 1971, continued to exhibit the lowest diversity and evenness indices in 1985 (Table 7). According to Moore and Rodger,[4] only station 5 retained any effect of sewage sludge input, with the benthic community being in essentially a transitional state.

Surface sediments at all 5 sampling stations in 1985 contained organic carbon concentrations between 2 and 3%; much higher values ranging from 3 to 8% were recorded in 1971. The decrease in organic carbon content is consistent with the termination of sewage sludge input, ongoing natural sedimentation, and the mineralization of sewage solids in bottom sediments. Accompanying the decline in organic carbon concentrations over time at the dumpsite has been an increase in the species

TABLE 2
Abundance of Common Benthic Epifaunal Species at Sewage Sludge Dumpsite 1 Compared to That at Control Areas Based on Trawl Sampling

Area	Trawl No.	Pandalus montagui Leach	Crangon allmani Kinahan	Buccinum undatum (L.)	Meganyctiphanes norvegica (M. Sars)	Chlamys septemradiata (Müller)	Nephrops norvegicus (L.)	Calocaris macandreae Bell
Disposal area	1	3	55	20	44	—	—	—
	2	32	106	12	52	—	—	—
	3	35	65	10	15	5	6	9
	4	25	21	20	—	—	1	—
	5	8	26	10	—	—	—	—
Central area	6	—	12	5	—	—	—	—
	7	11	8	1	—	—	—	6
	8	7	8	—	18	—	—	22
	9	—	—	—	150	—	2	18
Loch Fyne	10	—	5	—	80	—	2	40
	11	20	2	—	4	40	—	15
	12	20	10	1	8	25	—	35
	13	—	—	—	10	7	—	15
	14	8	14	—	3	9	—	15

From Halcrow, W., Mackay, D. W., and Thornton, L., *J. Mar. Biol. Assoc. U. K.*, 53, 721, 1973. With permission.

TABLE 3
Heavy Metal Concentrations in Selective Benthic Epifaunal Species Collected at Sewage Sludge Dumpsite 1 Compared to Those at Control Areas Based on Trawl Sampling

Species	Area	Cd (µg/g)	Co (µg/g)	Cu (µg/g)	Ca (%)	Mn (µg/g)	Ni (µg/g)	Pb (µg/g)	Zn (µg/g)
Buccinum	1	2.8	1.9	146	0.54	33	11	13	1520
undatum	2	1.1	1.6	142	0.67	46	3	7	254
	3	17.0	1.8	189	0.78	81	5	10	8520
Crangon allmani	1	1.2	3.0	112	1.33	36	7	17	121
	2	1.9	5.5	56	1.75	1280	13	23	166
	3	3.5	4.4	56	2.34	1670	10	16	158
Pandalus	1	1.2	2.9	113	1.42	37	7	19	119
montagui	2	1.5	5.0	86	1.24	1150	13	19	157
	3	3.1	3.3	77	1.29	1150	8	14	137
Meganyctiphanes	1	2.0	4.2	29	1.96	42	9	37	173
norvegica	2	0.9	3.6	19	0.77	523	10	19	84
	3	6.6	7.1	44	1.31	1410	18	19	234
Calocaris	1	1.3	0.9	95	0.17	18	20	2	223
macandreae	2	1.5	7.2	106	3.79	1350	12	49	141
	3	2.2	7.0	99	3.31	1260	10	35	117
Nephrops	1	0.6	2.3	49	1.42	37	5	11	75
norvegicus	2	0.9	3.3	30	0.95	43	7	16	82
Chlamys	1	9.5	5.4	33	0.84	117	13	23	792
septemradiata	3	17.8	4.5	28	1.06	798	12	15	781

Note: Area 1, sewage sludge dumpsite 1; area 2, control area; area 3, north of Arran.

From Halcrow, W., Mackay, D. W., and Thornton, I., *J. Mar. Biol. Assoc. U. K.*, 53, 721, 1973. With permission.

diversity of the benthic community and a return to environmental conditions typically encountered at unpolluted areas of the Firth of Clyde. The benthos displayed remarkable resiliency during the 11-year recovery phase following sewage sludge disposal.

d. Benthic Fauna: Sewage Sludge Dumpsite 2

Pearson et al.[16] concluded that the Clyde Sea area is enriched due to high nutrient inputs, especially to the Arran/Ayrshire basin. They also noted that the species composition of macrobenthic communities at localized waste disposal sites with high carbon inputs is indicative of organically enriched habitats. At sewage sludge dumpsite 2, Pearson[17] related the distribution of benthic fauna to a gradient of increasing organic carbon enrichment caused by disposal operations. He found active macrofaunal communities throughout the dumpsite, and, even in anoxic sediments at the center of the dumpsite, vast numbers of some populations were observed.

Pearson[17] documented changes in benthic fauna along an 11-km east-west transect across the center of the dumpsite. Three distinct groups of benthic organisms were indentified along the transect. Numerous small annelids and nematodes

TABLE 4
Summary of Abundance and Biomass of the Benthic Macrofauna at Garroch Head Sewage Sludge Dumpsite 1, Stations 1 to 5, in 1971

	Station				
	1	2	3	4	5
Abundance (ind./m²)					
Polychaeta	4220	9550	20,820	2610	9120
Oligochaeta	20	32,550	31,670	120	—
Mollusca	180	—	—	270	110
Crustacea	—	—	—	20	60
Echinodermata	—	10	—	10	—
Others	250	10	10	240	140
Total	4670	42,070	52,500	3270	9430
Wet weight biomass (g/m²)					
Polychaeta	15.419	40.091	102.157	82.583	37.922
Oligochaeta	0.010	56.683	56.000	0.073	—
Mollusca	12.030	—	—	12.365	12.434
Crustacea	—	—	—	0.060	0.110
Echinodermata	—	0.010	—	0.010	—
Others	2.679	0.010	4.550	2.679	1.445
Total	30.138	96.794	162.707	97.770	51.911

From Moore, D. C. and Rodger, G. K., *Mar. Ecol. Prog. Ser.*, 75, 301, 1991. With permission.

occupied the center of the dumpsite. The nematode *Pontenema* sp. dominated the very center of the dumpsite; *Scolelepis fuliginosa* also concentrated in this area. While *Pontonema* sp. decreased rapidly in abundance on either side of the center of the dumpsite, *Capitella capitata* and *Tubificoides benedeni* predominated throughout the central area. At the edge of the central area, a second group of species was recognized, being characterized primarily by the polychaetes *Chaetozone setosa*, *Mediomastus* sp., and *Notomastus latericeus*, as well as the mollusk *Thyasira flexuosa*. These species were either absent or present in very low abundances at the center of the dumpsite. At the eastern and western ends of the transect, species characteristic of unpolluted areas beyond the dumpsite perimeter were most abundant. This group was represented by the polychaetes *Lipobranchius jeffreysii*, *Lumbrinereis hibernica*, and *Spiophanes kroyeri* and the mollusks *Nucula sulcata* and *N. tenuis*. Species in this peripheral group tended to be larger than those at the center of the dumpsite.

Changes in species numbers, abundance, and biomass also were registered along the transect. Although the abundance and biomass of the fauna at the center of the dumpsite were high, species numbers were low. However, 1 km from the center, abundance and biomass decreased, whereas species numbers increased. A secondary peak in biomass occurred between 1 and 2 km from the center. Species numbers peaked 4 km from the center and then decreased toward either end of the transect. Abundance and biomass values also dropped to relatively low levels at the outer

TABLE 5

Summary of Abundance and Biomass of the Benthic Macrofauna at Garroch Head Sewage Sludge Dumpsite 1, Stations 1 to 5, in 1985

	Stn 1		Stn 2		Stn 3		Stn 4		Stn 5	
	1	2	1	2	1	2	1	2	1	2
Abundance (ind./m²)										
Polychaeta	880	100	1090	1290	540	570	770	680	13,550	4490
Oligochaeta	—	—	50	—	—	—	50	30	4330	980
Mollusca	420	270	490	490	130	120	150	130	880	560
Crustacea	30	20	40	40	20	10	160	30	120	10
Echinodermata	140	30	40	40	60	130	110	10	700	190
Others	190	160	210	210	80	160	190	90	5110	5370
Total	1660	1480	2320	2070	830	990	1430	970	24,690	11,600
Wet weight (g/m²)										
Polychaeta	28.102	47.441	13.967	22.761	16.550	10.941	20.303	16.616	36.516	17.127
Oligochaeta	—	—	0.025	—	—	—	0.011	0.002	0.867	0.162
Mollusca	39.574	36.115	23.320	17.771	13.850	5.979	1.185	6.202	48.841	57.925
Crustacea	0.233	0.156	0.149	0.043	9.143	0.002	10.290	0.149	0.536	0.001
Echinodermata	0.296	0.006	0.010	0.008	8.866	1.686	0.012	0.003	5.987	0.292
Others	0.395	0.421	0.098	7.822	0.070	0.451	0.940	0.173	33.450	14.469
Total	68.800	84.139	37.569	48.405	48.479	19.059	32.741	23.145	126.197	89.976

From Moore, D. C. and Rodger, G. K., Mar. Ecol. Prog. Ser., 75, 301, 1991. With permission.

TABLE 6
Top Five Species Dominance Lists for Garroch Head Sewage Sludge Dumpsite 1, Stations 1 to 5, 1971 and 1985

	1971			1985	
	Ind./m²	Cum. %		Ind./m²	Cum. %
Station 1					
Capitellidae	3170	67.88	Nemertea	160	10.03
Eunicidae (juv.)	680	82.44	*Chaetozone setosa*	130	18.18
Nemertea	240	87.58	*Ara alba*	115	25.39
Paraonis gracilis (*Levensenia*)[a]	80	89.39	*Spiophanes kroyeri*	105	31.97
Nucula turgida	70	90.78	*Prionospio malmgreni*	85	37.30
Station 2					
Peloscolex sp.	32,550	77.37	*Spiophanes kroyeri*	270	12.24
Capitellidae	6520	92.87	*Thyasira flexuosa*	235	22.90
Audouinia sp. (*Cirriformia*)[a]	970	95.18	*Chaetozone setosa*	140	29.25
Cirratulidae	820	97.12	Nemertea	130	35.15
Cirratulus sp.	570	98.48	*Nuculoma tenuis* *Nucula sulcata* *Mediomastus* sp. }	115	40.36
Station 3					
Peloscolex sp.	31,670	60.32	*Spiophanes kroyeri*	115	12.64
Capitellidae	17,740	94.11	Nemertea	95	23.08
Eunicidae (juv.)	1100	96.21	*Minuspio cirrifera*	80	31.87
Cirratulidae	820	97.77	Ophiuroidea	80	40.66
Cirratulus sp.	540	98.90	*Nephtys hystricis*	60	47.25
Station 4					
Chaetozone setosa	590	17.93	*Tharyx marioni*[a]	250	12.50
Capitellidae	350	28.93	*Minuspio cirrifera*	130	23.33
Diplocirrus glaucus	350	39.57	*Spiophanes kroyeri*	105	32.08
Paraonis gracilis (*Levensenia*)[a]	240	46.51	Nemertea	90	39.58
Eunicidae (juv.) Nemertea }	200	52.58	*Levensenia gracilis*	60	45.42
Station 5					
Capitellidae	8370	89.14	*Mediomastus* sp.	5185	28.58
Scolelepis fuliginosa	370	93.08	Nematoda	5185	57.06
Nemertea	130	94.46	Oligochaeta (indet.)	2335	69.90
Chaetozone setosa	100	95.53	*Capitella capitata*	1445	77.85
Lumbrinereis sp. *Paraonis gracilis* (*Levensenia*)[a] *Thyasira flexousa* }	50	96.06	*Ophryotrocha hartmanni*	610	81.21

Note: Cum. %: cumulative percentage of total abundance at each site.

[a] Synonym. (After Howson, C. M., Ed., Species directory to the British marine fauna and flora, Marine Conservation Society, Ross-on-Wye, 1987.

From Moore, D. C. and Rodger, G. K., *Mar. Ecol. Prog. Ser.*, 75, 301, 1991. With permission.

TABLE 7

Shannon-Wiener (SW) and Heip Indices of Diversity and
Evenness at Garroch Head Sewage Sludge Dumpsite 1,
Stations 1 to 5, in 1971 and 1985

	1971	1985		
		Sample 1	Sample 2	Total
Station 1				
No. species	23	42	34	51
SW index	1.300	3.315	3.127	3.412
Heip index	0.121	0.647	0.661	0.586
Station 2				
No. species	17	51	40	64
SW index	0.802	3.408	3.087	3.402
Heip index	0.077	0.584	0.536	0.461
Station 3				
No. species	13	24	31	44
SW index	0.932	2.793	3.037	3.242
Heip index	0.128	0.666	0.661	0.572
Station 4				
No. species	35	32	32	46
SW index	2.941	2.995	3.062	3.215
Heip	0.527	0.613	0.657	0.531
Station 5				
No. species	24	52	47	63
SW index	0.601	2.182	1.971	2.201
Heip index	0.036	0.154	0.134	0.130

From Moore, D. C. and Rodger, G. K., *Mar. Ecol. Prog. Ser.*, 75, 301, 1991. With
permission.

---

ends of the transect, being comparable to those of normal communities beyond the
dumpsite margins.

Pearson[17] attributed benthic community changes at the active Garroch Head
dumpsite to progressive maximization of carbon utilization as input levels increased.
At carbon input levels estimated to be 20 $gC/m^2/d$, the benthic system exhibited
remarkable stability over the 3-year study period. Hence, it appeared to have fully
adapted to continuously high organic carbon input. This was achieved at the community level through a progressive process involving continuous species replacement
to maximize density, biomass, and reproductive effort in order to utilize all available
carbon. Hence, at the center of the dumpsite a restricted community of continuously
breeding deposit feeders (e.g., *Capitella capitata, Scolelepis fuliginosa*) predominated, giving way along the gradient to a more diverse community of many trophic
types, mostly breeding only once a year (e.g., *Nucula sulcata, Nucula tenuis*, and

Melinna sp.) at the outer, lower end of the gradient. This highly evolved response in the benthic community is a strategy to maximize the utilization of available carbon dumped on the seafloor.

III. TEES RIVER ESTUARY

A. Historical Pollution

The Tees River estuary, located in northeast England, is a narrow stratified system about 30 km in length (Figure 5). As a highly industrialized estuary, the Tees has a long history of pollution problems. Because the river flows through densely populated urban and industrialized centers, biota in the estuary have been subject to serious anthropogenic impacts primarily due to sewage and industrial waste discharges, as well as habitat alteration related to land reclamation, dredging, and other activities. Pollution problems in the estuary have been chronicled as far back as the middle of the 19th Century when an iron and steel industry was established in the area.[18] Later, chemical and petrochemical industries sited along the river and in nearby watersheds contributed substantially to biochemical oxygen demand (BOD) loadings and the input of various chemical contaminants in the estuary. The Imperial Chemical Industries PLC provides an example.

During the first half of the 20th Century, population growth and development were rapid along the river, with contaminants discharged largely unabated. A once thriving fishing industry for flounders, salmon, sea trout, and eels began to decline in the 1930s, and by 1937 salmon were eliminated from the estuary. Alexander et al.[19] ascribed the demise of the fishery to diminishing dissolved oxygen concentrations associated with the discharge of untreated sewage and high amounts of cyanide derived from iron and steel manufacturing.[20] Gray[20] noted that development of the chemical and petrochemical industries and their waste discharges led to an increase in the 5-day BOD load from 14,000 kg/d in 1931 to 221,550 kg/d in 1966. By the early 1970s, water quality of the Tees River had deteriorated markedly, and at this time the estuary was considered to be among the most polluted in England.[21] Of the various pollution sources, industrial waste was deemed to be the most significant. The estuary received 1.37×10^6 m^3 of chemical contaminants and 1.1×10^5 m^3 of sewage per day.

Water quality in the Tees estuary began to improve in the 1970s after the Northumberland River Board and local industry agreed to reduce industrial discharges. Subsequent to this agreement, the BOD of discharges decreased by more than 50%.[22] Completion of the Portrack Sewage Works in 1985 decreased the input of untreated sewage sludge into the Tees, further improving water quality.[23] In addition, many industrial plants closed and were not replaced by other facilities.[24]

Biological surveys conducted in the estuary during the past two decades indicate gradual recovery of the flora and fauna from the severe pollution impacts incurred prior to the mid-1970s.[24] Between 1930 and 1970, macroalgae were eliminated from the most heavily polluted part of the estuary and major changes in the composition of brown and red macroalgae may have occurred elsewhere in the system.[23,25] Compared to the early survey by Alexander et al.[19] in the 1930s, Gray[20] found fewer

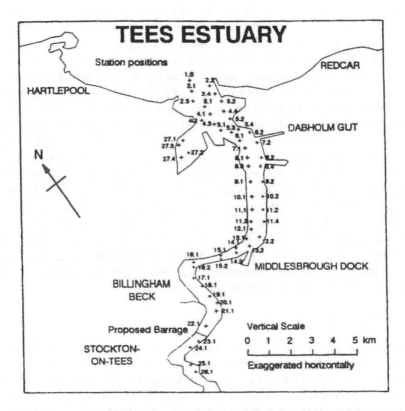

FIGURE 5   Map of the Tees River estuary. (From Tapp, J. F., Shillabeer, N., and Ashman, C. M., *J. Exp. Mar. Biol. Ecol.*, 172, 67, 1993. With permission.)

species of bivalves and lower abundances of bivalves and polychaetes in the early 1970s. Hardy et al.[22] subsequently disclosed that macroalgal species had recolonized parts of the Tees estuary devoid of seaweeds in the 1970s. Investigations of the benthos between 1979 and 1990 indicated an increase in both the abundance and the diversity of the fauna since Gray's[20] earlier work.[18,23,24] Pomfret et al.[26] likewise recorded a progressive increase in the abundance of fish during the 1980s. The following account provides an overview of the major changes in biotic communities of the Tees estuary associated with pollution inputs.

B. BENTHIC COMMUNITIES

1. Macroalgae

Hardy et al.[22] described the status of macroalgal populations in the Tees estuary during the early 1990s. They also compared their findings with those of Edwards[25] and Alexander et al.,[19] who conducted studies of the species composition and distribution of macroalgae in the estuary some 25 and 65 years ago, respectively. In the 1930s, Alexander et al.[19] reported that many species were distributed over extensive areas of the estuary. In studies by Edwards[25] conducted about 40 years later,

most macroalgal species were restricted to the estuarine mouth, and several prominent species (i.e., *Delesseria sanguinea, Membranoptera alata, Plumaria elegans,*
and *Ptilota plumosa*) were no longer present in the system.

Surveys by Hardy et al.[22] indicated that the estuary was relatively poor in macroalgal species, with the greatest number occurring at the mouth of the estuary and
progressively fewer species recovered upestuary. Hardy et al.[22] registered fewer species of brown (Phaeophyta) and red (Rhodophyta) seaweeds than did Edwards[25] and
Alexander et al.[19]; the largest decrease in the number of species took place between
1971 and 1991 when the number of species of Phaeophyta dropped from 12 to 7 and
that of Rhodophyta from 8 to 2. These losses of taxa were compensated in part by
increases in the number of species of green (Chlorophyta) and blue-green (Cyanophyta) seaweeds (Table 8). There is some evidence that the few species of Phaeophyta
remaining in the Tees are spreading into the higher reaches of the estuary. In addition,
in recent years species of Chlorophyta (e.g., *Entermorpha intestinalis, Rhizoclonium
tortuosum, Ulothrix flacca,* and *Urospora pencilliformis*) and Cyanophyta (*Microco-
leus lyngbyaceus*) have recolonized heavily polluted parts of the estuary in areas shown
by Edwards[25] to be devoid of macroalgal species. These changes infer recovery of
macroalgal populations following recent cleanup measures.

2. Benthic Fauna

Several important benthic faunal studies have been conducted in the Tees estuary
since the early 1930s, including those of Alexander et al.,[19] Gray,[20] Kendall,[27]
Shillabeer and Tapp,[18,23] and Tapp et al.[24] Alexander et al.[19] published an extensive
list of benthic fauna found throughout the length of the estuary in the 1930s. As in
the case of macroalgae, the numbers of benthic faunal species peaked at the estuarine
mouth, where 113 species were collected. Far fewer species were observed upestuary,
with only two species recovered in the area just above Billingham Beck about 15 km
from the estuarine mouth. By 1935, the distribution of the benthos appeared to be
strongly affected by pollution in the estuary.

TABLE 8
Number of Species of Macroalgae Recorded by Alexander et al.[19] (1929–1933), Edwards[21] (1970–1971), and Hardy et al.[22] (1991)

Years Surveyed Tees estuary	Number of species					
	Chlorophyta	Chrysophyta	Cyanophyta	Phaeophyta	Rhodophyta	Total
1929–1933	4	1	0	10	7	22
1970–1971	12	1	0	12	8	33
1991	15	1	2	7	2	27

From Hardy, F. T., Evans, S. M., and Tremayne, M. A., *J. Exp. Mar. Biol. Ecol.,* 172, 81, 1993. With
permission.

Gray[20] investigated the benthic community in summer 1971 and spring 1973 when the estuary was severely impacted by pollution. His work primarily focused on the spatial pattern of the fauna. The meiofauna exhibited no detectable effects of pollution, with abundance and biomass values for the group being comparable to those of other temperate estuaries not subjected to major pollution problems. Gastrotrichs and nematodes numerically dominated the meiofauna, with the highest density being 9×10^6 ind./m^2. At Seal Sand, the mean meiofaunal biomass in intertidal sediments amounted to 30 g/m^2 ash free dry weight, which was 50 times greater than that of the macrofauna. Gray[20] attributed this high biomass mainly to the presence of large numbers of oligochaetes and polychaetes (mean values of 2.07×10^5 and 8.01×10^4 ind./m^2, respectively). Much lower mean meiofaunal biomass values were recorded in the estuary along the open coast (1.12 g/m^2 ash free dry weight) and at Gare Sands (6.18 g/m^2 ash free dry weight). The total number of meiofauna at Gare Sands (2.5×10^6 ind./m^2) exceeded that at the open coast (0.9×10^6 ind./m^2).

Compared to the comprehensive list of macrofaunal species compiled by Alexander et al.[19] for the Tees estuary, the macrofauna chronicled by Gray[20] consisted of fewer species. The most dramatic reduction in the macrofauna occurred among the Bivalvia. For example, the 1935 survey list included a number of commonly occurring bivalves: *Mytilus edulis* L., *Macoma balthica* L., *Scrobicularia plana* (da Costa), *Tellina tenuis* da Costa, *Venerupis* (=*Paphia*) *pallustra* (Montagu), *Cerastoderma edule* (L.), and *Mya arenaria* L. In early 1970s surveys, only *M. balthica* and *M. edulis* remained, suggesting pollution impacts on this taxonomic group.[20]

In an 18-month investigation of the benthic fauna of the Seal Sands mudflat in the Tees River estuary, Kendall[27] uncovered a dense animal community dominated by small annelids in terms of both abundance and standing stock biomass. He stressed that the numerical structure of the fauna was highly stable. Both tube building and infaunal deposit feeding annelids existed in large numbers along the mudflat. The sabellid polychaete *Manayunkia aestuarina* Bourne was particularly numerous, reaching densities of ~1.3×10^6 ind./m^2. It numerically dominated the community for the duration of the sampling period in the Seal Sands area.

There is evidence of recovery of the benthic community from pollution impacts since 1979. For example, 33 benthic species were documented at standard stations in the estuary in 1979 compared to 78 species in 1985.[18] The most significant increase in benthic diversity between 1979 and 1990 developed at the seaward end of the estuary (Figure 6). However, diversity increased farther upestuary during the last two years. Abundance of the benthos also increased between 1979 and 1990, most conspicuously in the middle reaches of the estuary (Figure 7). These changes in the benthic community occurred concomitantly with large reductions in both industrial and domestic discharges into the estuary.

Table 9 lists the mean annual abundance of the 25 most common taxa at standard stations established in the Tees estuary during the 1979–1990 sampling period. Tapp et al.[24] identified four faunal zones based on species abundance and diversity data obtained during 1979, 1981, 1989, and 1990. These zones are defined as:

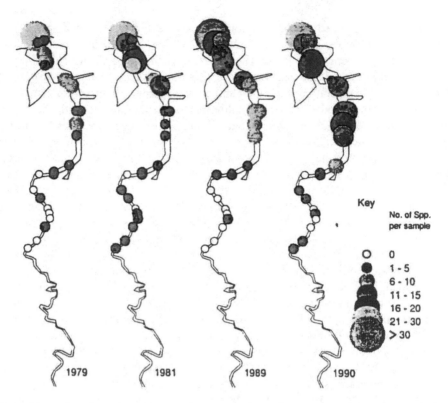

FIGURE 6 Number of benthic species per sample collected in the Tees estuary during 1979, 1981, 1989, and 1990. (From Tapp, J. F., Shillabeer, N., and Ashman, C. M., *J. Exp. Mar. Biol. Ecol.*, 172, 67, 1993. With permission.)

1. <u>Marine fauna</u> — that zone of the estuary characterized by taxa typical of the nearshore coastal waters. This zone contains the most diverse fauna of the estuary.
2. <u>Abundant fauna</u> — that zone of the estuary characterized by organic pollution tolerant species such as *Capitella capitata* and *Tubificoides benedeni*. This zone contains the highest abundance of benthic fauna.
3. <u>Sparse fauna</u> — that zone of the estuary limited to a few individuals of the pollution tolerant species.
4. <u>Extremely sparse</u> — that zone of the estuary devoid of macrobenthic life.

Figure 8 shows the distribution of the four faunal zones for the years 1979, 1981, 1989, and 1990. Most conspicuous is the shift in position of the marine zone progressively upestuary on the western side of the estuary during these four years. Table 10 provides the mean faunal statistics (i.e., mean numbers of species, mean abundances) for each faunal zone.

Improvements in water quality of the Tees River since the early 1970s have been responsible for the gradual recovery of the benthic community in the estuary. Since the late 1970s, there has been an increase in species abundance and diversity

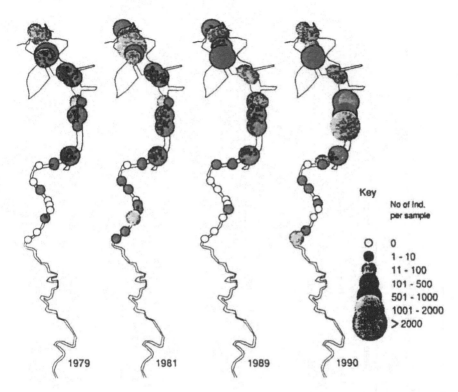

FIGURE 7 Abundance of benthic organisms per sample collected in the Tees estuary during 1979, 1981, 1989, and 1990. (From Tapp, J. F., Shillabeer, N., and Ashman, C. M., *J. Exp. Mar. Biol. Ecol.*, 172, 67, 1993. With permission.)

of the benthos. Improvement in the benthic biology of the estuary should continue to unfold during the 1990s as tighter controls are placed on industrial and sewage waste discharges.

C. Fish

Pomfret et al.[26] conducted demersal fish surveys in the Tees estuary between 1981 and 1988. Beam trawl samples were collected at eight stations upstream of the mudflats near the estuarine mouth. Three surveys were conducted each year between 1981 and 1983, and six surveys each year after 1983. Results of the beam trawl sampling were then compared to those of the Tyne and Wear estuaries.

Table 10 provides data on the number of species caught in the Tees, Tyne, and Wear estuaries for the 1982–1988 period. A simple method of comparing the data involved calculating the percentage of trawls in which each species had been recorded and summing these values. In the middle reaches, far fewer species were collected in the Tees than the Tyne and Wear estuaries. Pomfret et al.[26] ascribed these differences in trawl statistics to greater pollution in the middle reaches of the Tees estuary than in the moderately polluted Tyne and the relatively unpolluted Wear

TABLE 9
Characteristics of Four Faunal Zones Identified in the Tees Estuary During the 1979 to 1990 Sampling Period

Zone	1979		1981		1989		1990	
	Mean Species (1/10 m²)	Mean Individuals (1/10 m²)	Mean Species (1/10 m²)	Mean Individuals (1/10 m²)	Mean Species (1/10 m²)	Mean Individuals (1/10 m²)	Mean Species (1/10 m²)	Mean Individuals (1/10 m²)
Marine	7.4	67	10.7	285	14.9	142	9.9	81
Abundant	5.8	291	6.0	565	6.8	269	11.1	1390
Sparse	1.7	23	2.1	10	4.6	17	2.6	19
Extremely sparse	0	0	0.25	0.25	0.95	1.4	0.25	0.25

From Tapp. J. F., Shillabeer, N., and Ashman, C. M., *J. Exp. Mar. Biol. Ecol.*, 12, 67, 1993. With permission.

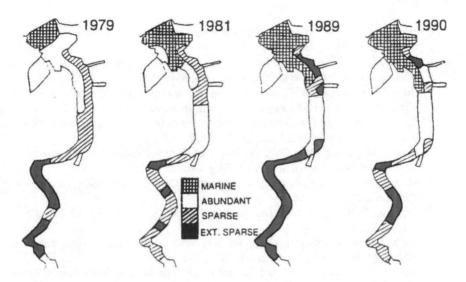

FIGURE 8 Characteristics of four faunal zones identified in the Tees estuary during the 1979–1990 sampling period. (From Hardy, F. T., Evans, S. M., and Tremayne, M. A., *J. Exp. Mar. Biol. Ecol.*, 172, 81, 1993. With permission.)

TABLE 10
Comparison of Beam Trawling Results from the Tees, Tyne, and Wear Estuaries and at Specific Mid-Estuary Sites of Similar Salinity Regime (Combined Data for 1982–1988)

Area/Station	Number of Species Recorded	Number of 1-km Trawls	Cumulative Percentage Occurrences
Tyne (all stations combined)	45	880	230
Wear (all stations combined)	24	69	368
Tees (all stations combined)	25	225	130
Tyne (Scotswood)	19	146	146
Wear (Hylton)	13	18	374
Tees (Stockton)	8	28	90

From Pomfret, J. R., Turner, G. S., and Phillips, S., *J. Fish. Biol.*, 33 (Suppl. A), 71, 1988. With permission.

estuaries. They noted, however, that abundance of demersal fish has increased progressively during the 1980s in the Tees estuary, presumably due to improvement in water quality of the system. Cleanup measures implemented in the estuary since the early 1970s appear to have contributed greatly to the upgrade of fish assemblages in the estuary.

REFERENCES

1. Dooley, H. D., Factors influencing water movements in the Firth of Clyde, *Est. Coastal Mar. Sci.*, 9, 631, 1979.
2. Steele, J. H., McIntyre, A. D., Johnston, R., Baxter, I. G., Topping, G., and Dooley, H. D., Pollution studies in the Clyde Sea area, *Mar. Pollut. Bull.*, 4, 153, 1973.
3. Harper, E. and Green, W. T., Marine disposal of sewage sludge by North West Water Authority and Strathclyde Regional Council, in *Marine Treatment of Sewage and Sludge*, Proceedings of a Conference held by the Institute of Civil Engineers, Brighton, 1987, Thames Telford, London, 1988, 137.
4. Moore, D. C. and Rodger, G. K., Recovery of a sewage sludge dumping ground. II. Macrobenthic community, *Mar. Ecol. Prog. Ser.*, 75, 301, 1991.
5. Mackay, D. W., Sludge dumping in the Firth of Clyde — a containment site, *Mar. Pollut. Bull.*, 17, 91, 1986.
6. Topping, G., Sewage and the sea, in *Marine Pollution*, Johnson, R., Ed., Academic Press, London, 1976, 303.
7. Mackay, D. W. and Topping, G., Preliminary report on the effects of sludge disposal at sea, *Effl. Wat. Treat. J.*, November, 1970, 641.
8. Mackay, D. W., Halcrow, W., and Thornton, I., Sludge dumping in the Firth of Clyde, *Mar. Pollut. Bull.*, 3, 7, 1971.
9. Halcrow, W., Mackay, D. W., and Thornton, I., The distribution of trace metals and fauna in the Firth of Clyde in relation to the disposal of sewage sludge, *J. Mar. Biol. Assoc. U.K.*, 53, 721, 1973.
10. Halcrow, W., Mackay, D. W., and Bogan, J., PCB levels in Clyde marine sediments and fauna, *Mar. Pollut. Bull.*, 5, 134, 1974.
11. Mackay, D. W., Tayler, W. K., and Henderson, A. R., The recovery of the polluted Clyde estuary, *Proc. Roy. Soc. Edinb.*, 76B, 135, 1978.
12. Rodger, G. K., Davies, I. M., and Moore, D. C., Recovery of a sewage sludge dumping ground. I. Trace metal concentrations in the sediment, *Mar. Ecol. Prog. Ser.*, 75, 293, 1991.
13. Henderson, A. R., Long-term monitoring of the macrobenthos of the upper Clyde estuary, *Wat. Sci. Technol.*, 16, 359, 1984.
14. Topping, G. and McIntyre, A. D., Benthic observations on a sewage sludge dumping ground, *Int. Counc. Explor. Sea Comm. Meet.*, ICES/E:30, 1972.
15. Jenkinson, I. R., Sludge dumping and benthic communities, *Mar. Pollut. Bull.*, 3, 102, 1972.
16. Pearson, T. H., Ansell, A. D., and Robb, L., The benthos of the deeper sediments of the Firth of Clyde, with particular reference to organic sediments, *Proc. R. Soc. Edinb.*, 90, 329, 1986.
17. Pearson, T. H., Benthic ecology in an accumulating sludge-disposal site, in *Oceanic Processes in Marine Pollution*, Vol. 1, *Biological Processes and Wastes in the Ocean*, Capuzzo, J. M. and Kester, D. R., Eds., Robert E. Krieger Publishing Company, Malabar, FL, 1987, 195.
18. Shillabeer, N. and Tapp, J. F., Long-term studies of the benthic biology of Tees Bay and the Tees estuary, *Hydrobiologia*, 195, 63, 1990.
19. Alexander, W. B., Southgate, B. A., and Bassindale, R., Survey of the River Tees estuary. II. The estuary chemical and biological, Technical Paper on Water Pollution Research No. 5, Her Majesty's Stationery Office, London, 1935.
20. Gray, J. S., The fauna of the polluted River Tees estuary, *Est. Coastal Mar. Sci.*, 4, 653, 1976.

21. Edwards, P., An assessment of possible pollution effects over a century on the benthic marine algae of Co. Durham, England, *Bot. J. Linn. Soc.*, 70, 269, 1975.
22. Hardy, F. G., Evans, S. M., and Tremayne, M. A., Long-term changes in the marine macroalgae of three polluted estuaries in northeast England, *J. Exp. Mar. Biol. Ecol.*, 172, 81, 1993.
23. Shillabeer, N. and Tapp, J. F., Improvements in the benthic fauna of the Tees estuary after a period of reduced pollution loadings, *Mar. Pollut. Bull.*, 20, 119, 1989.
24. Tapp, J. F., Shillabeer, N., and Ashman, C. M., Continued observations of the benthic fauna of the industrialized Tees estuary, 1979–1990, *J. Exp. Mar. Biol. Ecol.*, 172, 67, 1993.
25. Edwards, P., Benthic algae in polluted estuaries, *Mar. Pollut. Bull.*, 3, 55, 1972.
26. Pomfret, J. R., Turner, G. S., and Phillips, S., Beam trawl surveys as a monitoring tool in polluted estuaries in northeast England, *J. Fish. Biol.*, 33 (Suppl. A), 71, 1988.
27. Kendall, M. A., The stability of the deposit feeding community of a mud flat in the River Tees, *Est. Coastal Mar. Sci.*, 8, 15, 1979.

8 Case Study 7: Wadden Sea

I. INTRODUCTION

Increasing nutrient loads of the Rhine and Ems Rivers, as well as a number of smaller influent systems, have contributed to escalating eutrophication problems in Dutch coastal waters, notably the Wadden Sea. This highly dynamic estuarine system is bounded by one of the most important wetlands of the world, and it supports valuable commercial and recreational fisheries. However, due to a wide range of anthropogenic activities, these resources are being threatened. For example, bottom habitats and associated shellfish populations (e.g., cockles, mussels, and shrimp) are adversely affected by ongoing dredging and dredged-spoil disposal operations, as well as sand extraction operations. Overfishing has probably led to extinction of the oyster *Ostrea edulis* from the Dutch Wadden Sea. Closure of the Zuiderzee has been implicated in the demise of some finfish populations, such as the herring *Clupea harengus* and the anchovy *Engraulis enrasicolus*. Increased turbidity resulting from dredging and sand extraction may have impaired adult dab *Limanda limanda* and impacted critically important eelgrass habitat.[1]

Nutrient enrichment, accelerated contaminant discharges, and land reclamation and closures (e.g., Lauwerszee and Zuiderzee) are factors that may seriously alter biotic communities. Coastal engineering projects have significantly modified the morphology of the Wadden Sea. The loss of habitat due to embankments and the related changes in hydraulic processes has influenced much of the Wadden Sea environment. The creation of embanked areas separated from the sea — especially in the province of Groningen, south of the island of Ameland, and in the Balgzand region — has reduced the total area of the Wadden Sea and fringing salt marshes.[1,2] Because the Wadden Sea ecosystem is situated in the middle of the eastern Atlantic flyway and supports between 6 and 12 million birds, such modifications have presumably affected bird populations utilizing various terrestrial and aquatic habitats for feeding, breeding, or roosting.[3]

II. PHYSICAL DESCRIPTION

The Wadden Sea is a 600-km long, shallow physically controlled estuarine system in northwestern Europe consisting of a chain of tidal basins that are separated from the North Sea by a series of barrier islands and from the mainland by dikes (Figure 1). Extending from the Netherlands to Denmark, this highly dynamic system has a surface area of ~6000 km².

Sand flats comprise ~75% of the entire Dutch, German, and Danish Wadden Sea, and mud flats and mixed sediments, 7% and 18%, respectively.[3] Water exchange

271

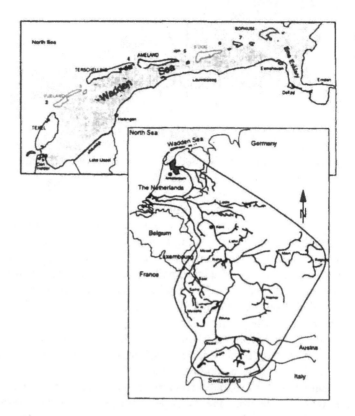

FIGURE 1   Map of the Wadden Sea showing different tidal basins (numbered, see Table 1 for description), the Rhine River (including the lower branches and Lake IJssel) and its drainage basin, and the coastal area of the North Sea. (From de Jonge, V. N. and van Raaphorst, W., in *Eutrophic Shallow Estuaries and Lagoons*, McComb, A. J., Ed., CRC Press, Boca Raton, FL, 1995, 129. With permission.)

with the North Sea occurs through several tidal inlets. However, water masses in the tidal basins are only interconnected during high tide, because the basins are separated from each other by elevated tidal flats (tidal watersheds). Seven tidal basins exist in the Dutch Wadden Sea; in the western part, these include the Marsdiep, Eijerlandsche Gat, and Vlie, and in the eastern part, the Borndiep, Friesche Zeegat, Lauwers, and Schild (Table 1). Along the mainland, freshwater discharges from the Rhine and Ems Rivers, as well as several smaller influent systems, have significantly influenced water quality in these basins. Riverine water quality deteriorated markedly in the 1960s and 1970s owing largely to industrial and municipal wastewater discharges.[4]

Freshwater inflow in the western Dutch Wadden Sea derives mainly from the Rhine River, which carries an average of 3.4×10^9 kg/yr of suspended matter,[5] and has a mean annual discharge of ~2200 m³/s. The Rhine, which is more than 1000 km long, splits into two main branches about 80 km from the North Sea. One branch, the IJssel River, flows northward into Lake IJssel, with some of the water sluiced out of the lake into the Wadden Sea. The other branch, the Rhine and Waal Rivers,

TABLE 1
Some Characteristics of the Different Tidal Basins of the
Dutch Wadden Sea

Number of the Tidal Basin	Name of the Tidal Basin	Total Surface Area (10⁶ m²)	Water Content at HW[a] (10⁶ m³)	Tidal Volume (10⁶ m³)	Turnover Time Water t (tides)	Surface area of Tidal Flats (10⁶ m²)
1	Marsdiep	712	3357	1140	17	107 (15%)
2	Eijerlandsche Gat	153	313	207	3	99 (65%)
3	Vlie	668	2248	1060	13	267 (40%)
4	Borndiep	309	812	478	10	139 (45%)
5	Friesche Zeegat	195	350	300	9	136 (70%)
6	Lauwers	200	300	240	8	130 (65%)
7	Schild	29	41	31	8	23 (80%)
	Total Dutch Wadden Sea	2266	7400	3388		901 (40%)

Note: See Figure 1 for location of basins.

[a] HW = High Water.

From de Jonge, V. N. and van Raaphorst, W., in *Eutrophic Shallow Estuaries and Lagoons*, McComb, A. J., Ed., CRC Press, Boca Raton, FL, 1995, 129. With permission.

flows westward into the North Sea.[6] This water then flows northward in a 50 to 70-km wide zone along the Dutch coast under the influence of residual currents and the wind, entering the Wadden Sea near Den Helder.[3,7] About 8–15% of the Rhine River runoff flows into the IJsselmeer via the northern branch of the Rhine. It ultimately enters the western Wadden Sea. More than 50% of Rhine River water flows via the Nieuwe Waterweg to the North Sea, and ~30% runs into the Haringvliet and also ultimately discharges into the North Sea.

The Ems River is about 200 km long and has a monthly discharge (at Pogum) of 25 to 390 m³/s.[8] The Ems River estuary forms part of the Wadden Sea, being situated at the border between the Netherlands and Germany. Excluding its outer delta, the Ems estuary has a surface area of ~500 km², with about half of this area covered by intertidal flats. The estuary lies in an agricultural area, and thus receives nutrient inputs and some contaminants from land runoff. A chlor-alkali plant and a pesticide factory caused mercury pollution in the Ems estuary near Delfzijl in the 1960s and 1970s. The Ems River and the Westerwoldsche Aa transport large amounts of nutrients both in dissolved and particulate forms. The effect of nutrient enrichment in this area of the Wadden Sea appears to be limited because of only moderate freshwater flow into the Ems estuary.[8]

III. POLLUTION SOURCES

Contaminants enter the Wadden Sea via several primary pathways, notably river inflow, wastewater discharges, land runoff, and atmospheric deposition. Both the

Rhine and Ems Rivers transport a variety of contaminants to the Wadden Sea, although reliable quantitative data on contaminants that reach the sea are not available, largely because the transport from Hoek van Holland northward along the Dutch coast has not been precisely determined.[1] Industrial development after World War II resulted in massive loading of the Rhine River with toxic organic chemicals, heavy metals, and other substances, culminating in rapid degradation of the system. High nutrient and organic loading effectively reduced dissolved oxygen levels and impacted biotic communities. By the 1960s, environmental conditions in the river had declined appreciably, falling to perhaps their lowest levels.[9] Remedial sanitation programs initiated in the 1970s led to improved water quality and partial recovery of biological communities in the early 1980s.[10,11] Today, the Rhine River carries reduced, albeit persistent loads of toxic chemicals, and it is a highly eutrophic system.[4]

The Rhine River has been a significant source of cadmium and mercury for the Wadden Sea; however, annual loads decreased substantially during the periods 1970–1975 (cadmium) and 1978–1983 (mercury).[1] Freshwater discharges from Lake IJssel represent a major source of PAHs for the western Dutch Wadden Sea.[12] Organic wastewater discharges are important in the Ems estuary.[13]

In the 1960s, several avian species breeding in the Wadden Sea suffered high mortality due to pesticide contamination. Reproductive failure among harbor seals was ascribed to PCB exposure. Other chemical contaminants (TBT, HCH, etc.) have been detected in sediment and biotic samples.

Currently, the most serious environmental concern in the Wadden Sea is eutrophication, which is the focus of this discussion. Nutrient enrichment and organic carbon loading have become a potentially serious problem in the Wadden Sea. According to Beukema,[14] the concentrations of nitrogen and phosphorus compounds, as well as phytoplankton, have roughly doubled in the western Wadden Sea since 1970. de Jonge[6] reported more than a 10-fold increase in phytoplankton production in the Marsdiep tidal channel since the early 1950s. Although dissolved oxygen levels are often depressed in summer, anoxic conditions rarely occur. When present, diminished oxygen concentrations tend to be localized because of strong tidal mixing.

Nutrient concentrations in the Wadden Sea depend on external inputs from river discharges, Lake IJssel, the North Sea, and atmospheric deposition. In addition, internal processes in the Wadden Sea may be as important as external inputs.[3] In the eastern Wadden Sea, allochthonous sources of nutrients and organic matter are primarily the Ems River estuary and the North Sea. Nondispersive tide-induced residual currents transport particulate nutrients from the North Sea into the Wadden Sea.[15,16] The mineralization of organic matter derived from the North Sea and/or the Wadden Sea tidal basins release dissolved nutrients. Allochthonous sources of organic carbon are significant to the overall carbon budget of the system.[17–19] During recent decades, the import of organic matter into the Wadden Sea has increased.[19–21]

In the western Wadden Sea, the Rhine River and Lake IJssel are important allochthonous sources of nutrients and, together with the North Sea, are significant allochthonous sources of organic matter. Internal processes — losses to sediments, uptake by primary producers, and mineralization of organic matter on the seafloor

— are major factors influencing the total concentrations of dissolved nutrients in the estuary. The effects of reduced or enhanced external loadings on the concentrations of dissolved inorganic nutrients in the system are difficult to determine because internal biological and physical-chemical transformations commonly dominate. Accurate quantitative data must continue to be collected on both allochthonous and autochthonous sources of nutrients and organic carbon, as well as internal processes, to provide greater understanding of eutrophication in the Wadden Sea.

A. NUTRIENT CONCENTRATIONS

Nienhuis[7] showed that the total nitrogen loadings in six Dutch estuaries ranged from 4–235 $gN/m^2/yr$ (Table 2). The total nitrogen loading in the Wadden Sea (50 $gN/m^2/yr$) was only slightly greater than that in North Sea coastal waters (40 $gN/m^2/yr$). Nitrogen loads of 40 to 50 $gN/m^2/yr$ usually do not result in extremely high summer chlorophyll concentrations.

TABLE 2
Nitrogen Loadings in the
Wadden Sea Compared to
Other Dutch Estuaries

Estuary	Load[a]
Wadden Sea	50
Dollard	61
Grevelingenmeer	4
Oosterschelde	5
Veerse Meer	34
Westerschelde	235

[a] $gN/m^2/yr.$

From Nienhuis, P. H., *Hydrobiologia*, 265, 15, 1993. With permission.

In the more turbid Dutch estuaries, light availability rather than nutrients is the primary limiting factor to phytoplankton production. This is true in the Ems-Dollard and Westerschelde estuaries which have extinction coefficients of 0.5 to 7.0/m. In contrast, the Grevelingen estuary with extinction coefficient values of 0.2 to 0.5/m is much clearer, and consequently has substantially higher chlorophyll concentrations. Intermediate extinction coefficients are evident in the Wadden Sea, Oosterschelde, and Veerse Meer systems.

In regard to nutrient levels in influent systems, the dissolved nitrogen (ammonium and nitrate) load of the Rhine and Meuse Rivers increased by a factor of 2 to 4 between 1950 and 1985. By comparison, the dissolved phosphorus load increased by a factor of 5 to 7 over the same period. These elevated nutrient loads resulted in a 3- to 5-fold increase in nitrogen and phosphorus concentrations in Dutch coastal waters.[7]

Figure 2 illustrates annual nutrient loads of the Rhine River and Lake IJssel between 1950 and 1992. Total phosphorus and phosphate loads increased greatly in the Rhine River after 1950 and then decreased significantly in the 1980s. The use of polyphosphates in detergents and artificial fertilizers in agriculture contributed substantially to the elevated levels.[1] Total nitrogen and dissolved nitrogen, in turn, began to increase significantly in the 1960s and have remained higher than total phosphorus and phosphate loads in recent years. Nitrate concentrations in the Rhine River have increased since the 1970s because of improved wastewater treatment.

In Lake IJssel, the loads of total phosphorus and phosphate increased acutely from 1971 to 1984 and then declined. A similar trend was uncovered for total nitrogen and dissolved nitrogen loads. However, the decrease in nitrogen loads during the 1980s was less clear than the decrease in phosphorus loads.

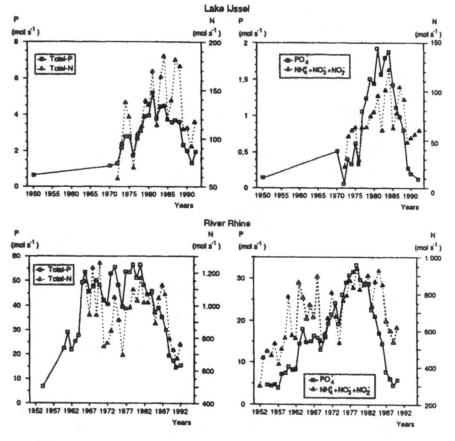

FIGURE 2   Annual nutrient loads of the Rhine River and Lake IJssel between 1950 and 1992. "Total" is the sum of dissolved and particulate nutrients. (From de Jonge, V. N. and van Raaphorst, W., in *Eutrophic Shallow Estuaries and Lagoons*, McComb, A. J., Ed., CRC Press, Boca Raton, FL, 1995, 129. With permission.)

Nitrogen concentrations vary seasonally in both the Rhine River and Lake IJssel (Figures 3 and 4). For example, total nitrogen concentrations followed a seasonal curve, being highest in winter and lowest in summer. A similar seasonal pattern has been shown for ammonium concentrations in the Rhine River, but this pattern is not evident in Lake IJssel. While phosphate concentrations peaked in Lake IJssel during the winter months and declined to their lowest levels in the spring, no such seasonal pattern in concentrations was apparent in the Rhine River (Figures 3 and 4).

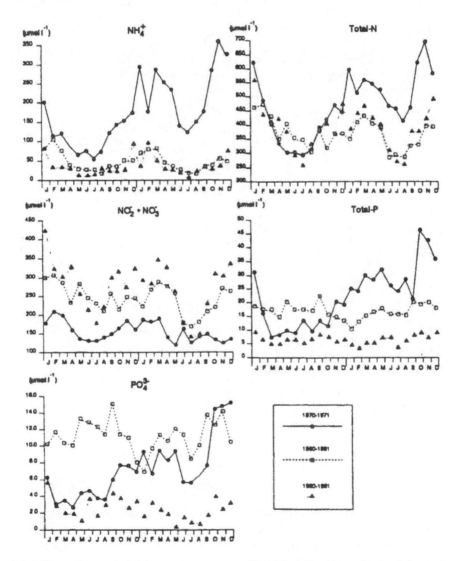

FIGURE 3 Seasonal variation in nutrient concentrations of the Rhine River when entering the Netherlands. (From de Jonge, V. N. and van Raaphorst, W., in *Eutrophic Shallow Estuaries and Lagoons*, McComb, A. J., Ed., CRC Press, Boca Raton, FL, 1995, 129. With permission.)

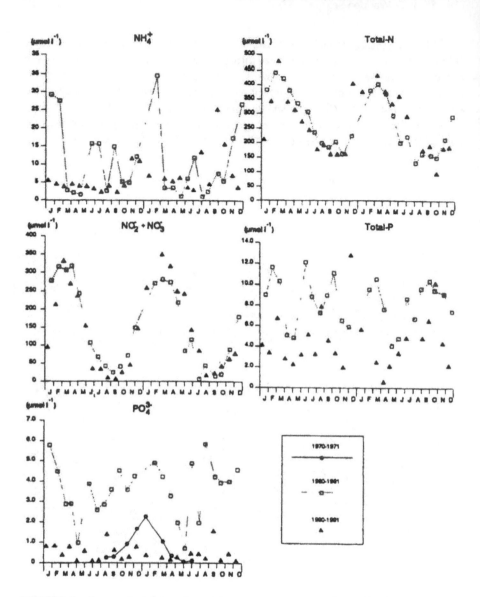

FIGURE 4 Seasonal variation in nutrient concentrations of Lake IJssel near one of its discharge points. (From de Jonge, V. N. and van Raaphorst, W., in *Eutrophic Shallow Estuaries and Lagoons*, McComb, A. J., Ed., CRC Press, Boca Raton, FL, 1995, 129. With permission.)

The mean annual concentrations of nutrients in the Marsdiep and Vlie Basins also varied greatly between 1950 and 1985 (Figure 5). In the early 1950s, the nutrient concentrations were relatively low in both basins. However, total phosphorus levels later increased substantially, peaking in the 1970s and early 1980s. Changes in total nitrogen levels were less dramatic.

In summary, important differences exist in nutrient supply in the eastern and western Wadden Sea. As noted above, nutrient supply in the eastern Wadden Sea is

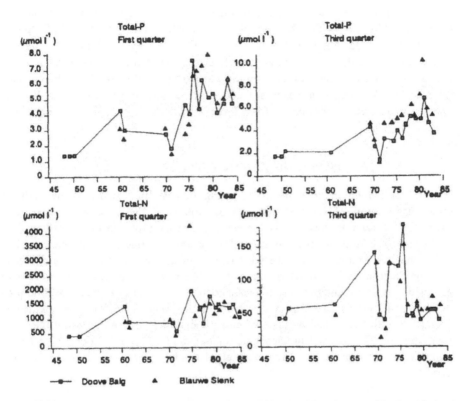

FIGURE 5 Concentrations of total-P and total-N in the Marsdiep and Vlie basins in the first and third quarter of the year, respectively, during the 1950–1985 period. (From van der Veer, H. W., van Raaphorst, W., and Bergman, M. J. N., *Helgol. Meeresunters.*, 43, 501, 1989. With permission.)

contingent upon the mineralization of organic matter from the North Sea and/or the adjacent Wadden Sea basins. Nutrients in the Ems estuary and adjacent Wadden Sea area originate from the Ems and Westerwoldsche Aa Rivers and the North Sea coastal zone.[3,8,22,23] In the western Wadden Sea, Lake IJssel is an important source of phosphorus,[6,16] and the Rhine River a significant source of phosphorus and nitrogen.[1,3]

Nutrient data are available for the entire Dutch Wadden Sea from 1970 onward. de Jonge and Postma[20] and Helder[22] provided an overview of nutrients throughout the Dutch Wadden Sea in the early 1970s. Inorganic nitrogen concentrations peaked in the western Wadden Sea. Phosphate concentrations were highest near freshwater sources, as well as tidal watersheds where enhanced mineralization of organic matter takes place. These distribution patterns of inorganic nutrients reflect the strong influence of inputs from the Rhine River and Lake IJssel.

Nitrogen and phosphorus concentrations increased sharply in the western Wadden Sea during the mid-1970s, reaching maxima between 1980 and 1986. Figure 6 shows concentrations of ammonium and nitrate as a function of salinity in the Marsdiep Basin, and Figure 7 depicts dissolved phosphorus and total phosphorus as a function of salinity in the basin. As is evident, both ammonium and nitrate concentrations were usually higher, regardless of salinity, in 1985–1986 than in the early 1960s and 1970s. The same was true for total phosphorus and phosphate concentrations. de Jonge and van Raaphorst[3] primarily attributed the largest increase in nutrient concentrations in the western Wadden Sea over these years to increases in nutrient concentrations in the northern part of Lake IJssel. Freshwater input to the Marsdiep tidal basin, situated in the westernmost part of the Wadden Sea, is dominated by discharges from Lake IJssel which contribute much of the nutrient supply.

IV. EUTROPHICATION EFFECTS

Nutrient enrichment in the Wadden Sea during the past few decades has had a direct effect on the biota of the tidal basins. Most notable is the increased primary and secondary production observed in the estuary. For example, in the Marsdiep tidal basin phytoplankton and benthic microalgal production has doubled since the late 1970s due to eutrophication in the western part of the Dutch Wadden Sea. Accelerated production and biomass of intertidal macrobenthos at this time were also ascribed to nutrient enrichment.[1] No such increases in intertidal macrobenthic production and biomass have been observed in the eastern Wadden Sea, which is consistent with the virtual absence of eutrophication in that area during the same time period.[23] The increase in phosphate loads from Lake IJssel is deemed to be the main factor responsible for stimulating algal productivity in the western Wadden Sea.[6,24]

A. Primary Producers

Primary production of phytoplankton in the western Wadden Sea (i.e., Marsdiep tidal basin) increased from 20–40 gC/m^2/yr in the early 1950s to 150–520 gC/m^2/yr in the early 1990s.[25–27] The highest annual primary production value (520 gC/m^2/yr) was measured in 1981/1982; since that time, primary production estimates have declined to ~300 gC/m^2/yr (Figure 8).[28] Hence, annual primary production of phytoplankton more than tripled from ~150 gC/m^2/yr around 1970 to >500 gC/m^2/yr in the early 1980s. Phytoplankton concentrations also increased. The lengthening of the blooming period of the flagellate *Phaeocystis pouchetii* (Haptophyceae) from 20 days in the early 1970s to more than 60 days in the mid-1980s contributed greatly to the higher phytoplankton numbers.[28–30]

Benthic microalgal production likewise increased substantially over the 1950–1990 period in the western Wadden Sea (Figure 8). Between 1970 and 1980, Cadee[21] found that the annual primary production of benthic microalgae more than doubled from <100 gC/m^2/yr to >200 gC/m^2/yr. He attributed this rising trend to escalating eutrophication of the western Wadden Sea. Benthic microalgal production was ~30–70% that of phytoplankton production.[7] The increased annual primary production of both phytoplankton and benthic microalgae was positively correlated with mean annual loads of orthophosphate from Lake IJssel.[6,23] According to de Jonge,[6] algal growth in the Marsdiep tidal basin is phosphate-limited.

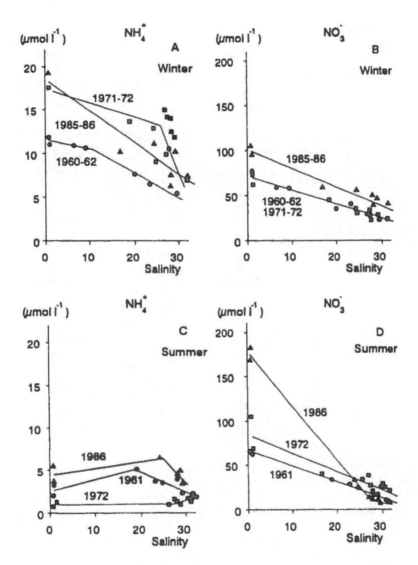

FIGURE 6 Ammonium (A, C) and nitrate (B, D) as a function of salinity in the Marsdiep basin during winter (December–February) and summer (May–July). (From de Jonge, V. N. and van Raaphorst, W., in *Eutrophic Shallow Estuaries and Lagoons*, McComb, A. J., Ed., CRC Press, Boca Raton, FL, 1995, 129. With permission.)

The biomass of both phytoplankton and benthic microalgae increased significantly in the Marsdiep tidal channel over the 40-year study period, although this increase was not established for the inner area.[3,23] Spring and summer phytoplankton chlorophyll *a* concentrations were about twice as high in the tidal channel area during the early 1970s than the mid-1970s.[21] In contrast, no statistically significant trend in phytoplankton chlorophyll *a* was detected in the tidal inlet or the inner area of the eastern Dutch Wadden Sea (Friesche Zeegat).[23] The low direct input of

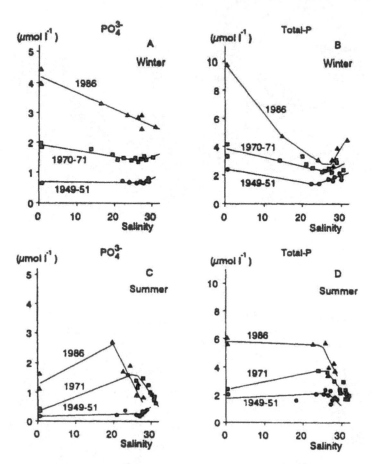

FIGURE 7 Dissolved inorganic P (A, C) and total phosphorus (B, D) as a function of salinity in the Marsdiep basin during winter (December–February) and summer (May–July). (From de Jonge, V. N. and van Raaphorst, W., in *Eutrophic Shallow Estuaries and Lagoons*, McComb, A. J., Ed., CRC Press, Boca Raton, FL, 1995, 129. With permission.)

nutrients relative to the residual inputs in the eastern Dutch Wadden Sea accounts for the differences in biomass trends observed between the two Wadden Sea areas.

Enteromorpha spp. and *Ulva* spp. are dominant benthic macroalgae in the Wadden Sea. Four species of *Ulva* occur there (i.e., *Ulva curvata, U. lactuca, U. rigida,* and *U. scandinavica*).[31] Peletier[32] indicated that the biomass of *Ulva* spp. in the Wadden Sea may vary by a factor of 15 over an annual period. During the past three decades, the mean annual biomass of *Ulva* spp. appears to have increased in the Dutch Wadden Sea, but conclusive data are lacking. In spring, *Enteromorpha radiata* occasionally forms extensive masses, completely covering sand flats. More commonly, much lower densities (~1 ind./m) are found.[33]

Phytoplankton and benthic macroalgae are the dominant primary producers in the Wadden Sea, together being responsible for more than 90% of the overall annual

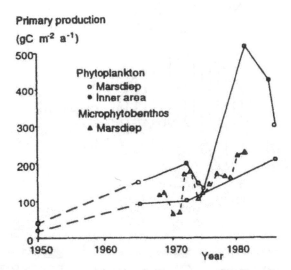

FIGURE 8   Annual primary production of phytoplankton and microphytobenthos at two locations in the Marsdiep tidal basin from 1950 to 1990. (From de Jonge, V. N. and van Raaphorst, W., in *Eutrophic Shallow Estuaries and Lagoons*, McComb, A. J., Ed., CRC Press, Boca Raton, FL, 1995, 129. With permission.)

production of organic carbon (Figure 9). However, both temporal and spatial variations of phytoplankton and benthic macroalgal populations may be considerable.[3,7] This variation complicates assessment of eutrophication problems in different areas of the estuary.

B. SECONDARY PRODUCERS

Because of extensive intertidal flats along the Dutch Wadden Sea, the macrobenthic community on mudflats and sandflats is an important component of the ecosystem. Beukema and Cadee[19] and Beukema[34] demonstrated that both annual production and biomass of the macrobenthic fauna inhabiting the intertidal flats in the western part of the Wadden Sea doubled during the 1970–1984 period. However, Nienhuis[7] could not delineate any increasing trend in production and biomass of this trophic group between 1970 and 1980, with most of the change being observed during the 1980s. The abundance of more than half of the macrobenthic species sampled at western Wadden Sea sites increased during the 1970s and 1980s.[34]

Figure 10 compares the biomass of the macrobenthos in the western (Balgzand) and eastern (Groningen) Dutch Wadden Sea. While the trend of increasing biomass is clearly evident at Balgzand, there is no indication of increasing standing stock at Groningen between 1970 and 1990.[23] Between 1970 and 1984, production of benthic macrofauna at Balgzand nearly doubled (Figure 11). Beukema[14] asserted that eutrophication in the western part of the Wadden Sea during most of the 1980s enhanced primary production and thus the food supply available for the macrobenthos, thereby

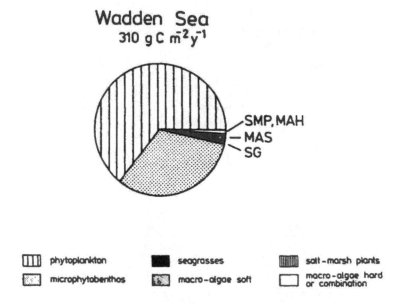

FIGURE 9   Carbon budget of primary producers in the Wadden Sea based on mean annual data. Note large contribution of phytoplankton and benthic macroalgae to the carbon budget. (From Nienhuis, P. H., *Hydrobiologia*, 265, 15, 1993. With permission.)

contributing to greater secondary production. de Jonge and van Raaphorst[3] and Beukema and Cadee[19] consider eutrophication the main factor responsible for the increased macrobenthic production in the western Dutch Wadden Sea.

In benthic communities of the entire system, meiofaunal biomass amounts to 0.5 gC/m², which is only 5% of the macrofaunal biomass (10–20 gC/m²). However, the meiofaunal P/B ratio is much greater than the macrofaunal turnover. Consequently, the contribution of meiofauna to the overall benthic metabolism may be substantial.

Aside from significant changes in species abundance and biomass, the species composition of the benthic community also varied greatly in the western Wadden Sea during the 1970–1990 period. Beukema[35] and Beukema and Essink[36] examined the dominant benthic macroinvertebrate species inhabiting Dutch Wadden Sea tidal flat areas. Only six or seven macrobenthic species of the tidal flats add an average of more than 1 g/m² each to the total biomass.[35] The succeeding seven species accounted for most of the total biomass in 1971 and 1972: the blue mussel *Mytilus edulis* (23% of the total biomass), the lugworm *Arenicola marina* (19%), the sandgaper *Mya arenaria* (17%), the cockle *Cerastoderma edule* (16%), the tellinid bivalve *Macoma balthica* (8%), and the polychaetes *Nereis diversicolor* (5%) and *Lanice conchilega* (3%). These seven species supplied more than 90% of the total biomass in a later study performed in 1977, when *C. edule* contributed the most biomass (26%). Other species, such as *Corophium volutator* and *Hydrobia ulvae*, have much higher numerical densities, but extremely low biomasses. A large fraction

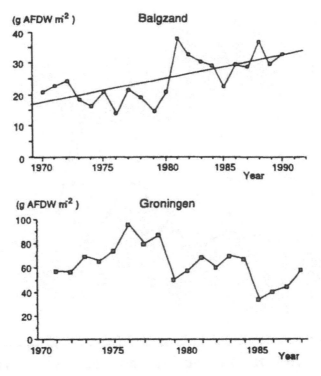

FIGURE 10 Time series of biomass of benthic macrofauna in the western (Balgzand) and in the eastern (Groningen) Dutch Wadden Sea. (Top figure from Beukema, J. J., *Mar. Biol.*, 111, 293, 1991. With permission. Bottom figure from de Jonge, V. N. and Essink, K., in *Estuaries and Coasts: Spatial and Temporal Intercomparisons*, Elliot, M. and Ducrotoy, J.-P., Eds., Olsen & Olsen, Fredensborg, Denmark, 1991, 307. With permission.)

of the macrobenthic invertebrate biomass (66%) is contained in the mollusks and polychaetes (about 33%), with coelenterates, crustaceans, echinoderms, and other groups comprising a minor fraction.

Beukema and Essink[36] compiled data on the macrobenthic fauna at Balgzand during 17 successive years (1969–1985). Fifteen stations were sampled at Balgzand, a 50-km tidal flat area in the western perimeter of the Wadden Sea, in an attempt to unravel the degree of similarity of population fluctuation patterns. Results of this long-term sampling program revealed that nearly 50% of population fluctuation patterns within the study area had a high degree of similarity. Table 3 lists important macrobenthic invertebrates registered at Balgzand.

Beukema and Cadee[19] followed the abundance patterns of 27 macrobenthic species at Balgzand over a 15-year period (1970–1984). The following 12 species exhibited significant increases in abundance (numbers ind./m²): *Nereis diversicolor, Heteromastus filiformis, Eteone longa, Scoloplos armiger, Scoloplos foliosa, Carcinus maenas, Littorina littorea, Arenicola marina, Scrobicularia plana, Macoma balthica, Mytilus edulis*, and *Anaitides* sp. The increase in abundance of *Abra tenuis*

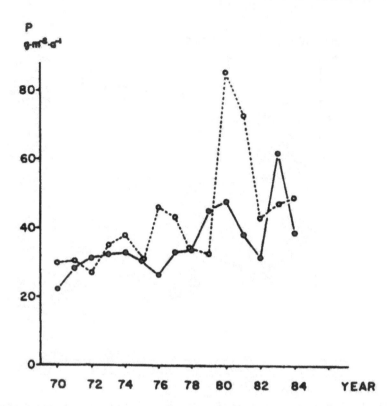

FIGURE 11 Changes in production of benthic macrofauna at Balgzand from 1970 to 1984. (From Beukema, J. J. and Cadee, G. C., *Ophelia*, 26, 55, 1986. With permission.)

and *Cerastoderm edule* was almost statistically significant. The remaining 13 species showed irregular fluctuations in numbers. No species decreased significantly in abundance during the study period.

Beukema[14] reported several significant changes in the species composition of the benthic community at Balgzand between 1970 and 1990. Over this 20-year period, the numerical proportion of polychaetes increased at the expense of crustaceans and mollusks. The overall mean weight per individual of the macrobenthos decreased (numbers of individuals of small-sized species increased more rapidly than those of large-sized species). The share of deposit feeders increased and that of carnivores decreased, although the absolute numbers and biomass of all feeding types increased. The proportion of suspension feeders was relatively constant. While numerical densities of many species increased over the study period, the total number of species fluctuated without a clear trend.

Reise[37] described the macrofaunal assemblages on the tidal flats of Konigshafen, a shallow sheltered bay on the island of Sylt in the northern part of the Wadden Sea. The dominant macrofaunal species formed dense assemblages on the tidal flats during the 1980s. This was the case for *Corophium volutator* and *Nereis diversicolor* on the upper tidal flats, *Arenicola marina* on the sandflats, and *Lanice conchilega* and *Mytilus edulis* near the low water line.

TABLE 3
Quantitatively Important Benthic
Macroinvertebrates Collected
Between 1969 and 1985 at
Balgzand, Dutch Wadden Sea

Abra tenuis (Montagu)
Anaitides spec. div.
Angulus tenuis (Da Costa)
Antinoella sarsi (Kinb.)
Arenicola marina (L.)
Bathyporeia spec.
Carcinus maenas (L.)
Cerastoderma edule (L.)
Corophium volutator (Pall.)
Crangon crangon (L.)
Eteone longa (Fabr.)
Heteromastus filiformis (Clap.)
Hydrobia ulvae (Penn.)
Lanice conchilega (Pall.)
Littorina littorea (L.)
Macoma balthica (L.)
Magelona papillicornis F. M.
Mya arenaria L.
Mytilus edulis L.
Nemertini spec. div.
Nephtys homberegii Sav.
Nereis diversicolor (O. F. M.)
Scolelepis foliosa (A. & M.-E.)
Scoloplos armiger (O. F. M.)
Scrobicularia plana (Da Costa)

From Beukema, J. J. and Essink, K., *Hydrobiologia*, 142, 199, 1986. With permission.

Polychaetes numerically dominated the benthic macrofauna in most areas of Konigshafen. *Corophium volutator* and *Hydrobia ulvae* appeared to be dominant in zones where they grew in dense aggregations. On sandflats, *Arenicola marina* had a mean density of ~40 ind./m. The cockle *Cerastoderma edule* also was a prominent member of the benthic community of sandy tidal flats. Varying greatly in abundance on muddy flats, the benthic macrofaunal community was frequently dominated by spionid and capitellid polychaetes (e.g., *Tharyx marioni*) and the oligochaete, *Tubificoides benedeni*. The macrofaunal biomass averaged about 15 gC/m^2 on the tidal flat, but where dense beds of mussels grew, the mean annual biomass approached 1000 gC/m^2.

The tidal flats of the Dutch Wadden Sea cover an area of 1300 km^2. The estimated mean macrofaunal biomass of the flats is 27 g/m^2 ash free dry weight, with a range of 19–34 g/m² ash free dry weight.[38] At the highest and lowest levels of the intertidal zone, where environmental factors are extreme, a low absolute abundance and biomass of species exist. Maximum macrofaunal biomass, exceeding 50 gC/m², takes place between the mid-tide level and 50 cm below the mid-tide level in sediments having a silt content of 10–20%. The highest species diversity values, >14/0.45 m², also are encountered at the mid-tide level, about 25 cm below the sediment-water interface in sediments having a silt content of 2–25%.[38,39]

C. Fish

The fish fauna of the Wadden Sea changed considerably-between 1950 and the early 1990s. Prior to 1962, dab (*Limanda limanda*) dominated the flatfish stock in the western Wadden Sea. However, abundances of this species declined-rapidly between 1962 and 1965, and since that time, plaice (*Pleuronectes platessa*) has become the dominant juvenile fish-species.[7,40] Eutrophication may have been a factor in causing this change.[3] Aside from plaice, herring (*Clupea harengus*) and sole (*Solea solea*) from North Sea waters utilize the Wadden Sea as a nursery area. Other species which may attain significant numbers in the Wadden Sea include eel (*Anguilla anguilla*), eelpout (*Zoarces viviparus*), bull rout (*Myoxocephalus scorpius*), and butterfish (*Pholis gunellus*). The abundance and distribution of these species are related to the presence of mussel beds which provide food and shelter.

Some species of fish once commonly found in the Wadden Sea are now either missing from the system or only rarely observed. For example, the disappearance of eelgrass beds resulted in the demise of the sea stickleback (*Spinachia spinachia*) and the deep-snouted pipefish (*Syngnathus typhle*) in the estuary.[1] The thornback ray (*Raja clavata*), a common species in the Wadden Sea prior to 1950, has decreased greatly in abundance in recent decades.[41] Its decline has not been linked to escalating eutrophication.

D. Birds

The tidal flats of the Wadden Sea are important feeding grounds for various water-fowl, including waders, ducks, geese, and other populations. Some waterfowl, such as eiders (*Somateria mollissima*) and gulls (*Larus argentatus*, *L. canus*, and *L. ridibundus*), breed on the Dutch Wadden Sea islands. The oystercatcher (*Haematopus ostralegus*) also breeds in areas of the Dutch Wadden Sea. Eiders and dark-bellied brent geese (*Branta bernicla bernicla*) overwinter in this area.

The main food chain in the Wadden Sea is dominated by phytoplankton and benthic filter feeders. Cockles (*Cerastoderma edule*), mussels (*Mytilus edulis*), clams (*Macoma balthica*), polychaetes (*Nereis diversicolor*), and other macrobenthic prey support huge numbers of birds frequenting the tidal flats. The diet of eiders consists of cockles (40%), mussels (40%), and other invertebrates (20%).[42] Cockles and mussels constitute a major portion of the diets of other waterfowl as well. Hence, the increase in secondary production of intertidal macrobenthos in the western Wadden Sea in recent decades runs parallel with the increase in the breeding population of some avian species (e.g., the oystercatcher). Therefore, eutrophication in the Wadden Sea may have affected major taxonomic groups at all trophic levels.

E. MAMMALS

de Jonge et al.[1] discussed changes in abundance of the common seal (*Phoca vitulina*) in the Dutch Wadden Sea during the 1950–1990 period. Population size of the common seal peaked in 1950 at more than 3000 animals. After 1950, the numbers decreased precipitously, falling to about 1200 animals in 1960, rebounding to approximately 1500 animals in 1964, and declining again to less than 500 animals during the 1970s. Between 1980 and 1987, the stock exhibited some degree of recovery, reaching a population size of about 1100 individuals. Reijnders[43] ascribed this recovery phase to increased reproductive success, increased numbers of rehabilitated young seals, and increased immigration of seals from the German Wadden Sea. Infection of the Dutch seal population in 1988 by *Canine distemper* virus caused a decline in abundance to only about 450 animals in 1989.[44]

V. CONCLUSIONS

A straightforward causal relationship exists between nutrient enrichment in the western Wadden Sea and increased primary and secondary production. Escalating algal biomass and primary production, followed by increases in stocks and production of herbivores, peaked during the 1970s and 1980s and appear to be related to the high nutrient inputs of influents systems, notably the Rhine River and Lake IJssel. However, alternative factors may have contributed to the observed trends. For example, decreases in the concentrations of heavy metals, halogenated hydrocarbons (e.g., PCBs, pesticides, etc.) and other toxic substances in the Rhine River[45] and the Dutch Wadden Sea[46] in recent decades could have resulted in the enhanced biotic production and biomass. Increasing suspended particulate matter,[47] rates of sedimentation,[19] and storm surges[48] must also be considered.

There is a growing awareness of the chronic eutrophication of the western Wadden Sea and its effect on living resources. The negative aspects of eutrophication are well chronicled in estuarine and coastal marine systems worldwide and include anaerobiosis, toxic algal blooms, mass kills of benthic and epibenthic organisms, and changes in species abundance and diversity patterns.[7] In the Wadden Sea, the mineralization of organic matter can lower oxygen concentrations; locally depleted oxygen levels may occasionally lead to increased, albeit spatially restricted, mortality of benthic organisms.[3] However, strong tidal currents and continuous flushing with water from the North Sea generally produce well-mixed conditions, and reduced oxygen levels are therefore not pervasive.[19] Consequently, eutrophication of the western Wadden Sea has not caused serious problems to date.

Despite its limited effects, eutrophication of the western Wadden Sea was significant during the 1970s and 1980s, when a doubling of nutrient concentrations occurred. In response, the rates of primary production roughly doubled. Functionally, community productivity of the benthic macrofauna also doubled.[19] While the species richness of the community remained more or less unchanged, the abundance and standing stock of individuals increased. The largest increase in abundance occurred among three deposit-feeding polychaete species: *Heteromastus filiformis*, *Nereis diversicolor*, and *Scoloplos armiger*.[14]

The species composition in the Wadden Sea did not change appreciably during the period of rising eutrophication. However, observed changes in the succession of different groups of algae may be related to nutrient enrichment in the tidal basins. Higher densities of *Phaeocystis pouchetii* temporarily generated excessive foam on beaches and created offensive odors.[3]

On the positive side, eutrophication has increased food availability for fish, birds, and mammals. Improved cockle fisheries, juvenile flatfish growth, and mussel culturing have been an economic benefit for man. There has been greater utilization of resources in the system.

The long-term effects of eutrophication on Wadden Sea ecosystem functioning remain unclear. Hence, it is imperative to pursue a strategy of avoiding excessive nutrient inputs and reducing future loads. An upgrading of water quality in the Wadden Sea depends on lowering pollutant loads in rivers that flow into the Wadden Sea and adjacent areas of the North Sea. Only then can the long-term eutrophication of the Wadden Sea be effectively controlled.

REFERENCES

1. de Jonge, V. N., Essink, K., and Boddeke, R., The Dutch Wadden Sea: a changed system, *Hydrobiologia*, 265, 45, 1993.
2. Dijkema, K. S., Changes in salt-marsh area in the Netherlands Wadden Sea after 1600, in *Vegetation Between Land and Sea*, Huiskes, A. H. L., Blom, C. W. P. M., and Rozema, J., Eds., Dr. W. Junk Publishers, Dordrecht, The Netherlands, 1987, 42.
3. de Jonge, V. N. and van Raaphorst, W., Eutrophication of the Dutch Wadden Sea (Western Europe), an estuarine area controlled by the River Rhine, in *Eutrophic Shallow Estuaries and Lagoons*, McComb, A. J., Ed., CRC Press, Boca Raton, FL, 1995, 129.
4. Admiraal, W., van der velde, G., Smit, H., and Cazemier, W. G., The Rivers Rhine and Meuse in the Netherlands: present state and signs of ecological recovery, *Hydrobiologia*, 265, 97, 1993.
5. van Urk, G. and Smit, H., The lower Rhine geomorphological changes, in *Historical Change of Large Alluvial Rivers: Western Europe*, Petts, G. E., Ed., John Wiley & Sons, New York, 1989, 167.
6. de Jonge, V. N., Response of the Dutch Wadden Sea ecosystem to phosphorus discharges from the River Rhine, *Hydrobiologia*, 195, 49, 1990.
7. Nienhuis, P. H., Nutrient cycling and food webs in Dutch estuaries, *Hydrobiologia*, 265, 15, 1993.
8. de Jonge, V. N., The Ems estuary, the Netherlands, in *Eutrophic Shallow Estuaries and Lagoons*, McComb, A. J., Ed., CRC Press, Boca Raton, FL, 1995, 81.
9. Wolfe, W. J., The degradation of ecosystems in the Rhine, in *The Breakdown and Restoration of Ecosystems*, Holdgate, M. W. and Woodman, M. J., Eds., Plenum Press, New York, 1978, 169.

10. Friedrich, G. and Muller, D., Rhine, in *Ecology of European Rivers*, Whitton, B. A., Ed., Blackwell, Oxford, 1984, 265.

11. van Urk, G., Lower Rhine-Meuse, in *Ecology of European Rivers*, Whitton, B. A., Ed., Blackwell, Oxford, 1984, 437.

12. van Meerendonk, J. H., Janssen, G. M., and Frederiks, B., De aanvoer van voeding-stoffen en microverontreingingen naar de Waddenzee en Elms-Dollard, Rijkswater-staat, Tidal Waters Division, Haren, Notanr., GWWS-88.002, 1988.

13. BOEDE, Biological research in the Ems-Dollard estuary, Rijkswaterstaat Communi-cations No. 40, The Hague, The Netherlands, 1985.

14. Beukema, J. J., Changes in composition of bottom fauna of tidal flat area during a period of eutrophication, *Mar. Biol.*, 111, 293, 1991.

15. Postma, H., Hydrography of the Wadden Sea: movements and properties of water and particulate matter, Wadden Sea Working Group, Report 2, Leiden, The Nether-lands, 1982.

16. van Raaphorst, W. and van der Veer, H. W., The phosphorus budget of the Marsdiep tidal basin (Dutch Wadden Sea) in the period 1959–1985: importance of the exchange with the North Sea, *Hydrobiologia*, 195, 21, 1990.

17. Postma, H., Introduction to the symposium on organic matter in the Wadden Sea, in *The Role of Organic Matter in the Wadden Sea*, Laane, R. W. P. M. and Wolff, W. J., Eds., Neth. Inst. Sea Res., Pub. Ser. 10, Texel, The Netherlands, 1984, 15.

18. de Wilde, P. A. W. J. and Beukema, J. J., The role of zoobenthos in the consumption of organic matter in the Dutch Wadden Sea, in *The Role of Organic Matter in the Wadden Sea*, Laane, R. W. P. M. and Wolff, W. J., Eds., Neth. Inst. Sea Res., Pub. Ser. 10, Texel, The Netherlands, 1984, 145.

19. Beukema, J. J. and Cadee, G. C., Zoobenthos responses to eutrophication of the Dutch Wadden Sea, *Ophelia*, 26, 55, 1986.

20. de Jonge, V. N. and Postma, H., Phosphorus compounds in the Dutch Wadden Sea, *Neth. J. Sea Res.*, 8, 139, 1974.

21. Cadee, G. C., Has input of organic matter into the western part of the Dutch Wadden Sea increased during the last decades?, in *The Role of Organic Matter in the Wadden Sea*, Laane, R. W. P. M. and Wolff, W. J., Eds., Neth. Inst. Sea Res., Pub. Ser. 10, Texel, The Netherlands, 1984, 71.

22. Helder, W., The cycling of dissolved inorganic nitrogen compounds in the Dutch Wadden Sea, *Neth. J. Sea Res.*, 8, 154, 1974.

23. de Jonge, V. N. and Essink, K., Long-term changes in nutrient loads and primary and secondary producers in the Dutch Wadden Sea, in *Estuaries and Coasts: Spatial and Temporal Intercomparisons*, Elliott, M. and Ducrotoy, J.-P., Eds., Olsen and Olsen, Fredensborg, Denmark, 1991, 307.

24. Brockmann, U., Billen, G., and Gieskes, W. W. C., North Sea nutrients and eutroph-ication, in *Pollution of the North Sea: An Assessment*, Salomons, W., Bayne, B. L., Duursma, E. K., and Forstner, U., Eds., Springer-Verlag, Berlin/Heidelberg, 1988, 348.

25. Postrha, H., Hydrography of the Dutch Wadden Sea, *Arch. Neerl. Zool.*, 10, 405, 1954.

26. Cadee, G. C., Increased phytoplankton primary production in the Marsdiep area (western Dutch Wadden Sea), *Neth. J. Sea Res.*, 20, 285, 1986.

27. Veldhuis, M. J. W., Colijn, F., Venekamp, L. A. H., and Villerius, L. A., Phytoplankton primary production and biomass in the western Dutch Wadden Sea (The Netherlands): a comparison with an ecosystem model, *Neth. J. Sea Res.*, 22, 37, 1988.

28. Cadee, G. C., Trends in Marsdiep phytoplankton, in *Proceedings of the 7th International Wadden Sea Symposium*, Dankers, N., Ed., Neth. Inst. Sea Res., Pub. Ser. 19, Texel, The Netherlands, 1991.
29. Cadee, G. C. and Hegeman, J., Seasonal and annual variation in *Phaeocystis pouchetii* (Haptophyceae) in the westernmost inlet of the Wadden Sea during the 1973 to 1985 period, *Neth. J. Sea Res.*, 20, 29, 1986.
30. Cadee, G. C. and Hegeman, J., Historical phytoplankton data of the Marsdiep, *Hydrobiol. Bull.*, 24, 111, 1991.
31. Koeman, R. P. T. and van den Hoek, C., The taxonomy of *Ulva* (Chlorophyceae) in the Netherlands, *Br. Phycol. J.*, 19, 9, 1981.
32. Peletier, H., Measurements of biomass, pigments, and photosynthesis of *Ulva* sp. in the western Wadden Sea in 1988–91, Report DGW-92.005, Tidal Waters Division, Rijkswaterstaat, The Netherlands, 1992.
33. van den Hoek, C., Admiraal, W., Colijn, F., and de Jonge, V. N., The role of algae and seagrasses in the ecosystem of the Wadden Sea: a review, in *Flora and Vegetation of the Wadden Sea*, Wolff, W. J., Ed., A. A. Balkema, Rotterdam, 1979, 9.
34. Beukema, J. J., Long-term changes in macrozoobenthic abundance on the tidal flats of the western part of the Dutch Wadden Sea, *Helgol. wiss. Meeresunters.*, 43, 405, 1989.
35. Beukema, J. J., Quantitative data on the benthos of the Wadden Sea proper, in *Ecology of the Wadden Sea*, Vol. 1, Wolff, W. J., Ed., A. A. Balkema, Rotterdam, 1983, 4/135.
36. Beukema, J. J. and Essink, K., Common patterns in the fluctuations of macrozoobenthic species living at different places on tidal flats in the Wadden Sea, *Hydrobiologia*, 142, 199, 1986.
37. Reise, K., Predator control in marine tidal sediments, in *Proceedings of the 19th European Marine Biology Symposium*, Gibbs, P. E., Ed., Cambridge University Press, Cambridge, 1985, 311.
38. McLusky, D. S., *The Estuarine Ecosystem*, Halsted Press, New York, 1981.
39. Beukema, J. J., de Bruin, W., and Jansen, J. J. M., Biomass and species richness of the macrobenthic animals living on the tidal flats of the Dutch Wadden Sea: long-term changes during a period with mild winters, *Neth. J. Sea Res.*, 12, 58, 1978.
40. Boddeke, R., Visserij-biologische veranderingen in de westelijke Waddenzee, *Visserij*, 20, 213, 1967.
41. Witte, J. Y. and Zijlstsra, J. J., The species of fish occurring in the Wadden Sea, in *Fishes and Fisheries of the Wadden Sea*, Dankers, N., Wolff, W. J., and Zijlstra, J. J., Eds., A. A. Balkema, Rotterdam, 1978, 10.
42. Swennen, C., Populatie-structuur en voedsel van de Eidereend *Somateria m. mollissima* in de Nederlandse Waddenzee, *Ardea*, 64, 311, 1976.
43. Reijnders, P. J. H., Waarom onze zeehondenpopulatie 1984 toch heeft gehaald, *Waddenbulletin*, 20, 79, 1985.
44. Osterhaus, A. D. M. and Vedder, E. J., Identification of virus causing recent sea death, *Nature*, 335, 20.
45. Dijkzeul, A., De waterkwaliteit van de Rijn in de periode 1970–1981, *Rijksinstituut voor Zuivering van Afvalwater Notanr.*, 82-061, 1, 1982.
46. Essink, K., Monitoring of mercury pollution in Dutch coastal waters by means of the teleostean fish *Zoarces viviparus*, *Neth. J. Sea Res.*, 19, 177, 1985.

47. de Wit, J. A. W., Schotel, F. M., and Bekkers, L. E. J., De waterkwaliteit van de Waddenzee 1971–1981, *Rijkinstituut voor Zuivering van Afvalwater Notanr.*, 82-065, 1, 1982.

48. Siefert, W., Bemerkenswerte veränderungen der wasserstände in den deutschen tideflüssen, *Die Küste*, 37, 1, 1982.

Index